Spectrophysics

Spectrophysics

ANNE P. THORNE

Senior Lecturer in Physics,
Imperial College,
University of London

CHAPMAN AND HALL
London

A HALSTED PRESS BOOK
JOHN WILEY & SONS
New York

First published 1974
by Chapman and Hall Ltd
11 New Fetter Lane, London EC4P 4EE

© 1974 Anne P. Thorne

Printed in Great Britain by
William Clowes & Sons Ltd.
London, Colchester and Beccles

Library of Congress Cataloging in Publication Data

Thorne, Anne P
    Spectrophysics.

    1. Spectrum analysis. I. Title.
QC451.T47   535'.84   74-1551
ISBN 0-470-86495-8 (Halsted Press)

To

L., A.L., and P.F.T.
but for whom this book would not have been
started or finished,

and to

Bridget, Meriel, Andrew and Janet
but for whom it would have been finished
long ago

Preface

This book describes the methods of experimental spectroscopy and their use in the study of physical phenomena. The applications of optical spectroscopy may be grouped under three broad headings: chemical analysis, elucidation of atomic and molecular structure, and investigations of the interactions of radiating atoms and molecules with their environment. I have used the word 'Spectrophysics' for the third of these by analogy with spectrochemistry for the first and in preference to 'quantitative spectroscopy'.

A number of textbooks treat atomic and molecular structure at varying levels of profundity, but elementary spectrophysics is not, so far as I am aware, covered in any one existing book. There is moreover a lack of up-to-date books on experimental techniques that treat in a fairly elementary fashion interferometric, Fourier transform and radiofrequency methods as well as prism and grating spectroscopy. In view of the importance of spectrophysics in astrophysics and plasma physics as well as in atomic and molecular spectroscopy there seemed a place for a book describing both the experimental methods and their spectrophysical applications.

This book is directed at students on a first degree course in physics and at postgraduates beginning research in the fields of experimental atomic or molecular spectroscopy, plasma physics or astrophysics. It is hoped this book will enable such students to set their own particular research problems in the context of similar problems and to compare the various methods of tackling them. The book is not, however, intended as a research manual, for it does not contain sufficient experimental detail for this purpose. In particular, Chapter 7 on microwave and radiofrequency spectroscopy describes only the principles of the various methods and the types of problem to which they can be applied.

The rather long chapter on atomic and molecular structure, Chapter 2, has been included to explain the concepts and terminology used in the later chapters for the benefit of any student who is not familiar with the subject. It is in no sense intended as a potted course on atomic structure. Some acquaintance with elementary wave mechanics is assumed in both this and later chapters.

A few remarks on the contents of Chapters 8 to 11 should help to indicate the scope of this book. Pressure broadening of spectral lines is discussed more fully than is usual in elementary textbooks, but from a physical rather than a mathematical point of view. Transition probabilities, line and oscillator strengths, absorption coefficients, and the relations between these quantities are treated at some length because they seem to engender considerable confusion among students. Moreover, topics such as optical depth, equivalent width, curves of growth and radiative transfer are often totally unfamiliar to students because they are usually to be found only in textbooks on astrophysics. The last chapter introduces some basic ideas of plasma physics, discusses ionization, continuous radiation and thermodynamic equilibrium, and shows how spectroscopy may be applied to problems in both laboratory and astrophysical plasmas.

This book has been completed at a time when the rapid development of tunable dye lasers is opening up new possibilities in all branches of spectroscopy. I have remarked on some of the new techniques at appropriate places in the text, but undoubtedly many more will pass into common usage in the next few years.

I owe my sincere thanks to a number of friends and colleagues who have helped me in writing this book. In particular, Dr H. G. Kuhn, FRS, of Oxford University most kindly read the whole manuscript and made numerous suggestions for its improvement. Dr R. C. M. Learner of Imperial College has provided a wealth of constructive criticism of most of the book, and Dr. D. D. Burgess, also of this college, has done likewise for the remainder. Finally, I must thank my husband and children for putting up, on the whole patiently, with a constantly pre-occupied wife and mother.

Contents

CHAPTER ONE

Introduction

1.1 The uses of spectroscopy

Spectroscopy can be used to determine the identity, the structure and the environment of atoms and molecules by analysis of the radiation emitted or absorbed by them. The light from a gaseous discharge when analysed by wavelength to form a spectrum is found to consist of discrete lines and bands, perhaps overlying a continuum. Each line or band is characteristic of a particular atom or molecule, and once the characteristic line pattern of an atom is known its appearance in the spectrum establishes the presence of that atom in the source. This aspect of spectroscopy is known as spectrochemical analysis. It may be made quantitative by measuring relative intensities as well as wavelengths. Secondly, one may deduce from the line or band pattern the characteristic energy levels, or stationary states, of the atom or molecule. This provides the experimental basis from which the theories of atomic and molecular structure have been developed and are still developing. Finally, the physical properties (temperature, pressure, etc.) of the gas or plasma containing the emitting or absorbing particles affect the intensity and wavelength distribution of the radiation in various ways. The study of these effects, with a view to establishing the physical parameters of the source, may be called, by analogy to spectrochemistry, spectrophysics.

For the first couple of hundred years after spectroscopy was first conceived by Newton it was put to none of these uses. Although a large number of spectral lines were observed and measured during the first half of the nineteenth century –

1

including the absorption lines in the sun's spectrum discovered by Fraunhofer and emission lines in the spectra of flames – it was not until about 1860 that Kirchhoff and Bunsen confirmed the link between particular wavelengths and particular atoms and so inaugurated spectrochemistry. In the next few years several elements were first discovered spectroscopically (rubidium, cesium, thallium and indium, for example). Apart from its laboratory uses, spectrochemical analysis was immediately applied to the identification of the Fraunhofer lines in the sun's spectrum. Helium was 'discovered' in the sun by Lockyer in 1868 many years before it was isolated in the laboratory. The identification of atoms and molecules in other stars, in nebulae and in inter-stellar space, together with estimates of their relative abundance, has continued to the present time and is of the utmost importance to astrophysics and cosmology.

The spectroscopic foundations of atomic structure were laid from 1885 onwards with the attempts by Balmer, Rydberg and others to group the observed spectral lines of a given atom in some meaningful fashion. The splitting of spectral lines in a magnetic field discovered in 1896 by Zeeman was interpreted by Lorentz in terms of oscillating charged particles whose charge/mass ratio identified them with the electrons discovered at about the same time by Thompson. However, it was not until 1913 that Bohr, using the atomic model recently proposed by Rutherford, produced the theory of the hydrogen atom that really established the link between spectra and structure. The part played by spectroscopy in atomic physics grew rapidly in scope in the 1920s and 1930s with the introduction of quantum mechanics and the discovery of electron spin and nuclear spin, and it is far from being played out now. The unravelling of the spectra of complex atoms (such as the rare earths) and of highly ionized species, and the study of hyperfine structure and isotope shift are examples of the continuing role of 'conventional' spectroscopy. Even the spectra of simple atoms, which had been considered well understood, have turned out on further investigation to possess curious features attributable to auto-ionization effects, the study of which has helped in the understanding of re-combination and scattering processes. The development of microwave and radio-frequency spectroscopy has allowed small energy differences such as those occurring in hyperfine structure and the Lamb shift to be measured far more accurately than is possible by optical spectroscopy. Theoretical atomic physics has made great advances since computers became readily available. A high degree of co-operation between theory and experiment is vital in the interpretation of the more complicated structures.

The third application of spectroscopy, which we call for convenience spectrophysics, consists of using the shifts, widths and intensities of spectral lines (and sometimes the intensity and spectral distribution of continuous radiation) to determine abundances, temperatures, pressures, velocities, electron densities and radiative transfer processes in the emitting or absorbing gas. This has been an important branch of astrophysics ever since the identification of the Fraunhofer lines. We shall be considering here the applications both to true astrophysics and to what is sometimes called laboratory astrophysics. The importance of quantitative spectroscopy in both these fields has increased rapidly in recent years for two main reasons. Until 1930 astrophysicists were limited in the range of wavelengths they could study by the 'optical window' in the earth's atmosphere, 3000 Å to 10 000 Å in round figures (there is actually partial transmission between molecular bands from 10 000 Å to 25 000 Å). The opening of the 'radio window' (roughly 1 cm to 10 m wavelength range) set off the rapidly growing and immensely important science of radio astronomy; and, of more direct relevance to the subject of this book, the development of rocket and satellite spectroscopy from 1945 onwards enabled 'optical' spectroscopists to emerge from their window and see for the first time the infra-red and the highly important region below 3000 Å. The second reason for spectrophysical advance has been the development of high temperature plasmas, initially for fusion work. The need for spectroscopic investigation of these plasmas stimulated lines of research that were immediately applicable to the study of astrophysical plasmas. Although laser diagnostic techniques have largely superseded spectroscopic methods in the case of laboratory plasmas, the availability of these sources has been of great value in developing theories of astrophysical processes and testing them under controlled laboratory conditions.

The general references [1-4], quoted at the end of this chapter give further details of historical development and traditional applications of spectroscopy.

1.2   Scope of this book

This book will be mainly concerned with those aspects of experimental spectroscopy that have definite physical applications – the third of the three uses described in the last section. Chapter 2 is a brief survey of atomic and molecular structure, designed to show the orders of magnitude of the effects to be expected in, for example, Zeeman splitting or rotational structure, and to introduce the quantum numbers and the necessary terminology. Since the applications of molecular spectroscopy tend to be chemical more often than physical, the molecular structure is even more perfunctorily dealt with in this chapter than is the atomic. Chapters 3, 4, 5 and 6 describe the experimental techniques of optical spectroscopy, by which is meant spectroscopy from the soft X-ray region to the sub-millimetre region – from a wavelength of, say, 10 Å (1 nm) to 1 mm in round numbers. The emphasis in these chapters is on high resolution and intensity measurements because these are necessary for the determinations of line shapes, transition probabilities, and so on. Absolute wavelength measurements and line identifications are of much greater importance in spectrochemistry and theoretical spectroscopy than in spectrophysics.

Chapter 7 surveys microwave and radiofrequency spectroscopy, but in very little experimental detail; the object is simply to point out alternative ways of measuring some of the quantities discussed in optical spectroscopy. Chapter 8 discusses the broadening of spectral lines by radiation damping, Doppler effect and interactions with other particles. Chapter 9 defines and relates transition probabilities, absorption coefficients, oscillator strengths and connected quantities and discusses such astrophysically important concepts as optical depth and curves of growth. Chapter 10 describes the various methods of measuring transition probabilities, partly because of their intrinsic importance in spectrophysics and partly because many such methods may equally well be used on lines of known transition probability to determine abundances and temperatures. Chapter 11 describes specifically some of the more important applications of spectroscopy to laboratory and astrophysical plasmas. In this chapter I have specifically excluded radio-astronomy on the experimental grounds that it is outside the wavelength region of this book (Chapter 7 notwithstanding) and on the theoretical grounds that it has closer links with true astronomy than with spectroscopy and cannot possibly be treated satisfactorily in a few paragraphs.

The book makes no attempt to cover X-ray, Mössbauer or Raman spectroscopy, although some brief comments on the latter are included at the end of Chapter 2.

1.3 The electromagnetic spectrum

The electromagnetic spectrum extends from the longest radio waves, whose wavelength is measured in kilometres, through the medium and short radio waves to the microwave, infra-red and visible regions. Beyond the visible, the ultra-violet merges into the X-ray region, and beyond these again come the γ and cosmic ray regions.

Fig. 1.1 is a diagram of the spectrum on which are marked the conventional spectral regions. The divisions are necessarily rather arbitrary, except for the closely defined visible region. The wavelength λ is marked in metres and is also given in microns or micrometres μm, Ångström units Å, and X-units XU where appropriate. With the introduction of the metre as the unit of length, optical wavelengths should properly be expressed in nanometres, nm, but the Ångström is still a recognized unit and the one most often found in spectroscopic literature. Since $1 \text{ Å} = 10^{-10}$ m, the nm and the Å are related by 1 nm = 10 Å. In the older literature the nanometre is often called a millimicron. The XU was originally defined independently in such a way that its relation to the Ångström unit turned out to be $1002.06 \text{ XU} = 1$ Å. It is now defined as exactly 10^{-3} Å or 10^{-13} m.

Despite the common classification of radiation by wavelength, the important quantity from the point of view of atomic and molecular structure is the frequency, $\nu = c/\lambda_{\text{vac}}$, because this is related to the difference in energy ΔE between two stationary states of the system by $h\nu = \Delta E$. Frequencies in Hz (sec^{-1}) are marked below wavelengths at the long wavelength end of Fig. 1.1. At shorter wavelengths it is more convenient to replace ν by a unit proportional to it: the wave-number $\sigma = \nu/c = 1/\lambda_{\text{vac}}$. Different authors use different symbols for wave-number, ν', $\bar{\nu}$, k and σ

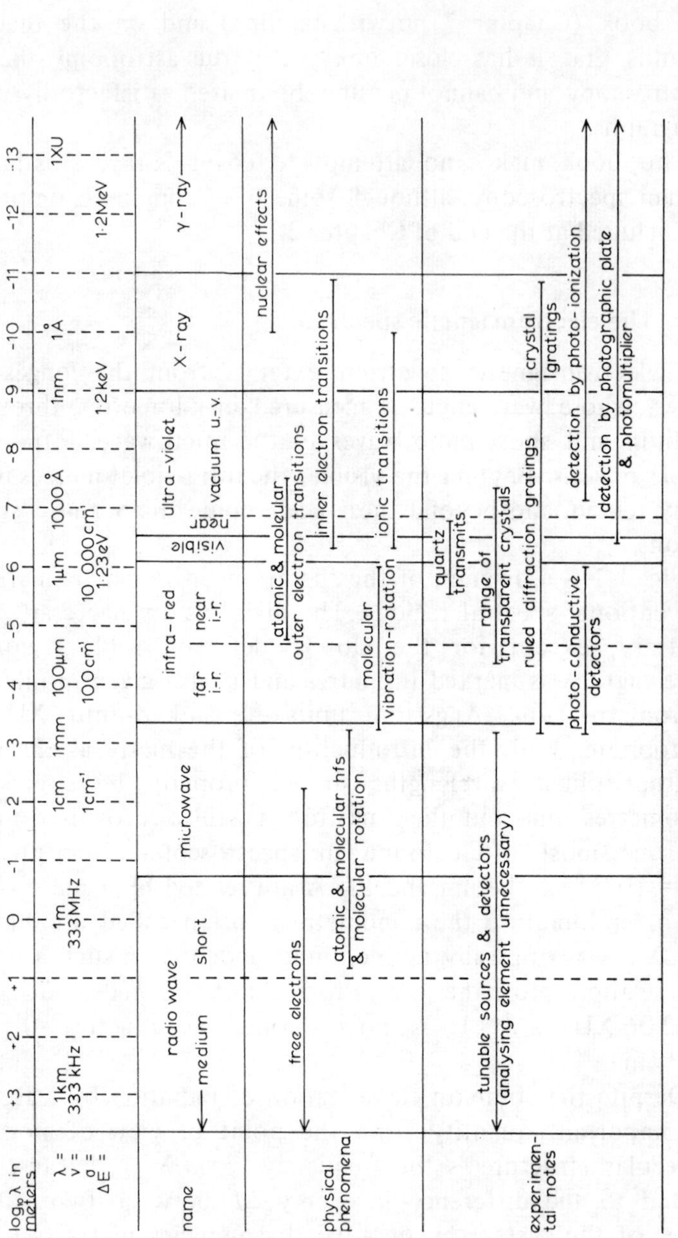

Fig. 1.1 The electromagnetic spectrum.

being all common; I use σ here because it seems less likely to be confused with ν itself or with other constants. The wave-number unit is the cm^{-1}, sometimes called the Kayser (K); its sub-unit, the millikayser (mK) is a convenient unit for hyperfine structure and line-width work, although radiofrequency spectroscopists use frequency units, $Mc\,sec^{-1}$ or MHz, for such quantities (1 mK = 29.979 MHz). Finally, very high photon energies (corresponding to very short wavelengths) are often expressed in electron volts (eV). 1 eV is equal to $1\cdot6 \times 10^{-19}$ J and corresponds to 8066 cm^{-1}. These numerical results are collected in the Appendix.

Atomic and molecular energy levels, like wave-numbers, are usually expressed in cm^{-1}. When converting from absolute energy units in the S.I. system, a factor 100 must be included to change m^{-1} to cm^{-1}:

$$\sigma\,(cm^{-1}) = 1/100\,\lambda_{vac}\,(m) = \nu/100\,c = \Delta E(\text{joule})/100\,hc$$

The use of λ_{vac} should be stressed. If the wavelength is measured in air, it must be multiplied by n, the refractive index of air, before conversion to frequency or wave number units:

$$\nu = \frac{c}{n\lambda_{air}}, \qquad \sigma = \frac{1}{n\lambda_{air}}$$

The correction is about 3 parts in 10^4 and is itself slightly wavelength-dependent. A set of tables for conversion of air wavelengths to wave numbers has been compiled by the National Bureau of Standards [5] that includes the refractive index correction in taking the reciprocal.

Returning to Fig. 1.1, it is seen that the visible region (7500 Å– 4000 Å) is bounded by the infra-red on the longwave side and the ultra-violet on the shortwave side. It is convenient to subdivide these regions into near and far infra-red, near and far ultra-violet, at limits somewhere around 40 μm and 2000 Å respectively, these being roughly the practical limits of transparency of optical materials suitable for prisms and lenses. Until fairly recently the 'optical' region from far infra-red to far ultra-violet was isolated from the radio and X-ray regions by gaps in which it had not been found possible to excite and detect radiation of the appropriate wavelength. The invention of radar in the 1939–45 war opened the door to the very short radio wave, or

microwave, techniques, while at the same time infra-red spectroscopists were pushing up to longer wavelengths. The two now
overlap in the sub-millimetre region. The short wavelength gap has
meanwhile been well filled, partly because of its interest to
plasma- and astro-physicists. Optical spectroscopy now reaches
wavelengths below 2 Å, while soft X-rays extend to 50 Å. The
distinction between 'optical' and 'X-ray' in the overlap region lies
in the origin of the spectra. Optical spectroscopy is concerned
with transitions of the outer or valence electron(s) and X-ray
spectroscopy with transitions of the inner electrons. The very
short wavelength optical spectra come from the outer electrons of
very highly ionized species.

A book on optical spectroscopy may therefore take as a logical
lower limit wavelengths of a few Å. The upper limit is conventionally set in the far infra-red. However, it will be seen that
the very smallest energy level splittings with which we shall be
concerned are of order 0.01 cm^{-1}, corresponding to wavelengths
of a metre or so. This is the reason for including a chapter on
microwave and radiofrequency spectroscopy. Apart from this one
chapter, however, we shall be concerned in this book with the
conventional 'optical' region below 100μm.

1.4   Term diagrams and line series

Spectroscopy is concerned with radiative transitions between
stationary states of an atom or molecule. In the conventional
energy level diagram these states are plotted vertically upwards in
order of increasing energy. Fig. 1.2 shows the simplest of all such
diagrams, that of the hydrogen atom with the fine structure
omitted. In this case the stationary states are defined by a single
quantum number n. In optical spectroscopy the numerical value of
the energy in units of cm^{-1} is known as the term value T.

Transitions between two stationary states i and j can be shown
on a term diagram by vertical arrows, pointing up for absorption
of radiation and down for emission. The wave-number of the
transition is simply the difference of the two term values: $\sigma_{ij} = T_i - T_j$. Atomic spectroscopists are in the habit of writing
transitions with the lower state first (regardless of whether they
are looking at emission or absorption lines) while molecular
spectroscopists, confusingly enough, write the upper state first.

The zero of energy is arbitrary because only energy differences are relevant. For the H atom the zero is conventionally defined for the electron and proton at rest at infinite separation. The bound states are then states of negative energy given by the Bohr relation $T = -R_H/n^2$ where R_H is the Rydberg constant for hydrogen ($109\ 678\ \text{cm}^{-1}$) and n is an integer. $n = 1$ is the state of lowest

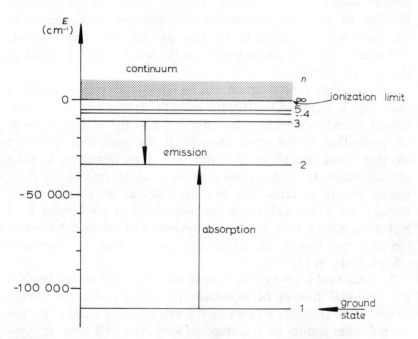

Fig. 1.2 Energy level diagram of the hydrogen atom. Fine structure is omitted. The energy levels are given by $E = -R/n^2$; for the ground state $E = -R$.

energy – i.e., the ground state. As $n \to \infty$ $T \to 0$, and the terms crowd together as they approach the ionization limit, corresponding to the complete removal of the electron. The energy difference between this limit and the ground state, $100\ Rhc$ joule, is the ionization energy of the atom.

For states of positive energy, that is for the system proton + electron + kinetic energy, the energy is no longer quantized, and there exists a continuum of possible states, as shown in Fig. 1.2. Transitions between the continuum and a bound state and transitions within the continuum are possible and are known as free-bound and free-free transitions respectively.

The hydrogen term diagram can equally well be taken to represent the energy of the electron or that of the atom as a whole. In a many-electron atom it is the atom as a whole that matters. One can construct a term diagram by starting with all the electrons in their lowest possible energy states – the ground state of the atom – and then assuming one electron (normally the most weakly bound) to be excited to successively higher energies. Because of the interactions between the electrons, the other electrons will be affected by this process, and the resulting stationary states are characteristic of the whole atom, not just of the excited electron.

The resulting term diagram will be considerably more complex than Fig. 1.2, but will retain the essential features of a series (or several series) of energy levels converging to the ionization limit, corresponding to complete removal of the most loosely bound electron from the atom, the other electrons remaining in their ground states. It is also quite possible to excite one of the more tightly bound electrons or two electrons at once, and obtain thereby an additional set of stationary states converging to a higher ionization limit in which the removal of one electron leaves the ion no longer in its ground state. This is illustrated schematically in Fig. 1.3.

The standard source of data for atomic and ionic energy levels is the National Bureau of Standards publication 'Atomic Energy Levels' [6]. This lists all measured levels of every element for the neutral atom and for several stages of ionization. All terms in these tables are referred to the ground state of the relevant atom or ion as zero so as to avoid inaccuracies from inexact values of ionization potentials. An N.B.S. bibliography [7] covers more recent work. The data for diatomic molecules are to be found in Herzberg's book [8] and in a more complete and more recent compilation 'Spectroscopic Data Relative to Diatomic Molecules' [9].

Transitions between terms may occur in emission or absorption, but emission spectra show many more lines than absorption spectra. In an ordinary discharge tube emission lines may start from any of the large number of levels populated by the excitation process, whereas in an absorbing gas or vapour at a moderate temperature absorption lines arise only from the ground state and possibly a few other levels sufficiently low-lying to have an appreciable population.

The transitions between any one term and a succession of higher terms form a line series converging to some limit. When the lower term is the ground state, the series limit is identical with the ionization limit; all other series connecting normal terms must have a series limit of smaller wavenumber (longer wavelength). This is shown schematically for H in Fig. 1.4.

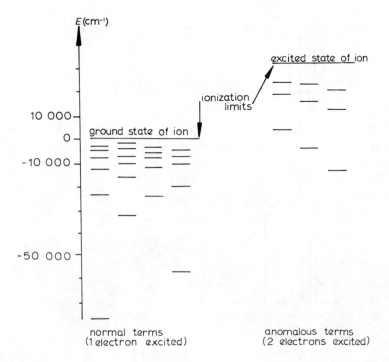

Fig. 1.3 Schematic energy level diagram of a many-electron atom. The diagram shows some of the energy levels of beryllium.

The Lyman series, for which the lower term has $n = 1$, converges to a limit in the far ultra-violet with $\sigma_\infty = R_H$ while the Balmer series, comprising transitions to $n = 2$, has its limit in the near ultra-violet ($\sigma_\infty = R_H/4$).

Beyond the series limit comes a region of continuous absorption or emission arising from the free-bound transitions. The relevant process in absorption (removal of an electron with finite kinetic energy) is known as photo-ionization; the inverse process (recombination of an ion and an electron with emission of radiation) is known as radiative recombination.

There exist a number of compilations of wavelengths making it possible to identify the atom responsible for a line of measured wavelength. The most widely used are probably the National

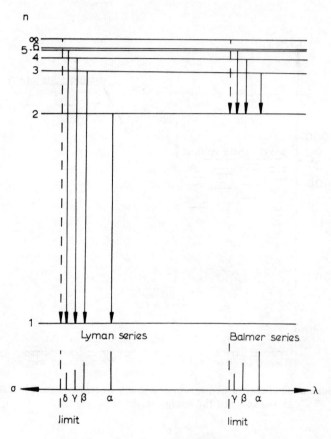

Fig. 1.4 Line series in the spectrum of hydrogen.

Bureau of Standards Multiplet Tables [10] and the M.I.T. Wavelength Tables [11]. Spectra of simple molecules may be identified from the book of Pearse and Gaydon [12].

The fact that energy levels (and the probability of transitions between them) are characteristic of atoms and molecules might appear at first sight to preclude the possibility of deducing anything from an observed spectrum about the environment of the atom or molecule responsible for it. The observed intensity of a line depends, however, on the distribution of population among the various excited states as well as on the intrinsic transition probabilities, and it is also affected by absorption and re-emission by neighbouring atoms. Moreover, the energy levels themselves, and hence the spectral lines, may be broadened, shifted or split by interactions with external magnetic and electric fields or with the micro-fields due to neighbouring particles. It is effects like these that allow one to deduce particle densities, temperatures, pressures and field strengths from precise measurements of line shapes, shifts, widths and intensities.

1.5 Sources, detectors and dispersers

Fig. 1.5 shows schematically the necessary elements of a spectroscopic system for emission (a) and for absorption (b). In emission the source is some form of arc, glow discharge, impulsive discharge, etc., in which the atom under study is excited, generally by electron collisions. In absorption the background source must have a more or less continuous spectrum over the relevant wavelength range, and it is followed by an absorption cell. The disperser, or analyser, spreads the light out spatially according to its wavelength. Finally, the detector is the photographic plate or photoelectric detector or other device that records the incident intensity as a function of position. Only in Fourier transform spectroscopy is the information on wavelengths presented and analysed in an essentially different way.

In the microwave and radiofrequency regions the disperser is redundant, because both the source and the detector can be tuned to any desired wavelength. The development of tunable dye lasers may soon reduce or eliminate the need for a disperser for absorption spectroscopy in the infra-red and visible regions, but meanwhile it is the discussion of dispersers that constitutes the greater part of any text on experimental spectroscopy. A few general points about them are worth making at this stage.

Dispersers may be classified, according to the type of dispersing element they use, as prism instruments (treated in Chapter 4), grating instruments (Chapter 5) and interferometers (Chapter 6). The only interferometers to be discussed in detail are the two types at present in general use for moderate- to high-resolution spectroscopy: the Fabry–Perot and the Michelson, the second of which is the instrument used in the important and relatively new technique of Fourier transform spectroscopy. An alternative classification takes account of the method of use – as spectrographs, scanning spectrometers or monochromators. Etymologically, one should look through a spectroscope, take pictures

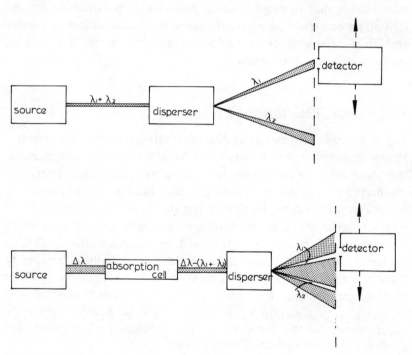

Fig. 1.5 Schematic spectroscopic system for (a) emission and (b) absorption.

with a spectrograph, measure (wavelengths or intensities) with a spectrometer, and isolate light of a given colour with a monochromator. However, the term 'spectroscope', far from being confined to the very narrow visible region, is often used as a generic name for all these instruments, and measurements are made with spectrographs as well as with spectrometers. In other respects these descriptions are correct.

Fig. 1.6 shows the essential features of a prism or grating instrument. Light from the slit S is collimated by the lens L_1 and dispersed by the prism or grating D. The camera lens L_2 forms a set of images of S in its focal plane P in light of different wavelengths, and this constitutes the spectrum. A narrow slit is essential if lines close in wavelength are to be separated. If the instrument is a spectrograph, the photographic plate is placed at P, and many wavelengths are recorded simultaneously. If the instrument is a spectrometer, an exit slit S' is placed somewhere along P with an appropriate detector (photomultiplier, Golay cell, etc.) behind it, and the spectrum is scanned either by tracking the exit slit and detector along P or by rotating the prism or grating. Obviously only one spectral element is recorded at a time and a

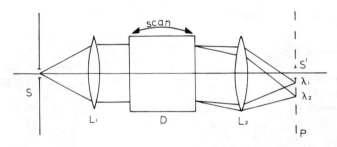

Fig. 1.6 Essential features of prism or grating spectrometer. For explanation, see text.

very accurate drive is needed for high resolution work. In a monochromator the exit slit is fixed, and the prism or grating mounting is chosen so that the emerging beam always comes out at the same angle, whatever its wavelength. A monochromator normally has a fairly wide band pass, so the rotation of the disperser does not have to be as accurately controlled as in a spectrometer. Fig. 1.6 is, of course, highly schematic, and it will be seen in Chapters 4 and 5 that dispersers may be mounted in many different ways. The concave grating, for example, acts as its own collimator and camera lens.

Fig. 1.7 shows similarly the essential elements of interferometric spectroscopy. The lens L_1 collimates the light from the *extended* source S, and L_2 focuses the interference fringes from the interferometer I in the plane P. The important point of distinction is that the 'spectrum' is a superposition of the interference patterns produced by each wavelength in the source, not a juxtaposition of geometric slit images. The interference pattern has circular symmetry, and the width of the fringes is determined by the properties of the interferometer. With the Fabry–Perot used photographically, the whole pattern, or some slice across it, is recorded on the plate at P. In use as a spectrometer, whether Fabry–Perot or Michelson, a circular aperture A isolates the central portion of the interference pattern, and the instrument is scanned by changing the optical path between the interferometer plates.

It should be pointed out that sources, dispersers and detectors are in fact interdependent: a disperser of high resolution may not

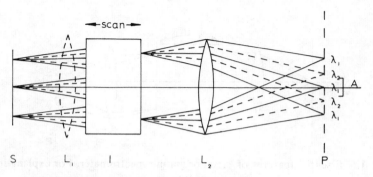

Fig. 1.7 Essential features of interferometric spectrometer. For explanation, see text.

be usable in practice if the source is weak and the detector insensitive. On the other hand, the high resolution may be wasted if the lines are intrinsically very broad or are photographed on a very grainy plate. This dichotomy of intensity and resolution is a constantly recurring one in spectroscopy, and it is important to think about the balance in any experiment. The far infra-red and to some extent the far ultra-violet are particularly crucial regions in the respect.

1.6 Survey of spectral regions

Let us, illogically but chronologically, start in the visible region and proceed outwards. If in fact we extend the 'visible' a little into both the infra-red and the ultra-violet we have a fairly well defined spectral region from 10 000 Å (1 μm) to 2000 Å which is by far the easiest to work in. Quartz is transparent throughout this region, and glass through most of it, so that there is a choice between prisms, gratings and interferometers as dispersers and there is no difficulty about windows or lenses. The spectrum can be recorded photographically or photoelectrically. A wide range of sources is available for both emission and absorption spectroscopy.

Going below 2000 Å in wavelength, first air (or oxygen, to be precise) and then quartz start to absorb. To overcome the first, the light path has to be evacuated – hence the name vacuum ultra-violet for this region. Quartz optics can be replaced by other crystals to extend the transmission range for a few hundred angstroms (as far as 1040 Å, the lithium fluoride limit), but only at low resolution. Interferometers run into additional problems with surface tolerances and low reflectivity. Below about 1800 Å gratings are the only available high resolution dispersers. Lenses and mirrors (which have low reflectivity in this region) are eliminated by using concave gratings. Below about 400 Å the gratings have to be used at grazing incidence to overcome the low reflectivity. On the other hand, detectors are no problem: either photographic or photo-electric methods may be used throughout the ultra-violet. The problems of running suitable light sources may become acute in the region below 1040 Å where windows cannot be used to contain or isolate different gases. The regions to the long and short wavelength side of 1040 Å are often called the Schumann and Lyman regions respectively, after the men who first explored them.

Going now the other way, into the infra-red, we find that the choice between prisms, gratings and interferometers remains open up to about 40 μm, the effective limit of crystal transmission. Interferometers made of thin films such as polythene can be used at still longer wavelengths, so that Fourier transform spectrometry is in competition with grating spectrometry throughout the infra-red region. With the advent of tunable dye lasers, spectroscopy without either gratings or interferometers becomes feasible in certain cases. Mirrors are used in preference to lenses through most of the infra-red; since reflection coefficients are high, several mirrors can be used without significant loss of intensity. The principal problem through most of the region is that of inadequate intensity. Most sources radiate relatively little energy in the infra-red, and detectors suffer from severe noise problems. It is often necessary to sacrifice resolution to obtain an adequate signal/noise ratio. Spectroscopy in the infra-red is usually done in absorption because of the lack of line sources of sufficient intensity. On the other hand, the vacuum requirements in the infra-red are not severe, because dry oxygen and nitrogen do not absorb, and it is only necessary to eliminate water vapour and carbon dioxide.

At wavelengths of a few tenths of a millimetre the infra-red overlaps with the microwave region, and there is an abrupt change of technique. Selective sources and detectors become easily available, first in the form of submillimetre masers at particular wavelengths and then as tunable klystron oscillators. Dispersers become entirely redundant, and absorption spectroscopy consists simply of observing the variation in signal as the source and detector together are swept over the required wavelength range.

Radiofrequency spectroscopy falls into two rather distinct categories. On one hand it is simply an extension of microwave spectroscopy to longer wavelengths; on the other hand it incorporates the numerous resonance methods that have been developed to study transitions between magnetic and/or hyperfine structure sub-levels. In these methods the transitions, although induced by the rf field, are usually detected not by the absorption of energy but by some other method, such as deflection due to a change in spin direction or a change in the polarization of resonance radiation.

1.7 Wavelength standards

The aim of early spectroscopists was to measure accurately the wavelengths of the lines investigated. Nowadays much of spectroscopy is concerned with line shapes, widths, splittings, and so on, in which absolute wavelength measurements are not necessary, but the establishment and maintenance of absolute standards is still an important part of spectroscopy.

Until 1960 the ultimate standard of length was the standard metre, the distance between two scratches on a platinum-iridium bar kept in Paris. There were obvious drawbacks to such a standard. The first was geographical and became evident as early as 1884, when it was shown that the great chart of absolute wavelengths compiled by Ångström some 16 years earlier in terms of the Uppsala metre would have to be corrected by 13 parts in

10^5, the discrepancy between the two metre bars. The second disadvantage became apparent with the development of high resolution spectroscopy at the end of the last century; the accuracy of wavelength measurement came to be limited not by the spectroscopic techniques but rather by the finite width of the scratches on the bar. To remedy this it was decided to adopt as an international standard of *wavelength* the red line of cadmium, emitted from a source in certain specified conditions. Once this had been determined as accurately as possible in terms of the Paris metre, all other wavelengths could be measured relative to it. But the standardization of the red cadmium line revealed the third disadvantage of the Paris metre: measurements over a period of years showed the bar to be actually shrinking slowly, presumably as the result of metal fatigue. (Similar measurements on the Imperial Yard in London revealed an even worse shrinkage.) All these difficulties led to the decision to adopt a wavelength as the ultimate standard of length. The red cadmium line was no longer the best choice, because line broadening arising from unresolved isotope shift in lines more suitable in other respects could now be eliminated by using separated isotopes. After some years of controversy over the relative merits of the green line of mercury and the orange line of krypton, the latter was chosen in 1960. The international standard of length is now defined to one part in 10^8 by taking one metre to be 1650 763·73 vacuum wavelengths of this line, emitted by the isotope Kr^{86} under certain specified conditions – i.e., $\lambda_{vac} = 6057\cdot80210 \times 10^{-10}$ m.

As a matter of historical interest, the mercury green line which was the runner-up in the contest was one of the lines investigated as a possible standard by Michelson in the 1890s, using visibility measurements with his own interferometer. He rejected it in favour of the red cadmium line because of its structure, which we now recognize as hyperfine structure and can eliminate by using a single even isotope of mercury.

It may be noted in passing that the fundamental standard of time or frequency is now also an atomic standard. It was agreed in 1967 that the reciprocal second, or Hz, should be defined as 9 192 631 770 times the frequency of the transition between the hyperfine structure levels of the ground state of Cs^{133}, a measurement that can be made to $1:10^{12}$. We shall meet this so-called atomic clock again in Chapter 7. In principle, since

frequency and wavelength are related by the velocity of light, only
two out of three of these quantities can be separately defined.

A large number of spectral lines, measured interferometrically
against the primary standard in different laboratories, have been
accepted as secondary standards. They are mostly lines of the rare
gases, with certain lines of iron, thorium, and so on. The
secondary standards are up-dated and added to every few years by
a commission of the International Astronomical Union [13]. For
most practical purposes the numerous tertiary standards now
available are quite accurate enough. Standards in the far ultra-
violet and infra-red regions are, however, far less accurate because
of the large number of intermediate steps from the primary
standard, as well as the difficulties of producing narrow lines and
measuring them precisely. The lack of good standards in the
vacuum ultra-violet is particularly inconvenient because they are
required for precise measurements of series limits and for the
analysis of iso-electronic sequences. It is often necessary to use
overlapping orders in a grating spectrum for the purpose, despite
the systematic errors to which this method is prone. Standards
established by the Ritz combination principle are more reliable:
the energy difference between two levels of a line in the vacuum
ultra-violet ν_{41} may be calculated by measuring accurate values
for the visible transitions $\nu_{43}, \nu_{32}, \nu_{21}$ [14, 15].

In addition to wavelength standards there is a need, particularly
in spectrophysics, for intensity standards. These will be discussed
in Chapter 3.

1.8   Units used in this book

I have used S.I. units in the belief that most physics courses are
now taught in these units. All formulae and expressions are in the
form appropriate to S.I. units, but I have not stuck rigorously to
them in the text when other units are more convenient or more
widely used in practice. Practising spectroscopists, as already
remarked, express energy differences in cm^{-1} (or Hz in the rf
region); the joule is an enormously large unit on the atomic scale
and even the eV is inconveniently large except in the far
ultra-violet. For wavelengths most spectroscopists still use Å in
preference to nm, and it is certainly much less confusing to
express the reciprocal dispersion of a spectrograph in Å/mm rather

than in nm mm^{-1}. In the literature of plasma physics and astrophysics number densities are still given in cm^{-3} rather than in m^{-3}. I have in fact often quoted number densities in both units, not because I believe the reader to be incapable of multiplying by 10^6 but because I think it useful to get a feeling for the orders of magnitude met in practice. For the same reason, although the magnetic induction B appears in weber m^{-2}, I have occasionally also quoted values in gauss for the benefit of those accustomed to c.g.s. units. Where possible I have tried to keep $(e^2/4\pi\epsilon_0)$ bracketed as a unit, so that anyone working in c.g.s. units can simply replace it by e^2. In any case, the conversion can easily be made simply by replacing ϵ_0 wherever it appears by $1/4\pi$. Expressions for the Rydberg constant, the first Bohr radius, the Bohr magneton, etc., are to be found in c.g.s. units as well as S.I. units in the Appendix.

References

General

1. Harrison, G. R., Lord, R. C. and Loofbourow, J. R. 'Practical Spectroscopy', Prentice-Hall, 1948
2. Kuhn, H. G. 'Atomic Spectra', Longmans, 1969
3. Sawyer, R. A. 'Experimental Spectroscopy', Dover, 1961
4. Walker, S. and Straw, H. 'Spectroscopy', Vols. 1 and 2, Chapman and Hall, 1961/2

Wavelength conversion

5. Coleman, C. D., Bozmann, W. R. and Meggers, W. F. 'Table of Wavenumbers', N.B.S. Monograph no. 3, 1960

Energy levels and wavelengths

6. Moore, C. E. 'Atomic Energy Levels', N.B.S. Circ. 467, Vols. I, II, III, 1949-58
7. Hagan, L. and Martin, W. C. 'Bibliography on Atomic Energy Levels and Spectra', N.B.S. Special Publication 363, 1972
8. Herzberg, G. 'Spectra of Diatomic Molecules', Van Nostrand, 1950
9. Rosen, B. 'Spectroscopic Data relative to Diatomic Molecules', Vol. 17 of 'International Tables of Selected Constants', Pergamon, 1970
10. Moore, C. E. N.B.S. Multiplet Tables: 'Multiplets of Astrophysical Interest', N.B.S. Tech. Note 36, 1959; 'Ultra-violet Multiplet Tables', Circ. 488, 1950-62
11. Harrison, G. R. 'M.I.T. Wavelength Tables', Wiley, 1959
12. Pearse, R. W. B. and Gaydon, A. G. 'The Identification of Molecular Spectra', Chapman and Hall, 1963

Wavelength standards

13. *Transactions of the International Astronomical Union,* **12A**, 137, 1964, and earlier volumes
14. Samson, J. R. 'Techniques of Vacuum Ultra-violet Spectroscopy', Wiley, 1967
15. Edlen, B. Wavelength Measurements in the Vacuum Ultra-violet, *Rep. Prog. Phys.* **26**, 181, 1963

CHAPTER TWO

Summary of atomic and molecular structure

Before discussing the methods of spectroscopy, it is worth summarizing the main features of atomic and molecular spectra in order to see what effects one is likely to be investigating in any given spectral region. Some understanding of structure and of the terminology used to describe it is also a necessary prerequisite to the later chapters on spectrophysics. Any one-chapter survey is bound to be superficial, but should serve at least to define some of the terms used in later chapters. A selection of books covering atomic and molecular structure in proper detail, at varying degrees of rigour, is given at the end of the chapter.

The schematic spectra below diagrams illustrating multiplet structure, Zeeman splitting, etc., are intended to show the relative spacings, but not the relative intensities, of the components.

2.1 The hydrogen atom

It was stated in Chapter 1 that the energy levels of the hydrogen atom, neglecting fine structure, are given by the simple relation $T = -R_H/n^2$ where R_H is a constant and n a positive integer. Bohr derived this relation by assuming the electron to follow a circular orbit around the nucleus, obeying the laws of classical mechanics with two exceptions: the orbital angular momentum must be an integral multiple of $\hbar$ $(=h/2\pi)$, and in such an allowed orbit the electron does not radiate. It is easily shown that for this system the energy of a state having angular momentum $n\hbar$ is

$$E = -\frac{\mu e^4 Z^2}{2\hbar^2 n^2 (4\pi\epsilon_0)^2},$$
(2.1)

23

where μ is the reduced mass of the electron

$$\frac{mM_{\mathrm{H}}}{m + M_{\mathrm{H}}},$$

Z is the charge on the nucleus (unity for the H atom), and the constants are in S.I. units, giving E in joules; dividing by $100ch$ to convert to wavenumbers (Section 1.3), one obtains for the Rydberg constant

$$R = \frac{2\pi^2 \mu e^4}{100ch^3 (4\pi\epsilon_0)^2} \ \mathrm{cm}^{-1} \qquad (2.2)$$

which is numerically equal to 109 678 cm^{-1} for hydrogen. For an infinitely heavy nucleus $\mu \to m$, and $R = 109\ 737{\cdot}31$ cm^{-1}.

The radii of the Bohr orbits are given by

$$a_n = \frac{\epsilon_0 h^2 n^2}{\pi\mu e^2 Z} \equiv \frac{n^2}{Z} a_0 \qquad (2.3)$$

where a_0, the radius of the first Bohr orbit, is treated as the 'atomic unit' of length. It is numerically equal to 0·529 Å, or 5·29 x 10^{-11} m.

Rather than discuss the modifications and additions to Bohr's theory, we shall move straight on to the wave mechanical theory. The stationary states of the atom are described by Schrödinger's time-independent equation

$$\frac{\hbar^2}{2\mu} \nabla^2 \psi + (E - V)\psi = 0 \qquad (2.4)$$

(see, for example, [1], [2], [3]). Here ∇^2 is the Laplacian operator and V is the potential energy of the electron. ψ, the wave function, is to be interpreted as a 'probability amplitude' such that $\psi^2 (r, \theta, \varphi)$ – or, more precisely, $\psi^* \psi$ – is the probability of finding the electron at the point in space defined by the co-ordinates r, θ, φ (Fig. 2.1). The nucleus is taken as the origin of coordinates. If V is a function of r only, as in the case of the hydrogen atom for which $V = -Ze^2 (4\pi\epsilon_0 r)$, the equation can be separated into three independent differential equations by writing

$$\psi(r, \theta, \varphi) \equiv R(r)\, \Theta(\theta)\, \Phi(\varphi) \qquad (2.5)$$

These three equations can then be solved for R, Θ, and Φ respectively. Certain restrictions are imposed on the mathematically possible solutions in order to make ψ behave in a physically sensible fashion: it must be single valued, continuous and finite everywhere. These restrictions limit the possible values of the energy E to the Bohr values $-RZ^2/n^2$ when E is negative. (For positive E the values are unrestricted.) For a given n they also limit the orbital angular momentum to the values $\sqrt{[l(l+1)]}\hbar$, where l, the orbital angular momentum quantum number, can take any of

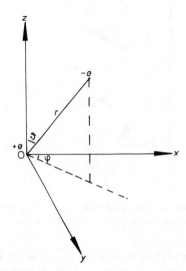

Fig. 2.1 Spherical polar co-ordinates.

the n integral values from 0 to $n-1$. Finally, the component of the angular momentum in a fixed direction in space (conventionally, the z-axis) is restricted to the values $m_l \hbar$ where m_l may take any integral value between $+l$ and $-l$ ($2l+1$ values in all). It should be made clear that, although the orbital angular momentum *vector* is written as $l\hbar$, its actual magnitude is $\sqrt{[l(l+1)]}\hbar$, sometimes written $l^*\hbar$ as short-hand. For large l obviously $l^* \to l$. Since the maximum component of $l\hbar$ along the z-axis is only $l\hbar$, not $l^*\hbar$, l can never point exactly in the z-direction. This is illustrated in Fig. 2.2 for $l = 2$.

The wave function ψ depends on all three of the quantum numbers n, l, m_l – that is, for fixed n each l, m_l combination describes a different electron distribution (on the old Bohr theory, a different shape and orientation of electron orbit). More specifically, $\Phi(\varphi)$ depends only on m_l, $\Theta(\theta)$ depends on l and m_l, and $R(r)$ depends on n and l. Φ has the explicit form $\Phi = e^{-im\varphi}$, so $\Phi^*\Phi$ is always 1; consequently, the electron distribution has

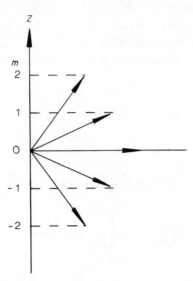

Fig. 2.2 Relation between the quantum numbers l and m_l. $l = 2$ in this diagram, and its magnitude is therefore $\sqrt{6}$ or 2·45.

cylindrical symmetry about the z-axis. The wave function for $l = 0$ has two important characteristics: first, $\Theta \equiv \Theta^* = 1$, so that the electron distribution is spherically symmetric, and, secondly, $R_{n,0}$ does not vanish at the origin, so that there is a finite probability of finding the electron near the nucleus.

2.2 Degeneracy and spin

For a given n we have seen that there are n possible values of the quantum number l, and for each of these there are $2l + 1$ possible values of m_l and hence $2l + 1$ different electron probability distributions. Since the energy depends only on n, all of these have the same energy and are said to be degenerate. The degeneracy of the state n is

$$\sum_{l=0}^{n-1} (2l + 1) = n^2 .$$

The m_l degeneracy is easily understood, for in the absence of an external field there is no reason for singling out any one direction in space. The l degeneracy is a direct result of the Coulomb potential and is peculiar to hydrogen.

A fourth quantum number is required to specify the state of the electron completely. Certain effects in atomic spectra – notably, fine structure and Zeeman effect – can be explained by attributing to the electron an intrinsic spin angular momentum $s\hbar$, of magnitude $\sqrt{[s(s + 1)]}\hbar$ (or $s^{\star}\hbar$), where s always has the value $\frac{1}{2}$. The component of spin in the z-direction, m_s, is also quantized: m_s can take only the values $+\frac{1}{2}$ (spin 'up') or $-\frac{1}{2}$ (spin 'down'). m_s is the fourth quantum number. In the absence of an external field the energy does not depend on m_s so that (still ignoring fine structure) the energy levels of the H atom are in fact $2n^2$-fold degenerate. Electron spin is a natural consequence of the full relativistic treatment of the electron first formulated by Dirac, but not of the Schrödinger treatment. It can be included formally in the wave function by writing the latter as $\psi(r, \theta, \varphi, s)$. When spin-orbit interaction is neglected this can be re-written as the product of the spatial function $\psi(r, \theta, \varphi)$ and a spin function $\sigma(s)$.

$$\text{i.e. } \psi(r, \theta, \varphi, s) \approx \psi(r, \theta, \varphi)\sigma(s) \qquad (2.6)$$

2.3 Many-electron atoms: the central field approximation

The exact form of Schrödinger's equation for an atom with more than one electron includes in the potential energy V not only the attractive interaction between each electron and the nucleus, $-(Ze^2)/(4\pi\epsilon_0 r_i)$, but also the repulsive interactions $+(e^2)/(4\pi\epsilon_0 r_{ij})$ between each pair of electrons. These cross-terms make an exact solution impossible, and some form of approximation is essential. As an obvious first approximation one assumes each electron to move independently in some field whose potential V represents the combined effect of the nucleus and all the other electrons. If, further, one takes a spherical average over the latter, so that V is a function of r only and not of angle, one arrives at the central field approximation. This has the enormous advantage that the variables may be separated as in the case of hydrogen. Moreover, the Θ and Φ equations are identical with those of hydrogen, since V does not appear in them. Only the R equation is different: $V(r)$ no longer has the simple Coulomb form $-Ze^2/4\pi\epsilon_0 r$. An electron moving in a central field may therefore still be described by the quantum numbers n, l, m_l (to which, as with hydrogen, we tack on m_s). Its energy, however, is determined by the form of $V(r)$ and now depends on l as well as n. Since everything is still spherically symmetric there is no reason, to this approximation, for the energy to depend on m_l or m_s.

2.4 The Pauli exclusion principle and shell structure

Were it not for the Pauli Exclusion Principle, all the electrons of every atom in its ground state would have $n = 1$, corresponding to minimum energy, and the table of the elements would not be a periodic table. The exclusion principle forbids any two electrons to have all four of their quantum numbers identical. This limits the number of electrons that can have the same principal quantum number n. For each of the n possible values of l there are $2l + 1$ different values of m_l, and for each of these two possible values of m_s, totalling $2n^2$, as has already been seen. Table 2.1 shows the vacancies available for the first three values of n. Each cell in the penultimate line may contain two, and only two, electrons. The nomenclature in the bottom line should be noted: electrons having $l = 0, 1, 2, 3$ are known as s, p, d, f electrons respectively. $2p$, for example, denotes an electron having $n = 2, l = 1$. The origin of this nomenclature is purely historical, deriving from the sharp, principal, diffuse and fundamental series of lines in the spectra of the alkali metals. It is unfortunate in that the s has nothing whatever to do with the spin angular momentum vector s. After f the order becomes alphabetical.

Table 2.1

n	1	2			3								
l	0	0	1		0	1			2				
m_l	0	0	+1	0	−1	0	+1	0	−1	+2	+1	0	−1	−2
m_s	↑↓	↑↓	↑↓	↑↓	↑↓	↑↓	↑↓	↑↓	↑↓	↑↓	↑↓	↑↓	↑↓	↑↓
Nomen-clature	1s	2s	2p			3s	3p			3d				

In the Bohr theory of the hydrogen atom the radius of the electron orbit increases as n^2. In wave mechanics one is concerned with a probability distribution rather than an orbit, but the expectation value for r still depends to a first approximation on n^2. All electrons with the same n may therefore be regarded as forming a sort of 'shell' around the nucleus. This concept has even greater validity when applied to the 'sub-shells' formed by electrons having identical values of n *and* l. There are $2(2l + 1)$ electrons in each sub-shell, as seen from the table. A full sub-shell has several important features. First, the pairing of positive and negative values of m_l and m_s means that there is no resultant angular momentum. Secondly, the wave functions are such that the resultant electron distribution is spherically symmetric. Thirdly, the structure is a very stable one. This last follows from the fact that all the electrons in the sub-shell have the same radial charge distribution (the function $R(r)$ depends only on n and l, not on m_l). Consequently the electrons screen each other only partially from the nuclear charge. As one goes along a row of the periodic table, adding one electron and one proton per element, the effective positive charge increases as the sub-shell fills, and the electrons become more strongly bound. This is shown experimentally by the steady increase in ionization potential up to that of the inert gas that corresponds to the filling of the last vacancy.

Closed shells and sub-shells may therefore be thought of as forming an inert 'core' surrounding the nucleus, and the behaviour of an atom is determined mainly by the electrons in unfilled sub-shells, the valence electrons. The core screens most of the nuclear charge from the valence electrons, so long as the latter do not have an appreciable probability of being found within it. It is this spherically symmetric core that justifies the central field approximation. That part of the interaction between the valence electrons that is not included in the central field will be discussed later. Insofar as it can be neglected, the energy of the atom is determined solely by the n and l values of each of the valence electrons – that is, by the electron configuration. In describing the configuration, the number of electrons in each sub-shell is shown by a superscript. For example, four electrons in the $2p$ sub-shell ($n = 2$, $l = 1$) is written as $2p^4$. If one of these is excited to a $3s$ state the configuration becomes $2p^3 3s$.

2.5   Features of the central field model

The assumption that each electron moves independently in a central field allows one to use hydrogen-like wave functions for the angular part of each wave function, $\Theta(\theta)\Phi(\varphi)$. The radial part $R(r)$ depends on the form of the potential $V(r)$ for each electron, and this in turn depends on the details of the radial electron charge distribution and hence on $R(r)$ for all the other electrons.

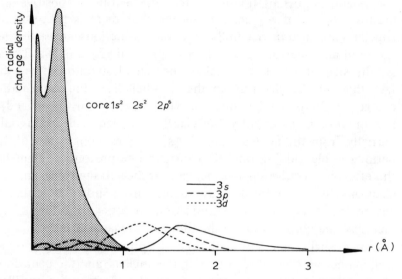

Fig. 2.3 Dependence of radial charge density on l. The radial charge density, $4\pi r^2 R^2$, is plotted against r for $n = 3, l = 0, 1, 2$.

This vicious circle is usually tackled by the self-consistent field approach. One guesses a plausible wave function for each electron, computes the corresponding $V(r)$ from Poisson's equation, and solves Schrödinger's equation to obtain a new set of R's. The cycle is repeated until V and R become consistent. Without going into these calculations one can still understand why the energy should depend on l as well as n, which is the most important distinction between the central field atom and the hydrogen atom. Regardless of the exact form of V, the behaviour of the radial wave functions is such that only an s-electron ($l = 0$) has a finite probability of being found close to the nucleus. Fig. 2.3 shows how the radial probability depends on l in a typical case for $n = 3$, $l = 0$, 1, 2. The probability distribution for the core electrons, $n = 1$ and $n = 2$, is also shown on the diagram. It is the s-electrons that have the greatest probability of penetrating the core; they therefore 'see' on average a higher positive charge and so are more closely bound. Electrons with the maximum value of l, d-electrons in this case, are mostly well outside the core and have almost hydrogen-like

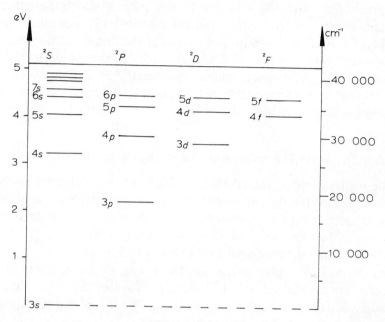

Fig. 2.4 Dependence of energy levels on l. The energy level diagram is that of sodium with fine structure omitted.

energies and wave functions. Fig. 2.4 shows a typical form of energy level diagram for a single valence electron. The different l-values are conventionally displaced sideways in order of increasing l.

Since the field acting on a given electron in the central field model is obtained by taking a smeared-out spherical average over the charge distributions of all other electrons, the model obviously works best when there is only one valence electron, as in the alkali metals. Atoms with two or more valence electrons are found to have several different possible energy states, or terms, for any given electron configuration. To account for these one must look more specifically at the direct electrostatic repulsion between the valence electrons. There is also another interaction that has so far been ignored, in hydrogen as well as in the central field model: this is the magnetic interaction between the orbital angular momentum and the intrinsic electron spin angular momentum. In considering these effects one usually starts with the electrostatic interaction because in many cases of practical interest it is the greater of the two. We may justify the order in which these three interactions – central field, residual electrostatic, magnetic – are taken by quoting values typical of one of the first excited states of a simple atom: central field energy ~ 5 eV, residual electrostatic energy ~ 1- 2 eV, magnetic energy from $\sim 10^{-3}$ eV in light atoms to $\sim 10^{-1}$ eV in medium atoms (often comparable with electrostatic energy in heavy atoms).

2.6   Electrostatic interactions: the quantum numbers L and S

The central field potential $V(r)$ in which any outer electron moves comprises not only the contributions of the nucleus and core electrons but also the greater part of the potential due to other electrons in the same sub-shell. The residual electrostatic inter-action arising from departures from spherical symmetry in the distribution of these other electrons has to be applied as a correction to the central field energy. Provided the correction is small, which is often the case, it can be calculated from the unperturbed central field wave functions.

Let us see what happens with two electrons. The magnitude of the electrostatic repulsion between them depends on an integral of the form $\int \psi^2(1)\,(e^2/4\pi\epsilon_0 r_{12})\,\psi^2(2)\,\mathrm{d}\tau_1\,\mathrm{d}\tau_2$ where $\psi(1)$ and $\psi(2)$ are the wave functions of the two electrons and the integral is taken over all space. For a given electron configuration $(n,\,l)_1$ and $(n,\,l)_2$ are fixed, by definition, but the value of the integral is different for different values of $(m_l)_1$, $(m_l)_2$ – that is, for the $(2l_2 + 1)$ possible angular distributions of the second electron relative to the first. In other words, the interaction depends on the relative orientation of the l vectors and hence on the magnitude of the resultant orbital angular momentum $L\hbar$ formed by the vector addition $L = l_1 + l_2$ (see Fig. 2.5). L acts as an additional quantum number for the two-electron system, taking integral values $l_1 + l_2$,

$l_1 + l_2 - 1, \cdots |l_1 - l_2|$. As with l, the actual magnitude of the total orbital angular momentum is $\sqrt{[L(L + 1)]}\hbar$, or $L^*\hbar$. Its component in the z-direction, M_L, can take the $2L + 1$ integral values from $+L$ to $-L$. States with different L-values are called terms and are labelled with the capital letters S, P, D, $\cdots$ for $L = 0, 1, 2, \cdots$ by analogy with the single electron terminology. For example, two p electrons can give rise to S, P and D terms ($L = 0, 1, 2$). If one of the electrons is an s-electron ($l_1 = 0$), then $L \equiv l_2$, and there is no splitting into terms of different L value.

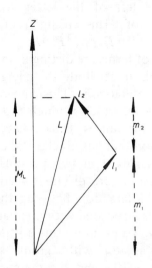

Fig. 2.5 Vector model for the quantum number L.

This is consistent with the spherical symmetry of the charge distribution of an s-electron.

In the case of three or more electrons, the total orbital angular momentum is defined in the same way by $L = l_1 + l_2 + l_3 + \cdots$ and may in general take any integral value from the maximum, $|l_1| + |l_2| + |l_3| + \cdots$, down to zero.

The splitting of a configuration into different terms in accordance with the relative orientation of the ls is essentially a classical effect, although to evaluate the energies one must treat the electrons as smeared-out charge distributions rather than point charges. It is found, however, that a given configuration splits into more terms than can be accounted for by the range of possible

L-values. In particular, an atom with two valence electrons has two distinct sets of terms, called singlets and triplets from the way in which they behave when magnetic interactions are taken into account. These two sets are distinguished by an additional quantum number S, defined by $S = s_1 + s_2$ and having the two possible values $S = 1$ and $S = 0$. $S\hbar$ is the total spin angular momentum vector of the system, of magnitude $\sqrt{[S(S + 1)]}\hbar$ (or $S^{\star}\hbar$). The multiplicity of a term is given by $2S + 1$, for reasons that will appear in the next section, so $S = 1$ corresponds to the triplet terms and $S = 0$ to the singlet. The multiplicity is shown as a superscript to the left of the letter giving the *L*-value; for example, the full list of terms arising from the configuration of two *p*-electrons is $^1S, \, ^1P, \, ^1D, \, ^3S, \, ^3P, \, ^3D$.

Since the two sets of terms are distinguished by their *S*-values, it might seem reasonable to attribute the energy difference between them to spin-spin interaction. This, however, is a magnetic dipole-dipole interaction, which is much too weak to account for the observed energy differences. The explanation is to be found in an exchange interaction that has no classical counterpart. Its effect is that the system can exist in two possible states in which the motions of the electrons are differently correlated, so that the average electrostatic interaction between them is also different. Each of these states is associated with one particular value of S for reasons of symmetry. To examine this a little further, let us take the particular case of an atom with a ground-state configuration s^2 (helium, the alkaline earths and the mercury group). When one electron is raised to a state of arbitrary l, the excited state configuration is sl, which has only one possible value of L, $L = l$. To be specific, suppose $l = 1$, so that we are concerned with the ^3P and ^1P terms arising from the configuration sp. In the zero-order approximation, in which the non-central part of the interaction between the two electrons is neglected, the wave function for the system can be written as the product of the two independent electron wave functions $\psi_s(1) \, \psi_p(2)$. But this wave function is not physically satisfactory, because if we interchange the electrons we get a different wave function (and hence a different electron distribution) $\psi_p(1) \, \psi_s(2)$ for an identical system – identical because the electrons are indistinguishable. A physically acceptable wave function should change either not at all or only in sign when the electrons are interchanged, the change in sign being permissible because it is the square of the wave function that determines the electron distribution. The two linear combinations

$$\psi_+ = \psi_s(1)\,\psi_p(2) + \psi_p(1)\,\psi_s(2)$$

$$\psi_- = \psi_s(1)\,\psi_p(2) - \psi_p(1)\,\psi_s(2)$$

(2.7)

satisfy this requirement: the first is symmetric and does not change, and the second is anti-symmetric and changes sign. Both may be thought of as describing a situation in which the excitation energy is shared between the two electrons. When one comes to evaluate the electrostatic repulsive energy

$$\int \frac{e^2}{4\pi\epsilon_0 r_{12}}\,\psi_+^2\,d\tau_1\,d\tau_2$$

and

$$\int \frac{e^2}{4\pi\epsilon_0 r_{12}}\,\psi_-^2\,d\tau_1\,d\tau_2$$

it is easily seen that in addition to straightforward Coulomb integrals of the form

$$\int \frac{e^2}{4\pi\epsilon_0 r_{12}}\,\psi_s^2(1)\psi_p^2(2)\,d\tau_1\,d\tau_2,$$

such as we have already discussed, there must be cross-terms of the form

$$\int \frac{e^2}{4\pi\epsilon_0 r_{12}}\,\psi_s(1)\psi_p(2)\psi_p(1)\psi_s(2)\,d\tau_1\,d\tau_2$$

which have no classical interpretation. These are known as exchange integrals. Setting the Coulomb integral equal to J and the exchange integral to K, one finds for the interaction energies of the ψ_+ and the ψ_- state respectively:

$$E_+ = J + K$$

$$E_- = J - K$$

(2.8)

The exchange interaction therefore splits the configuration into two terms differing in energy by $2K$. Both J and K are necessarily positive, so the ψ_+ state lies above the ψ_- state. One can interpret this as a correlation in the electron motions. In the ψ_- state the electrons are on average further apart. This can be proved explicitly, but it is obvious from inspection of ψ_- that the probability of finding them close together in this state is vanishingly small: $\psi_- \to 0$ as co-ordinates (1) $\to$ co-ordinates (2).

It remains to show the connection of the spin with all this. The Pauli principle can be re-stated in the form that the total wave

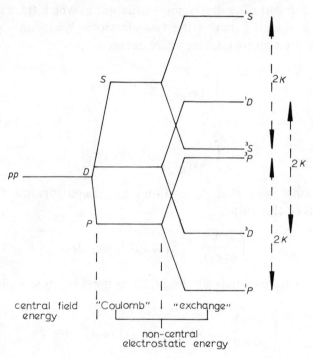

Fig. 2.6 Terms arising from the configuration *pp*.

function must be antisymmetric. To the extent that spin-orbit interaction is a second-order effect, the total wave function can be written as the product of an orbital part and a spin part $\psi\sigma$ (Equation 2.6). The Pauli principle requires an antisymmetric spin function σ_- for ψ_+ and a symmetric spin function σ_+ for ψ_-. Table 2.2 shows schematically how the four possible spin combinations give rise to one antisymmetric and three symmetric spin functions. The pairs (b) and (c) of column 1 have to be combined as in column 2 to form spin functions that are invariant or simply change sign when the electrons are interchanged. Column 3 gives the symmetry of each function for this interchange: clearly (c) is antisymmetric and the rest symmetric. The component of total spin in the z-direction, M_S of column 4, is obvious from inspection if the upward-pointing arrows are taken to define the positive z-direction. The three σ_+ functions can be attributed to the three possible orientations of a total spin $S = 1$ and must be combined with the antisymmetric orbital function; ψ_- therefore represents the triplet term. The one σ_- function is associated with $S = 0$ and is combined with ψ_+ to give the singlet term. In this approximation the ^3P term is triply degenerate, because in the absence of

Table 2.2

1	2	3	4	5	6
Elect Spins	Spin function	Symmetry	M_s	S	Orbital function
(a) ↑↑	↑↑	σ_+	$+1$		
(b) ↑↓	↑↓ + ↓↑	σ_+	0	1 (triplets) ψ_-	
(c) ↓↑	↑↓ − ↓↑	σ_-	0	0 (singlets) ψ_+	
(d) ↓↓	↓↓	σ_+	-1		

some field there is nothing to distinguish the three M_S values. The degeneracy is removed when the spin-orbit interaction is introduced as will be seen in the next section.

Fig. 2.6 summarizes the results of this section for a pp configuration. The Coulomb J-integral has different values for each of the possible L-values representing S, P and D terms. Each of these terms is split by the appropriate K-integral (which in this particular case has the same value in each term) into a singlet and a triplet term. It may be noted that the singlet-triplet order in the P-term is the reverse of that found for the sp configuration (Equation 2.8). This follows from evaluation of the integrals and can best be explained by the fact that 36 different $(m_l, m_s)_1$ $(m_l, m_s)_2$ combinations go to form 6 terms. The energy of each term is a linear combination of several different integrals and simple correlation arguments are inapplicable. In the case of equivalent electrons – that is, electrons having the same values of n and l – some terms are forbidden by the Pauli principle. p^2, for example, gives rise only to ^3P, ^1S and ^1D.

When there are three or more valence electrons, S is defined by $S = s_1 + s_2 + s_3 + \cdots$ and takes integral or half-integral values according to whether the number of electrons is even or odd. An even number of electrons leads to odd multiplicity – e.g., singlets, triplets and quintets for four electrons ($S = 0, 1, 2$); and an odd number leads to even multiplicity – e.g., doublets and quartets for three electrons ($S = \frac{1}{2}, \frac{3}{2}$). As in the case of two electrons, the terms that can be formed from *equivalent* electrons are limited by the Pauli principle.

2.7   Magnetic interactions: the quantum numbers j and J

The electron has an intrinsic magnetic moment μ_s by virtue of its charge and spin angular momentum $s\hbar$. The classical value of the gyromagnetic ratio would suggest the relation

$$\mu_s = -\frac{e}{2m}\, s\hbar,$$

but the experimental value for μ_s, confirmed theoretically by Dirac's relativistic treatment, is twice this:

$$\mu_s = -g_s\mu_B s \tag{2.9}$$

where

$$\mu_B = \frac{e\hbar}{2m}$$

μ_B is the Bohr magneton and is numerically equal to

$$0.927 \times 10^{-23} \begin{cases} \text{J per (wb m}^{-2}) \\ \text{amp m}^2 \end{cases}$$

or 0.467 cm^{-1} per (wb m^{-2}). g_s can be taken as exactly 2 for most purposes, although recent very accurate experiments have shown it to be greater by about $1:10^3$.

This magnetic moment is the cause of the fine structure observed in atomic spectra. Consider first an atom with one valence electron. As a result of its motion in the central electric field, the electron experiences a magnetic field B_l proportional to its angular momentum $l\hbar$. The resulting energy is given by

$$E_{ls} = - \mu_s . B_l$$

or $$E_{ls} = a s . l \qquad (2.10)$$

where a is a constant that can be calculated if the central field potential is known. The angular momenta $l\hbar$ and $s\hbar$ are coupled together by this interaction to form a new vector $j\hbar$, the total angular momentum, given by $j = l + s$ (Fig. 2.7). Like the other

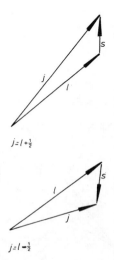

$$j = l + \tfrac{1}{2}$$

$$j = l - \tfrac{1}{2}$$

Fig. 2.7 Vector model for the quantum number j.

angular momenta, $j\hbar$ has a magnitude of $\sqrt{[j(j + 1)]}\hbar$, or $j^*\hbar$, and is quantized: the quantum number j is restricted to the values $l + \tfrac{1}{2}$ and $l - \tfrac{1}{2}$, corresponding to l and s almost parallel and almost anti-parallel. As seen from Fig. 2.7 the scalar product $l . s$ can be expressed in terms of j, l and s by means of the cosine rule: $l . s = \tfrac{1}{2}(j^{*2} - l^{*2} - s^{*2})$. There are thus two possible values of E_{ls} corresponding to the two possible j-values. This is the cause of the doublet structure of the spectra of atoms with one valence electron – for example, the alkali metals. The fine structure levels are labelled with the j-values as a subscript: $^2P_{1/2}$ and $^2P_{3/2}$ from a p-electron, $^2D_{3/2}$ and $^2D_{5/2}$ from a d-electron, etc. For an s-electron there is no splitting, and $^2S_{1/2}$ is the only level. The fine structure splitting is always small compared to the difference between levels of different l, with the single exception of the hydrogen atom, where levels of different l are degenerate. In this case, additional relativistic effects have to be taken into account in determining the fine structure. One of the textbooks listed should be referred to on this subject.

When l and s are coupled in this way, the quantum numbers m_l and m_s cease to be meaningful. This is because l and s have to be regarded as precessing around their resultant j, so their projections on the z-axis are no longer constant (Fig. 2.8). The state of the

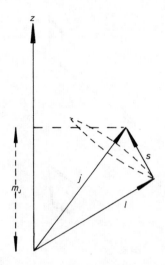

Fig. 2.8 Relation between the quantum number j and m_j.

electron is now described by j and the projection of j on the z-axis, m_j. The quantum numbers n, l, j, m_j form an alternative set to n, l, m_l, m_s. A level of given j still has a degeneracy of $2j + 1$ corresponding to the possible values of m_j from $+j$ to $-j$, and this degeneracy is removed only by an external field. It can be easily verified that the total number of states of given l in this representation, which is $2(l + \frac{1}{2}) + 1$ from $j = l + \frac{1}{2}$ plus $2(l - \frac{1}{2}) + 1$ from $j = l - \frac{1}{2}$, is identical with the number $2(2l + 1)$ in the m_l, m_s representation.

In an atom with several valence electrons the total angular momentum $J\hbar$ can be defined in a similar way by $J = L + S$, provided the magnetic interaction is small compared to the electrostatic. The quantum number J can take any of the $(2S + 1)$

integral or half-integral values $L + S$, $L + S - 1$, $\cdots$ $L - S$. (If $L < S$, there are $(2L + 1)$ possible values $S + L \cdots S - L$). It is now evident why S determines the multiplicity of a term. However, the superscript $2S + 1$ is still used even if the actual number of levels into which the term splits is less than $2S + 1$, as happens when $L < S$. In particular, S terms $(L = 0)$ are unsplit whatever their multiplicity. As in the one-electron case the spin-orbit interaction energy is given by

$$E_{LS} = A\,L.S$$

where

$$L.S = \tfrac{1}{2}\,(J^{*2} - L^{*2} - S^{*2}) \tag{2.11}$$

The constant A can be determined in the simpler cases from the as of the individual electrons. As in the single-electron case, the levels are labelled by their J values. For example, a ^4D term $(L = 2$, $S = 3/2)$ is split into four levels $^4D_{1/2}$, $^4D_{3/2}$, $^4D_{5/2}$, $^4D_{7/2}$. These levels are not equally spaced: it is easily shown from Equation 2.11 that the interval between two levels of a given term

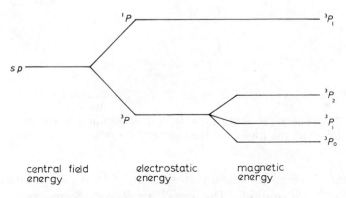

Fig. 2.9 Energy levels arising from the configuration *sp*.

is proportional to the larger of the two J-values concerned (the Landé Interval Rule). Fig. 2.9 shows the complete set of energy levels derived from the *sp* configuration considered in the last section. The threefold degeneracy of the ^3P term remarked on there is removed by the L–S interaction because the L effectively defines a direction in space to differentiate between the three M_S

values. Fig. 2.10 shows the lower few terms in the energy level diagrams of an atom with one valence electron (potassium) and an atom with two valence electrons (calcium).

The fine structure shown in Figs. 2.9 and 2.10 is small (a few tens of cm^{-1}) compared to the term separation, which is typically of the order of 1 eV, or several thousand cm^{-1}. The heavier the element, however, the greater the fine structure; for example, the

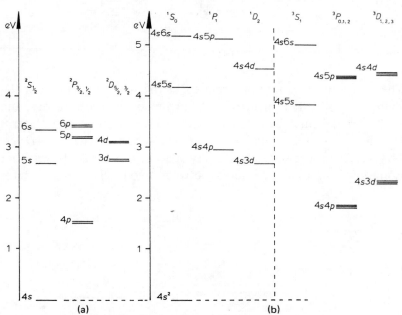

Fig. 2.10 Energy level diagrams for (a) potassium and (b) calcium. The doublet and triplet splittings on these diagrams are not to scale.

lowest P term of cesium is split by 554 cm^{-1}, compared to 17 cm^{-1} for sodium. The $L - S$, or Russell–Saunders, coupling scheme that has been used in this section is valid only when the magnetic interaction is much smaller than the electrostatic interactions; otherwise L and S cease to be meaningful quantities. It is sometimes possible to describe the system by another form of coupling, 'j-j' for example, but sometimes J is the only quantum number that can be defined in addition to the n, l of the individual electrons. Whatever the interactions, the total number of levels and all the possible J-values are fixed by the configuration; it is the arrangement, grouping and labelling of the levels that depends on the coupling. Thus J is always a 'good quantum number', and so also is its projection on the z-axis, M_J. $M_J\hbar$ is, of course, the component of total angular momentum in the z-direction. M_J can take the $2J + 1$ values from J, $J - 1$, $\cdots$ to $-J$, so that in the absence of an external field each J-level has a degeneracy of $2J + 1$. As is the case for the other angular momentum vectors, J can never point directly along the z-axis because its magnitude is $\sqrt{[J(J + 1)]}$ and its maximum component in that direction is only J.

2.8 Transitions and selection rules

Radiative transitions are concerned with the interaction between an atom and a radiation field, and a full treatment is possible only with the use of quantum electrodynamics. However, most of the essential results for our purpose can be obtained by simpler methods, in which the electromagnetic field is introduced as a time-dependent perturbation (see, for example, [6]). The perturbation produces a certain probability of finding the electron in a different state ψ_f from the one it started in ψ_i, provided that the frequency of the oscillating field obeys the relation $h\nu = |E_f - E_i|$. The probability per unit time is known as the transition probability and may refer either to upward transitions (absorption) or to downward (emission). The relations between the emission and absorption probabilities are discussed in Chapter 3. Since they are proportional to one another, we shall not differentiate between them in this section, which is concerned only with the rules determining what transitions may actually occur and what sort of line spectrum results.

By expanding the interaction between the electron and the electromagnetic field as a power series the transition probability can also be expressed as a series, of which the first term is proportional to the square of the so-called electric dipole strength or transition moment for the two states concerned, $|R_{if}(er)|^2$, where

$$|R_{if}(er)|^2 = |R_{if}(ex)|^2 + |R_{if}(ey)|^2 + |R_{if}(ez)|^2 \qquad (2.12)$$

and

$$R_{if}(ex) = \int \psi_i \, ex \, \psi_f \, d\tau$$

with similar expressions for the y and z components. R_{if} obviously has the dimensions of an electric dipole moment. By analogy with the wave-mechanical expression $\int \psi_i^2 x \, d\tau$ for the expectation value of x in the stationary state i, we can interpret $\int \psi_i ex \, \psi_f \, d\tau$ as the expectation value for the dipole evaluated for the mixed states i and f. This integral is frequently written in the form $|ex_{if}|$:

$$\text{i.e., } |R_{if}|^2 = |ex_{if}|^2 + |ey_{if}|^2 + |ez_{if}|^2$$

A transition for which R_{if} is not zero is 'allowed' and is called an electric dipole transition. If R_{if} does vanish, one may be concerned with the next term in the expansion, which corresponds to magnetic dipole and electric quadrupole transitions, and so on to higher orders. The actual value of the transition probability for an allowed transition – that is, the magnitude of R_{if} – is a subject to which we shall return in Chapters 9 and 10.

The rules determining allowed transitions are known as selection rules. We discuss them with reference to the x-component; similar considerations apply to the other two components. First, it is clear that when $i \equiv f$ the integral $\int \psi_i^2 x \, d\tau$ vanishes identically, because ψ^2 does not change sign when $x \to -x$, and the positive and negative contributions therefore cancel. This confirms that an atom in a stationary state does not radiate. Secondly, by writing $x = r \sin \theta \cos \varphi$ and $\psi = R\Theta\Phi$, the integral can be expressed as the product of three integrals, each involving only one of the variables r, θ, φ. The Θ and Φ functions for any central field electron are identical with those of hydrogen, as we have already seen. They have the property that $\Theta\Phi \to (-1)^l \Theta\Phi$ under inversion of the co-ordinates – that is, $x \to -x$, etc., which in polar coordinates corresponds to r, θ, $\varphi \to r$, $\pi - \theta$, $\pi + \varphi$. Since r, and hence $R(r)$, is unchanged by inversion, the effect on ψ is determined by $\Theta\Phi$ and we have

$$\psi \to + \psi \qquad \text{for even } l$$

$$\psi \to - \psi \qquad \text{for odd } l$$

By the argument already used, the integral $\int \psi_i \psi_f x \, d\tau$ must vanish unless the product $\psi_i \psi_f$ changes sign on inversion. This requires either ψ_i or ψ_f to have odd l; in other words, their $l-$ values must differ by $1, 3, 5 \cdots$. In fact it can be shown that the Θ integral vanishes except when the difference is 1, and the first selection rule therefore becomes

$$\Delta l = \pm 1 \qquad (2.13)$$

This selection rule can be generalized by introducing the concept of parity. A term is said to have odd or even parity according to whether its wave-function does or does not change sign on inversion. The function for several independent electrons is simply the product of the individual functions, and the parity is therefore determined by $(-1)^{\Sigma l}$ and is odd or even as the algebraic sum Σl is odd or even. Obviously all terms rising from a given configuration have the same parity. However, the parity of a term is well defined even when configuration mixing makes it impossible to define the ls of the individual electrons. Transitions are then governed by Laporte's generalization of the l selection rule: electric dipole transitions occur only between terms of opposite parity.

We have seen that the J-value of any level is also well defined, and there is a corresponding selection rule

$$\Delta J = 0, \pm 1 \qquad (2.14)$$

with $J = 0 \rightarrow J = 0$ forbidden

This rule cannot be derived from the Schrödinger wave functions, which do not include spin. A textbook such as Kuhn's or Woodgate's should be consulted for an explanation. In so far as L and S are good quantum numbers, there are additional selection rules

$$\Delta L = 0, \pm 1 \text{ with } L = 0 \rightarrow L = 0 \text{ forbidden}$$

and

$$\Delta S = 0$$

The ΔL and ΔS rules are frequently broken, particularly in the heavier elements where L-S coupling is not a good description. The singlet-triplet intercombination line $6s^2 \; ^1S_0 - 6s6p \; ^3P_1$ is one of the strongest lines in the mercury spectrum (2537 Å).

Transitions forbidden by electric dipole radiation may occur by magnetic dipole or electric quadrupole radiation, for which the selection rules are different. In particular, the parity rule for these transitions is even → even or odd → odd. This allows transitions between terms in the same configuration ($\Delta l = 0$) which are forbidden for electric dipole radiation. Their probabilities are, however, far smaller, by factors of the order of 10^{-5} and 10^{-8} respectively in the visible region. The upper levels of such 'forbidden' lines are described as metastable. The lines are important in astrophysics because at sufficiently low gas densities a forbidden transition has a good chance of occurring before the atom is knocked out of its excited state by a collision. The frequency dependence of electric quadrupole transitions makes them increasingly probable at very short wavelengths. Induced magnetic dipole transitions, on the other hand, are important at radio frequencies, and most of Chapter 7 is concerned with magnetic dipole transitions between different levels of the same term.

Electric dipole transitions are evidently restricted by the l selection rule to transitions between adjacent columns in simple energy level diagrams such as those of Fig. 2.10, together with the singlet-triplet intercombination lines in the two-electron system. Transitions from the ground state of potassium (Fig. 2.10a) are limited to the resonance line(s) $4s-4p$, the higher members of the so-called principal series $4s-np$ and the continuum. The other important series are the sharp, $4p-ns$, the diffuse, $4p-nd$, and, to a lesser extent, the fundamental, $3d-nf$. In the two-electron spectrum (Fig. 2.10b) each of these series appears twice, as a singlet series and a triplet series with different limits. In a complex spectrum the number of possible series is much greater; the transition $pp-pd$, for example, is made up of the lines $^1S-^1P$, $^1P-^1P$, $^1P-^1D$, $^1D-^1D$ and $^1D-^1F$, repeated for the triplet terms, quite apart from any intercombination lines.

In general each 'line' is split into a number of components by the fine structure. The potassium resonance line, for example, has two components $4s\,^2S_{1/2}-4p\,^2P_{1/2}$ and $4s\,^2S_{1/2}-4p\,^2P_{3/2}$, separated by the width of the $4p\,^2P$ fine structure, 58 cm^{-1} or 34 Å. The group of lines arising from transitions between two particular terms is known as a multiplet. Fig. 2.11 shows an example of a complex multiplet (both terms split) in the calcium spectrum of Fig. 2.10b. In this case the total spread of the line pattern is about 160 cm^{-1}, and the splitting of ^3D is much smaller than that of ^3P, giving the appearance of a simple triplet. Except in the very light atoms the resolution of the multiplet structure does not present a particularly difficult problem, at any rate in transitions between states of moderate excitation (fine structure diminishes rapidly with increasing n as well as with increasing l in any given atom). In heavy atoms the problem is rather to link together components of the same multiplet that may be separated by many hundreds of Ångströms. In mercury the ^3P–^3D multiplet analogous to that in Fig. 2.11 extends over 6200 cm^{-1}. In

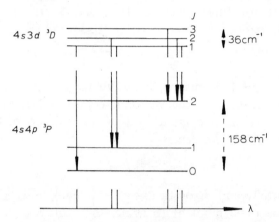

Fig. 2.11 Complex triplet in the spectrum of calcium. Six components are allowed by the selection rule $\Delta J = 0, \pm 1$.

complex spectra, particularly those involving partly filled d and f shells, such as the transition elements and the rare earths, the number of energy levels is so large and the density of lines so high that an instrument of high resolution may be necessary to sort out lines falling close to one another by pure chance.

2.9 Zeeman and Stark effects

These effects describe the splitting of spectral lines in magnetic and electric fields respectively. The Zeeman effect is much the easier to interpret and has been very widely used to help analyse spectra in the optical region. It plays a leading part in microwave and radiofrequency spectroscopy, as will be seen in Chapter 7. Historically, it was the anomalous Zeeman effect that led Goudsmit and Uhlenbeck to postulate the existence of electron spin. The Stark effect is of little use in the interpretation of spectra, but it plays an important part in molecular theory. It is responsible for the pressure broadening by charged particles discussed in Chapter 8, and it is used for scanning in microwave spectroscopy.

The Zeeman effect can be explained in terms of a semi-classical model, describing the interaction of an electronic magnetic moment with the external magnetic field. The intrinsic magnetic moment associated with electron spin has already been introduced to account for fine structure. If L and S are good quantum numbers Equation 2.9 can be written for several electrons as

$$\mu_s = - g_s \mu_B S^\star \tag{2.15}$$

where μ_B is the Bohr magneton $e\hbar/2m$ and $g_s = 2\,S^\star$ is a reminder that the magnitude of the vector is $\sqrt{[S(S + 1)]}$. Similarly the magnetic moment associated with the orbital part of the angular momentum can be written as

$$\mu_L = - g_L \mu_B L^\star \tag{2.16}$$

Experimentally and theoretically g_L is actually unity – that is, the ratio of magnetic moment to associated *orbital* angular momentum is equal to the classical gyromagnetic ratio for a particle of charge e, $e/2m$. L and S both precess about their resultant J (see Section 2.7 and Fig. 2.8 for the one-electron case), so the components of μ_S and μ_L perpendicular to J average to zero. Only the components in the J-direction contribute to the resultant magnetic moment μ_J: $\mu_J = \mu_L \cos(L, J) + \mu_S \cos(S, J)$. This can be expressed as

$$\mu_J = - g_J \mu_B J^\star \tag{2.17}$$

where g_J is known as the Landé g-factor. If L-S coupling is valid, g_J can be evaluated in a straightforward fashion by applying the cosine rule to the angles (L, J) and (S, J) with the result

$$g_J = 1 + \frac{J^{\star 2} + S^{\star 2} - L^{\star 2}}{2 J^{\star 2}} \tag{2.18}$$

It is easily seen that for a singlet term $(S = 0, J = L)\, g_J = 1$, while for an S term $(L = 0)$, in which there is no orbital contribution to the magnetic moment, $g_J = 2$.

The energy of the magnetic moment μ_J in the field B is

$$E_B = -\mu_J . B = g_J\mu_B J^* . B$$

Now $J^* . B$ is simply B times the component of J in the direction of B, which is taken to define the z-axis. This component is M_J (Section 2.7) and is restricted to the values $J, J - 1, \cdots -J$

$$\therefore E_B = g_J\mu_B B M_J \qquad (2.19)$$

The magnetic field splits each level into $2J + 1$ sub-levels of equal separation $g_J \mu_B B$, as shown in Fig. 2.12 for $J = 3/2$. This removes

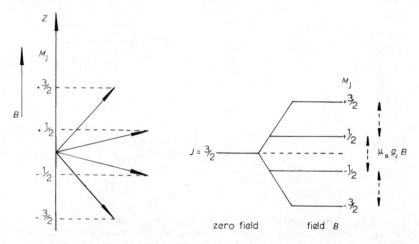

Fig. 2.12 Splitting of energy levels in a magnetic field. The M_J values and energy levels are shown for the case $J = 3/2$.

the last degeneracy in the wave function: each magnetic sub-level corresponds to one and only one set of quantum numbers.

The selection rule for electric dipole radiation is

$$\Delta M_J = 0, \pm 1 \qquad (2.20)$$

$\Delta M_J = 0$ gives π-components, polarized with the electric vector parallel to B, when viewed at right angles to the field, and $\Delta M_J = \pm 1$ gives σ-components, polarized with the electric vector perpendicular to B. Viewed along the field direction, the σ-components are circularly polarized, and the π-components do not appear at all. Fig. 2.13a shows the Zeeman pattern for a $^2S_{1/2} - ^2P_{3/2}$

transition (e.g. one of the sodium yellow lines) and Fig. 2.13(b) shows it for a $J = 2 \leftarrow J = 1$ transition (e.g., one component of the $^3P-^3D$ multiplet of Fig. 2.11). If the g_J factors of the two levels are exactly equal – in particular, if they are both 1 as in a singlet-singlet transition – the spacings in the upper and lower levels are equal also. All the $\Delta M_J = +1$ transitions then coincide, as do the $\Delta M_J = 0$ and $\Delta M_J = -1$ transitions, giving only three Zeeman components separated by $\mu_B B$. This is the so-called 'normal' Zeeman effect; it is strictly abnormal in that it occurs

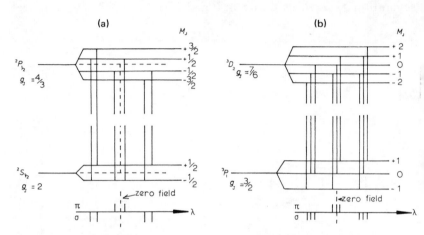

Fig. 2.13 Zeeman pattern for (a) a $^2S_{1/2}-^2P_{3/2}$ and (b) a $^3P_1-^3D_2$ transition. In both cases the g-factors of the upper and lower terms are not very different, and the components form three groups rather similar to a 'normal' Zeeman pattern. σ components obey $\Delta M_J = \pm 1$ and π components $\Delta M_J = 0$.

only for singlet states. If the g-factors are not very different, as in Fig. 2.13, the components still form $\sigma-\pi-\sigma$ groups, but if the gs are very different the pattern bears little resemblance to the 'normal' triplet and may be very complicated for large J. It is always symmetric about the field-free line.

As g_J is not very different from unity, the order of magnitude of the splitting is determined by $\mu_B B$, which is $0\cdot046$ cm^{-1} for a field of $0\cdot1$ wb m^{-2} (1 kilogauss). This is comparable with the Doppler line width for a light element (see Section 8.3), and it is necessary to use high fields, high resolution and light sources with narrow lines to resolve a complex Zeeman pattern. Magnetic dipole transitions between the sub-levels of the same J-level (see

Chapter 7) provide the most accurate determinations of the splittings.

In this treatment it has been implicitly assumed that the interaction with B is small compared with the internal interactions in the atom so that spin-orbit coupling is undisturbed and J remains a good quantum number. If the field is high enough for the Zeeman splitting to compare with the fine structure, which may be only 1 cm^{-1} or so for a light element, we have to think about the validity of this assumption. To go to the opposite extreme, if it were possible to make $\mu_B B$ much greater than the L-S interaction, L and S would interact directly with the field, with $(M_L + 2M_S) \mu_B B$ as the principal term in the energy and the L-S interaction energy as a small correction. This situation is known as the Paschen-Back effect. It is not of much practical importance as far as fine structure is concerned because such large fields are required, but the analogous effect in hyperfine structure *is* important and is studied mainly by the methods of Chapter 7. The intermediate field case, in which the internal and external magnetic interactions are of comparable size is important for both fine and hyperfine structure and is discussed in the books of Kuhn and Woodgate.

The Stark effect differs from the Zeeman effect because atoms have no permanent electric dipole moment to interact with the electric field. It is a good deal more difficult to treat in an elementary fashion, and we shall discuss it here in a qualitative way only, aiming mainly at providing a base for the description of pressure broadening in Chapter 8.

In all atoms except hydrogen, which must be treated separately, the effect of lowest order is quadratic: the interaction energy is proportional to the square of the electric field, F^2. One can think of the electric field as distorting the electron distribution so as to induce a dipole αF, where α is the polarizability. The interaction energy between the induced dipole and the field is αF^2. α may reasonably be expected to depend on the angular distribution of the electron charge and thus on M_J. The distributions represented by $+M_J$ and $-M_J$ have the same symmetry with respect to the direction of F (which we take to be along the z-axis), so the Stark effect depends only on $|M_J|$, and each sub-level except $M_J = 0$ remains doubly degenerate. The polarizability and hence the magnitude of the splitting cannot, however, be related to J and M_J

by a simple semi-classical approach as in the Zeeman effect. The main features of the treatment by perturbation theory are as follows.

The potential energy of an electric dipole $-er$ in an electric field F in the z-direction is $V_F = er \cdot F = ezF$. But the expectation value for V_F evaluated for a pure stationary state ψ_0 is zero, because $\int \psi_0^2 z \, d\tau = 0$, as pointed out in Section 2.8. This is just another way of saying that the atom has no permanent electric dipole moment. It is therefore necessary to go on to the second order perturbation, in which V_F is evaluated for a perturbed wave function representing the distorted charge distribution. This perturbed function is constructed by adding to ψ_0 a linear combination of the wavefunctions representing all the other states of the system,

$$\psi_{\text{pert}} = \psi_0 + \sum_n a_n \psi_n \qquad (2.21)$$

where a_n determines the amount of each state to be included. The values of the coefficients – that is, the amount of the distortion – are found from the first order perturbation:

$$a_n = \frac{\int \psi_n ezF\psi_0 \, d\tau}{E_0 - E_n} \equiv F \frac{|ez|_{n,0}}{E_0 - E_n} \qquad (2.22)$$

This integral is identical with the z-component of the electric dipole strength (Equation 2.12) that determines the transition probability between the states ψ_0 and ψ_n. The selection rules that decide which states n are to be mixed with ψ_0 are therefore identical with the selection rules for electric dipole transitions. The contribution to the interaction energy E_F made by each state for which a_n does not vanish is found from the second order perturbation theory to be

$$(E_F)_n = a_n \int \psi_n eFz\psi_0 \, d\tau$$

$$= F^2 \frac{|\int \psi_n ez\psi_0 \, d\tau|^2}{E_0 - E_n} \equiv F^2 \frac{|ez|_{n,0}^2}{E_0 - E_n} \qquad (2.23)$$

by using Equation 2.22. The state ψ_0 is effectively pushed away from any perturbing state ψ_n lying near it. From the symmetry of the expression, ψ_n is pushed away an equal amount by ψ_0. In other words, the effect of the electric field is that the two states repel one another by an amount proportional to F^2 and inversely proportional to their separation. The repulsion occurs only between configurations of opposity parity and, within these configurations, between sub-levels of the same M_J satisfying $\Delta J = 0, \pm 1$. Its magnitude is found by evaluation of the integrals to depend on M_J^2. The selection rule $\Delta M_J = \pm 1$ is not appropriate to this case because the electric field is necessarily along the z-axis (π-component). There are no x- and y-contributions to the interaction energy.

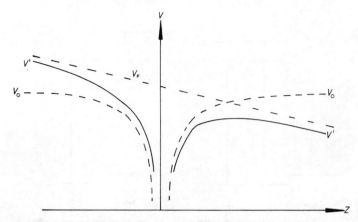

Fig. 2.14 Distortion of central field potential by an external electric field. The full curves V' show the result of superimposing the external potential V_F (dashed line) on the central field potential V_0 (dashed curves).

Fig. 2.14 shows diagrammatically the physical basis of the perturbation. The potential $V_F(z)$ due to the external field is superposed on the symmetric central field potential $V_0(r)$ to give a resultant V' that is asymmetric in the z-direction. The symmetric unperturbed wave function is given the necessary asymmetry to fit V' by an admixture of functions of opposite parity. Put another way, the θ-dependence of the modified potential means that it is no longer possible to separate the R and Θ parts of the Schrödinger equation, and the quantum number l loses its original meaning. M remains a good quantum number because the cylindrical symmetry leaves the Φ function unaffected.

Fig. 2.15 shows the Stark splitting of the alkali resonance lines as a typical example of the quadratic effect. The ground level is not much perturbed because the first excited configuration is rather a long way above it. Because of the crowding together of excited levels as the excitation increases, the excited levels tend on balance to be shifted down, and sub-levels of different M_J are shifted by different amounts. The result is an asymmetric line splitting together with a shift, usually towards the red, of the centre of gravity of the line pattern. Fields in the range 10^6 - 10^7 volts m^{-1} are usually required to produce observable Stark

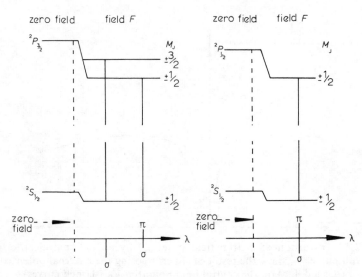

Fig. 2.15 Quadratic Stark effect for the alkali resonance lines. The $^2S_{1/2}$ - $^2P_{1/2}$ line is shifted to the red; the $^2S_{1/2}$ - $^2P_{3/2}$ line is split into two components, both red-shifted.

splitting – say a few tenths of a wavenumber. The splitting is largest for light atoms and increases with the excitation because of the smaller energy differences between the perturbing states. At values of the quantum number n so high that the Stark splitting becomes comparable with the energy differences between states of different l, the latter become effectively degenerate, the second order perturbation treatment breaks down, and the Stark effect becomes linear in F, as with hydrogen.

Before we pass on to the linear Stark effect, another result of the electric field perturbation should be mentioned briefly. The dipole radiation selection rule $\Delta l = \pm 1$ may be broken because of the configuration mixing in the perturbed wave functions. For example, the admixture of p- in the s-wave function allows the 'forbidden' transitions $s-d$ and $s-s$ to appear in the alkali spectra.

The hydrogen atom shows a linear Stark effect because of its l-degeneracy. The fine structure from spin-orbit and relativistic effects is of order 0.1 cm^{-1}, which is comparable with the Doppler line-width at room temperature, so it may usually be ignored in comparison with a Stark splitting large enough to be readily measurable. All states of the same n are thus effectively degenerate. In applying perturbation theory to degenerate states one immediately runs into the difficulty that the a_n in Equation 2.22 and the E_n in Equation 2.23 go infinite for two states of equal energy. For a sensible solution the numerator, if not already zero, must be made to vanish for such states,

$$\text{i.e., } \int \psi_{01} \, ez\psi_{02} \, d\tau = 0 \text{ when } E_{01} = E_{02} \qquad (2.24)$$

If ψ_{01} and ψ_{02} have $\Delta l = \pm 1$, $\Delta m = 0$, Equation 2.24 as it stands will not be satisfied. We can, however, arrange for the integral to vanish by using for the zero-order wave functions some linear combination of ψ_{01} and ψ_{02}, say u_{01} and u_{02}. The new functions satisfy the unperturbed Schrödinger equation with the same energy E_0 since linear combinations of solutions are themselves solutions. But the expectation values of ezF evaluated for u_{01} and u_{02}, $\int u_{01}^2 ezF \, d\tau$ and $\int u_{02}^2 ezF \, d\tau$, do *not* vanish because the u's are necessarily aysmmetric in the z-direction. There is therefore a first order splitting proportional to F. The important distinction from the quadratic effect is that the linear combination of *degenerate* functions does not contain coefficients proportional to F. One set of zero-order functions is just as permissible as another until some perturbation forces one to choose a physically sensible combination. A similar situation has already been discussed in the case of two valence electrons, when the introduction of the electrostatic repulsion necessitated a choice of wave functions compatible with the indistinguishability of the electrons (Equation 2.7).

The result of the full treatment for hydrogen is that a state of given n splits symmetrically into $2n - 1$ sub-levels, whose separation is proportional to F and to n. The line pattern is therefore also symmetric about the field-free line. Fig. 2.16 shows the effect for H_α, $n = 2 \leftrightarrow n = 3$. The splitting is a few cm^{-1} in a field of order 10^5 V m^{-1}, which is very much larger than the quadratic effect shown by other atoms. The Stark splitting of hydrogen is of enormous importance in plasma spectroscopy because hydrogen is a common constituent of most plasmas and its Stark pattern can

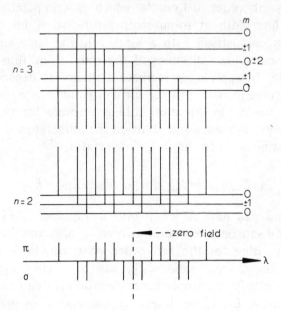

Fig. 2.16 Linear Stark effect for the H_α line. The level splitting is greater for $n = 3$ than for $n = 2$. The resulting line pattern is symmetric about the field-free line.

be calculated with great reliability. This allows accurate comparison of theoretical and experimental line shapes, as a result of which electron densities can be deduced from measured hydrogen line profiles.

2.10   Configuration interaction and auto-ionization

The interactions discussed so far for an atom in zero external field have all been based on the concept of single well-defined configurations, that is, the assumption that a definite (n, l) value can be allocated to each electron. A comparison of theoretical with experimental values of energy levels and transition probabilities shows that this assumption is not always justified. It is often necessary to mix in to the basic configuration contributions from one or more other configurations in order to describe the system properly. With no external field perturbation to affect the symmetry, configurations can only interact if they have the *same* parity, in contrast to the Stark effect. The other rule is that within two such configurations only terms of the same J can interact with one another. The strength of the interaction, or the amount of mixing, falls off inversely as the energy difference between the terms, so that for most simple spectra the effect is relatively unimportant so long as we are concerned with configurations representing excitation of a single valence electron. In complex spectra, where there are several valence electrons, several excited states may easily have comparable energies; in the transition elements, for example, the excitation energies of the $3d$ and $4s$ electrons are very similar. Even in simple spectra configuration interaction can be significant if two valence electrons are excited. In Be, for which the ground state is $2s^2$ $^1S_0$, the 'anomalous' term $2p^2$ ^3P has about the same energy as the terms arising from the $2s\,3s$ and $2s\,3d$ configurations of the normal spectrum. Perturbations are therefore to be expected between the levels of the same J. The $2s\,3p$ configuration does not interact with $2p^2$ because it has the opposite parity.

Of particular interest is the configuration mixing between a bound state and an adjacent ionization continuum that gives rise to auto-ionization. In a simple term diagram, such as those of Fig. 2.10, only one valence electron is excited, and all terms converge to the same limit, corresponding to the ion in its ground state plus the electron with zero kinetic energy. In the continuum states beyond the series limit the electron has finite kinetic energy. But there exist bound states equal in energy to the continuum states due either to the simultaneous excitation of two valence electrons or to the excitation of an electron from an inner shell – e.g., one of the p electrons from the p^6 closed shell in an alkali atom. Fig. 2.17 shows as an example of the former the energy level diagram for calcium, with the singlet-triplet term splitting omitted for simplicity. On the left hand side are the 'normal' terms $4s\,nl$, converging to the $4s$ $^2S_{1/2}$ ground state of Ca^+. On the right hand side are the 'anomalous' terms, in which one of the valence electrons is in either a $3d$ or a $4p$ state while the other one is excited to various higher states. Removal of this second electron from the atom leaves the calcium ion in either a $3d\ ^2D$ or a $4p\ ^2P$ state. All of the bound states converging to either of these two limits, except the lowest, have greater energy than the normal ionization energy. The interaction between such a high bound state and the continuum state of the same energy leads to a mixing of the wave functions and hence of the properties of the two

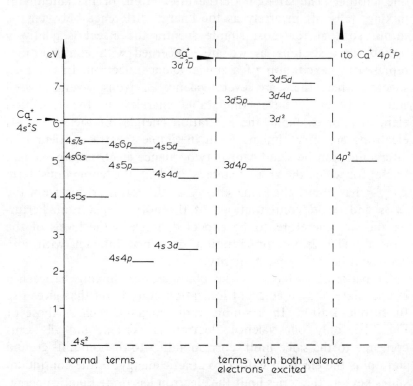

Fig. 2.17 Energy level diagram of calcium showing 'anomalous' terms. The 'normal' terms on the left converge to the ground state of Ca$^+$; the 'anomalous' terms on the right converge to excited states of Ca$^+$. For simplicity only the centre of gravity of each configuration is shown, the term structure being omitted.

states. The bound state becomes in some measure a piece of continuum and is thereby broadened. One can also think of the mixing as denoting a certain probability of finding the system in the ionized state – in other words, of undergoing a radiationless transition, as shown schematically in Fig. 2.18. This process is known as auto-ionization. Its probability may be as high as 10^{13}

sec^{-1}, in contrast to a typical value of 10^8 sec^{-1} for an allowed radiative transition. It will be seen in Section 8.2 that the natural width of a line depends on the lifetimes of the levels it connects according to the uncertainty principle $\Delta v \cdot \Delta t \sim 1$. Normally $\Delta t = 10^{-6}$ to 10^{-9} sec, but for an auto-ionizing level Δt may well be shorter by a factor of 10^5. Any spectral line starting or finishing on such a level will be extremely diffuse, possibly as much as a few thousand wavenumbers wide. It must be remembered that not all

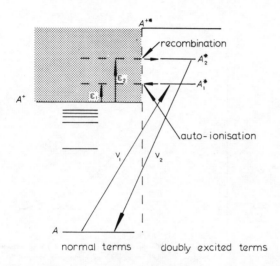

Fig. 2.18 Auto-ionization shown schematically. The process of absorption and auto-ionization is represented by: $A + h\nu_1 \rightarrow A_1^* \rightarrow A^+ + \epsilon_1$; radiative recombination is: $A^+ + \epsilon_2 \rightarrow A_2^* \rightarrow A + h\nu_2$.

anomalous levels can undergo auto-ionizing transitions, since the parity and J selection rules, plus additional rules for L and S when L-S coupling is valid, must be obeyed. The non-auto-ionizing levels have normal widths.

Most auto-ionizing levels are studied by means of absorption from the ground or a low excited state, and as they lie beyond the normal series limit the relevant transitions are usually in the vacuum ultra-violet region. In most sources the lines are hard to detect in emission. Apart from their great widths, auto-ionizing lines can be very asymmetric. The profiles were first calculated by Fano, whose results agree well with the experimental shapes. An example of an asymmetric absorption line is shown in Fig. 2.19. The absorption at some distance from the line is the ordinary continuous absorption from the ground state to the continuum above the normal series limit. The auto-ionizing transition forms a kind of resonance in the continuum, rising steeply on one side to a peak value and dropping on the other to zero before going back to the steady level. The absorption on this side of the line is thus *less* than that of the unperturbed continuum and appears as a window in the latter.

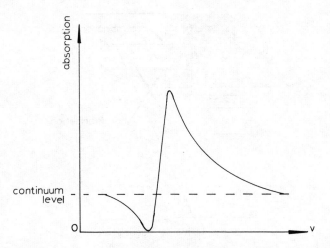

Fig. 2.19 Example of asymmetric auto-ionizing line profile. The dip on the left appears as a window in the continuous absorption.

Inverse auto-ionization, or radiative recombination, is often an important astrophysical process and is also shown schematically in Fig. 2.18. It produces a resonance hump in the recombination continuum; Fig. 2.19 may equally well be taken as a plot of recombination cross-section against frequency.

2.11   Ionic Spectra

If one electron is removed from an atom, the binding energy of the remaining electrons is increased because of the extra un-balanced nuclear charge. The ordinates of the energy level diagrams therefore have to be scaled up, transitions between different levels require larger energy jumps, and the spectra fall further into the ultra-violet region. In spectroscopic terminology the spectrum of a neutral atom is designated I, the spectrum of the singly ionized atom II, of the doubly ionized atom III, and so on. A set of spectra in which the degree of ionization increases by one for each element along a row of the periodic table is known as an iso-electronic sequence becasue the number of electrons remains constant along it. The study of the systematic differences in such a sequence is particularly valuable in spectral analysis, energy level calculations, and the prediction of transitions in spectra that have not been observed or directly identified. In the hydrogen-like ions He II, Li III, Be IV $\cdots$, the energy of a state of given n is proportional to Z^2, where $Z = 2, 3, 4 \cdots$. For these ions the hydrogen energy level diagram is simply scaled up by Z^2 and the wavelengths decreased by the same factor – e.g., the Lyman α line $n = 2 \leftrightarrow n = 1$ which lies at 1216 Å in H I occurs at 304 Å in He II, 135 Å in Li III, and so on. In all other iso-electronic sequences this simple relation is modified by the screening effects of the other electrons, but it remains approximately true that the scaling factors along an iso-electronic sequence are $4, 9, \cdots$.

It is clear that after a few stages of ionization the transitions among the lower levels will fall in the extreme ultra-violet, from a few hundred Å to a few Å. The highly stripped ions of N, O, C, Si, Fe, etc., in the sun's corona radiate in this wavelength region. Much work has recently been done on these spectra, partly because of the increasing numbers of observations from rockets and satellites and partly because of the development of suitable laboratory sources for their excitation.

2.12 X-ray spectra

X-ray spectra arise from transitions between the innermost electron shells, where there is relatively little screening of nuclear charge. The binding energies are therefore much greater, and the wavelengths correspondingly shorter, than for optical transitions. The heavier the atom, the shorter the wavelength, a relation given quantitatively by Moseley's law $\nu \propto Z^2$ (later modified to $\nu \propto (Z - s)^2$, where s represents the screening effect of the electrons in the same or any lower shells). The energy needed to remove an electron from an inner shell of a given atom is similar to that needed to excite an ion that has lost nearly all its electrons, and so the X-ray spectra of the light elements overlap in wavelength (in the region below 100 Å) the spectra of the highly stripped ions of the same elements. Moreover, for the very light elements there is no hard and fast distinction between X-ray spectra on the one hand and the inner-shell excitation of Section 2.10 on the other. For spectral analysis in this overlap region, often known as the XUV, it is common to use both optical (ruled) diffraction gratings and Bragg (crystal) spectrometers for both 'optical' and X-ray spectroscopy.

X-ray spectroscopy has grown up with its own notation: shells with $n = 1, 2, 3, \cdots$ are designated K, L, M, $\cdots$. For $n = 1$ there is only one value of l ($l = 0$), but for $n = 2$ the two sub-shells $l = 0$ and $l = 1$ are denoted by L_I and L_{II}. Similarly, there are three M sub-shells M_I, M_{II}, M_{III}. X-ray spectra of all elements are qualitatively similar and are notably less complex than optical spectra, because the energy of the atom when one electron is missing from an inner shell depends predominantly on the effective nuclear charge and not on the number or arrangement of the valence electrons. However, an X-ray transition that removes an electron from an inner shell to an energy level right outside the atom does show fine structure because the final level is affected by the other electrons and by the crystal fields. As a result, X-ray spectroscopy is a valuable tool in solid-state physics. However, since both the applications and the experimental methods differ entirely from those of optical and radiofrequency spectroscopy, we shall not be further concerned with X-ray spectroscopy in this book.

2.13　Hyperfine structure and isotope shift

In the discussion of atomic structure so far the only property of the nucleus that has been brought in is its charge Z. This section will show how other properties of the nucleus affect the electronic energy levels and hence the spectrum. First, the energy levels depend on the mass of the nucleus, so that analogous transitions in different isotopes have slightly different frequencies. Secondly, any magnetic moment possessed by the nucleus interacts with the magnetic field produced at the nucleus by the motion of the electrons, and any nuclear electric quadrupole moment interacts with the gradient of the electronic electric field at the nucleus. All these effects are of the same order of magnitude, and the term hyperfine structure (hfs) is sometimes used for any of them. More often it is taken to include magnetic dipole and electric quadrupole interactions but to exclude isotope shift.

Let us consider first the nuclear magnetic moment μ_I associated with a nuclear spin angular momentum $I^* \hbar$. I can take small integral or half-integral values, according as the number of protons and neutrons (the mass number) is even or odd. μ_I is proportional to I, and the relation between them can be expressed in the same way as that for electron spin, Equation 2.15:

$$\mu_I = g_I \mu_N I^* \qquad (2.25)$$

μ_N is the *nuclear* magneton $e\hbar/2M_p$ where M_p is the proton mass. The nuclear g-factor g_I defined by this equation is, like the Landé g-factor, of order unity, but may be positive or negative. The nuclear magneton is smaller than the Bohr magneton by a factor m/M_p, or $1/1836$, and μ_I is correspondingly three orders of magnitude smaller than μ_S. The energy level splitting is caused by the interaction of this magnetic moment with the time-averaged magnetic field B_J produced at the nucleus by the orbital motion and spin of the electrons. B_J is in the direction of J and has the same order of magnitude as the internal field B_l that we have already used to explain fine structure (spin-orbit interaction). The ratio of hyperfine to fine structure splitting is therefore determined essentially by the ratio of the magnetic moments μ_I/μ_S and is of order 10^{-3}. Even the largest hyperfine splittings are only a few hundredths of a wavenumber in the light elements, rising to a wave number or so for heavy elements.

The coupling between I and J brings in another quantum number F determining the total angular momentum of the atom according to $F = I + J$. Just as in the case of the L–S interaction, F can take the $2I + 1$ values $J + I, J + I - 1, \cdots J - I$ (or $2J + 1$ values from $I + J$ to $I - J$ if $J < I$). The interaction energy is

where
$$E_{IJ} = \mu_I . B_J \propto g_I \mu_N I . J = A I . J$$
$$I . J = \tfrac{1}{2} (F^{*2} - I^{*2} - J^{*2}) \qquad (2.26)$$

A level of given J is split into $2I + 1$ (or $2J + 1$) hyperfine levels whose separation is proportional to the hyperfine structure constant A. The analogy with fine structure is obvious: Equation

2.26 can be obtained from Equation 2.11 if *J, L, S* are replaced by
F, J, I respectively, and the relative separations obey the Landé
interval rule with *F* replacing *J*. Conventionally the same letter *A*
is used for both the fine and hyperfine structure constants in spite
of the difference of three orders of magnitude between them. The
actual value of the hyperfine *A* depends on g_I and on the
electronic wave functions that determine the relation between B_J
and *J. A* is at its largest for an unpaired *s*-electron, may be

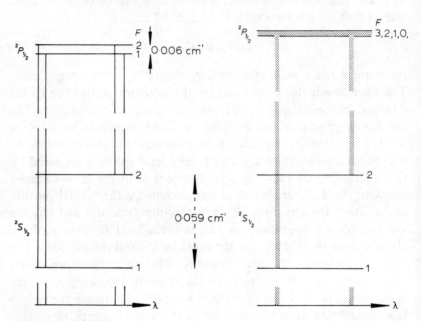

Fig. 2.20 Hyperfine structure in the resonance lines of sodium. The
hyperfine components are determined by the selection rules $\Delta F = 0, \pm 1$. The
splitting of the $^2P_{3/2}$ level is too small to show on the scale of the diagram.

appreciable for a $p_{1/2}$ electron, and is usually negligibly small for
$p_{3/2}$ and *d, f, · · ·* electrons. Fig. 2.20 shows the hfs of the
resonance lines of sodium, for which $I = 3/2$. The selection rule for
F needed to determine the line pattern is the same as for *J*:
$\Delta F = 0, \pm 1$ with $F = 0 \leftrightarrow F = 0$ forbidden.

Line splittings of the order 0·1 to 0·01 cm^{-1} are very difficult
to measure accurately in the optical region, mainly because of the
finite width of the lines themselves (Chapter 8). Direct transitions
between the hyperfine levels, however, fall in the radiofrequency

or microwave regions of the spectrum, and very many such structures have been measured with great accuracy by magnetic resonance and more recently by level crossing and double resonance methods. The behaviour of the hyperfine levels in an *external* magnetic field, or the Zeeman effect of hfs, is fundamental to these methods, which are described in Chapter 7. The rf methods are not, however, universally applicable, and high resolution optical spectroscopy has not been entirely superseded in hfs work.

The nuclear electric quadrupole interaction contributes to the hfs when neither the quadrupole moment Q nor the gradient of the electric field produced at the nucleus by the electrons is zero, conditions which require $I \geqslant 1$ and $J \geqslant 1$ respectively. The electric field gradient is a measure of the departure of the electronic charge distribution from spherical symmetry. Assuming axial symmetry in the z-direction it is described by $\partial^2 V_e / \partial z^2$, and the interaction energy can be written as

$$E_Q = eQ \, \frac{\partial^2 V_e}{\partial z^2} \, f(F, I, J) = Bf(F, I, J) \qquad (2.27)$$

$f(F, I, J)$ is a somewhat lengthy function of the three quantum numbers which is given in the text books (e.g. [5], [6]). In free atoms departures from spherical symmetry are not large, so B is generally smaller than the magnetic hyperfine structure constant A, and quadrupole effects usually show up only as small corrections on magnetic effects. In molecules and solids they are much larger, because of the strong asymmetry of molecular and crystalline electric fields (Sections 7.4 and 7.7).

Isotope shift is the name given to the very small differences in energy found between corresponding electronic states of different isotopes. Since different levels are shifted by different amounts, the corresponding lines of two isotopes are slightly displaced from one another. In the lighter elements the effect is due to the difference in mass itself, which affects the Rydberg constant directly (Equation 2.2). It can easily be verified from Equation 2.1 that the shift $\Delta\sigma$ in a line of wavenumber σ for two isotopes differing in mass by ΔM is

$$\frac{\Delta\sigma}{\sigma} = \frac{\Delta E}{E} = \frac{\Delta\mu}{\mu} = \frac{m\Delta M}{M^2}$$

In the case of hydrogen/deuterium this is about 10^{-3}. In all atoms except hydrogen and the hydrogen-like ions there may be in addition to this 'normal' shift a 'specific mass effect' that depends in a rather complicated way on the correlation of the electron motions and cannot be reliably calculated. Generally, although seemingly not always, the normal and specific effects are of comparable size. Because of the rapid decrease with increasing M, both effects can usually be ignored in the last third of the periodic table, but in the middle third the specific effect may well be large enough to introduce significant uncertainties to the evaluation of the volume shift, which is the dominant effect for the heavier elements.

The volume or field effect is a result of the finite spread of the nucleus. Adding neutrons alters the size and sometimes the shape of the nuclear charge distribution and changes the electrostatic interaction with any electron charge distribution overlapping it. Only s-electrons, and to a lesser extent $p_{1/2}$ electrons, have an appreciable probability of being found within the nucleus, and isotope shifts are confined to transitions in which the number of such electrons changes. Comparisons of theoretical with experimental absolute values are bedevilled by uncertainties in electron wave functions, screening constants and specific mass shifts, but relative shifts – the relative spacing between different isotopes of the same element – have contributed much useful data to work on nuclear models. The shift between adjacent isotopes increases up the periodic table, rising to around a wavenumber with the heaviest elements. Isotope shift must necessarily be studied by high resolution optical spectroscopy, but the task has been made much easier by the availability of very pure samples of separated isotopes.

2.14 Molecular structure: the Born-Oppenheimer approximation

In molecules one has to consider not only the electronic energy but also the vibrational energy from the relative motion of the nuclei and the rotational energy of the molecule as a whole. The complete wave equation is far too complex to solve. The first simplification is the Born-Oppenheimer approximation, in which the wave function is written as the product of three independent wave functions $\psi_e \psi_v \psi_r$ depending on the electron co-ordinates, the inter-nuclear separation and the rotational angle respectively. To the same approximation the energy is the sum of the electronic, vibrational and rotational energies: $E = E_e + E_v + E_r$.

2.15 Electronic energy of diatomic molecules

Suppose we first ignore vibration and rotation and attempt to solve Schrödinger's equation for the electrons in the field of the two nuclei at a fixed separation r. The central field method cannot be applied when there are two centres of force, and the approximations that can be used are notably less satisfactory than in the atomic case. The general method of tackling the problem depends on the principle that, because any system exists in the state of lowest possible energy, the best wave function is the one that gives the lowest energy. The technique is therefore to guess at some plausible wave function incorporating one or more parameters, insert it in Schrödinger's equation, calculate the corresponding electronic energy, and then adjust the parameters to minimize the energy. Because of the compactness of the closed shells in the separate atoms, it is sufficient to consider only the valence electrons. As in the atomic case these are treated to a first approximation as independent particles.

There are two general approaches to the initial wave function. The molecular orbital approach treats the valence electrons as belonging to both nuclei, usually by writing their wave functions as linear combinations of the atomic wave functions. The Heitler-London approach starts with the separated atoms and allows for an exchange of valence electrons. Both approaches yield both 'bonding orbitals', representing a high probability of finding the electrons in the region of low potential between the nuclei, and 'anti-bonding orbitals', representing a low probability. For a bonding orbital the electronic energy is less than for the atoms at infinite separation, and the molecular state is stable; for an anti-bonding orbital it is greater, giving an unstable or repulsive molecular state.

Suppose we now carry this process through for different values of the inter-nuclear separation r. With decreasing r and increasing overlap of electronic wave functions, the bonding and anti-bonding states diverge more and more, as shown by the dotted curves of Fig. 2.21. But these curves represent only the electronic energy, and for the molecule as a whole, while still ignoring the motion of the nuclei, it is necessary to add in the repulsive energy $Z_1 Z_2 e^2 / 4\pi\epsilon_0 r$ between them (dashed curve). At sufficiently small r this must eventually dominate the energy. The effect of super-posing it on the electronic energy is shown by the solid curves. The lower curve now passes through a minimum, the position and depth of which determine the equilibirium inter-nuclear distance r_0 and the binding energy D_e respectively. r_0 is

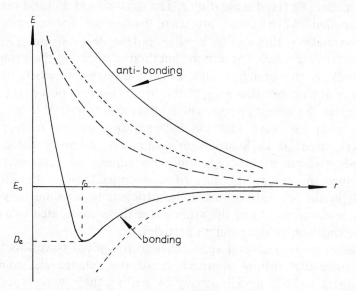

Fig. 2.21 'Electronic' energy of diatomic molecule as a function of inter-nuclear distance. The dotted curves give the electronic energy only, and the dashed curve is the nuclear repulsion term $Z_1 Z_2 e^2 / 4\pi\epsilon_0 r$. The full curves are the result of superposing these two.

typically between 1 and 3 Å and D_e ranges up to 5 eV or, in a few molecules, a little higher.

The electron distribution represented by the equilibrium state is necessarily symmetric with respect to the two nuclei if one is dealing with a homonuclear molecule. Its centroid must be at the mid-point between the two nuclei, coinciding with the centroid of the nuclear charge, and so the molecule has no permanent electric dipole moment. In the case of a heteronuclear molecule, however, the energy balance is generally favoured by a concentration of electron density near one nucleus. In NaCl, for example, the $3s$

electron contributed by the sodium atom and the unpaired $3p$ electron from the chlorine atom both go into a molecular orbital whose probability density is much greater near the chlorine atom than near the sodium atom. As a result, the centroid of the electronic charge distribution is shifted towards the chlorine atom. and the molecule *does* possess a permanent electric dipole moment. Although to some extent such a molecule may be regarded as if it were formed of two well-defined ions, Na^+ and Cl^-, the electron transfer is by no means complete. There is an infinite gradation possible between pure covalent and pure ionic bonding, and the

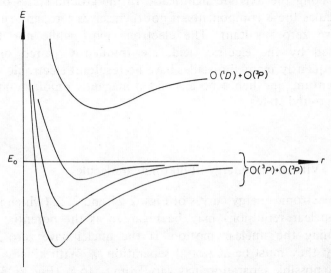

Fig. 2.22 Electronic states of O_2. Only a few of the known states are shown.

electric dipole moment of any molecule is a measure of its degree of ionic bonding. As will be seen in Section 2.18 the appearance of vibrational and rotational spectra depends on the existence of this dipole moment.

The curves of Fig. 2.21 are assumed to be the result of combining two atoms in their ground states. If one starts with one or both atoms in an excited state, one can construct similar sets of curves representing electronically excited states of the molecule. The excitation energy of these states is comparable with that of the atomic states; were it not for the vibration and rotation, transitions between them would give rise to a line spectrum in the visible and ultra-violet. Fig. 2.22 shows some of the lower states of

O_2. It can be seen that two ground-state atoms can form several stable molecular states. This is because their angular momenta can be coupled in several different ways, just as in an atom a given electron configuration can give rise to several different terms. A description of the coupling schemes and the relevant terminology may be found in the textbooks (refs [7-11], for example), but would take too much space here. One particular point is, however, worth making. Because of the strong electric field along the inter-nuclear axis, the orbital quantum numbers l_i of the individual electrons are not 'good quantum numbers': only the components of l_i along the axis are significant. In the ground states of many molecules these components, if not themselves zero, are paired off to give zero resultant. The electron spins, while not directly affected by the electric field, are frequently paired off also. Consequently many molecules have no resultant electronic angular momentum, and hence no associated magnetic dipole moment, in their ground states.

2.16   Vibrational energy of diatomic molecules

The electronic energy curves of Figs. 2.21 and 2.22 (which include the nuclear repulsion) may be regarded as the potential curves governing the nuclear motion. If the nuclei have zero kinetic energy they must be at rest at separation r_0. With kinetic energy E_v the possible separation may vary from r_1 to r_2 (Fig. 2.23).

The shape of the potential curve depends, as we have seen, on the r-dependence of the electronic wave functions ψ_e, but it can often be closely described by an empirical expression known as the Morse potential:

$$V = D_e (1 - e^{-\beta x})^2 \qquad (2.28)$$

V is the energy referred to D_e, the minimum of the curve, β is a constant and x ($\equiv r - r_0$) is the displacement of the nuclei from their equilibrium separation. For small displacements this can be expanded as

$$V = D_e \beta^2 x^2 (1 - \frac{\beta x}{2} + \cdots)$$

To the first order, the restoring force $-\partial V/\partial x$ is $-2 D_e \beta^2 x$, so the nuclear motion is simple harmonic. The classical vibrational frequency is

$$\nu_{\text{osc}} = \frac{1}{2\pi} \sqrt{\frac{2D_e}{\mu}} \beta$$

where μ, the reduced mass of the nuclei, is given by

$$\mu = \frac{M_A M_B}{M_A + M_B} \tag{2.29}$$

Classically, the vibrational energy is determined by the amplitude of the vibrations and may have any arbitrary value. Quantum-mechanically, the energy is restricted to certain discrete values,

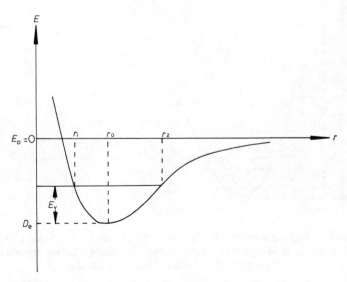

Fig. 2.23 Vibrational energy in a diatomic molecule. r_1 and r_2 are the 'classical turning points' for vibrational energy E_v.

which are obtained by solving Schrödinger's equation (Equation 2.4) in one dimension for the potential $V = D_e \beta^2 x^2$. Physically well-behaved solutions (finite, single-valued, etc.) are possible only for values of E given by

$$E_v = h\nu_{\text{osc}}(v + \tfrac{1}{2}) \tag{2.30a}$$

where the vibrational quantum number v is limited to $v = 0, 1, 2,$
$\cdots$. For practical purposes the vibrational energy is usually
expressed in wavenumbers and designated by the letter G:

$$G(v) = \frac{E_v}{100ch} = \omega_e(v + \tfrac{1}{2}) \qquad (2.30b)$$

The vibrational constant ω_e is in units of cm^{-1} and depends on
the particular electronic term considered. The vibrational levels are
evidently equally spaced; the first few are shown in Fig. 2.24. An

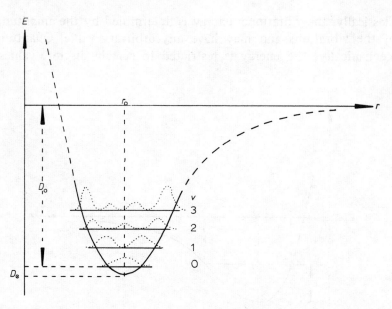

Fig. 2.24 Vibrational energy levels and wave functions. The solid part of the
potential curve is assumed to be parabolic. The dotted curves represent ψ_v^2,
the square of the vibrational wave function.

important difference between the quantum and the classical
oscillator is the existence in the former of the so-called zero-point
energy, the vibrational energy $\tfrac{1}{2}\,\omega_e$ associated with the lowest
level $v = 0$. ω_e is usually in the range 300-3000 cm^{-1}, so the
zero-point energy is of order 0.1 eV. The dissociation energy of
the molecule D_0 is measured from the lowest vibrational level and
is therefore less than D_e, the minimum of the curve, by about
0.1 eV, a not insignificant difference in a total of 2 or 3 eV.

When the full Morse potential (Equation 2.28) is inserted in the Schrödinger equation, instead of just the first term in the expansion, the allowed energy levels are found to be

$$G(v) = \omega_e (v + \tfrac{1}{2}) - x_e \omega_e (v + \tfrac{1}{2})^2 \qquad (2.31)$$

where x_e the anharmonicity constant is small and positive (~ 0.01). The vibrational levels therefore crowd together more closely with increasing v, but they remain finite in number. They do *not* converge towards the dissociation limit as atomic energy levels converge towards the ionization limit. For large values of v, terms of yet higher order in v may be required to describe the energy correctly, since the Morse potential itself is not always a good approximation for large displacements.

The vibrational wave function ψ_v, or rather its square ψ_v^2, determines the probable separation of the nuclei for any given energy E_v. The ψ_v^2 are shown as dotted curves in Fig. 2.24. It can be seen that as v increases the nuclei are increasingly likely to be found near their extreme positions, the turning points of the classical motion.

2.17   Rotational energy of diatomic molecules

Let us now regard the molecule as a rigid dumbell – masses M_A and M_B joined by a rigid bar of length r_0 – rotating about an axis through its centre of mass (Fig. 2.25). The moment of inertia I

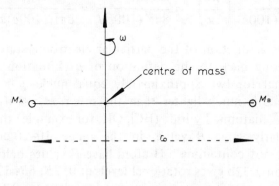

Fig. 2.25 Rotation in a diatomic molecule.

about this axis is easily found to be $I = \mu r_0^2$ in terms of the reduced mass μ (Equation 2.29). Classically, for an angular velocity ω the system has angular momentum $P = I\omega$ and energy $\tfrac{1}{2} I\omega^2 =$

$P^2/2I$. Quantum-mechanically, one must solve Schrödinger's equation (Equation 2.4) for r fixed and $V = 0$. Lacking any dependence on r, ψ is a function of angle only – $\psi = \Theta(\theta)\,\Phi(\varphi)$ – and moreover the Θ and Φ functions are identical with those of the hydrogen atom since the potential $V(r)$ in the hydrogen equation affects only the radial part of the wave function. The usual requirements that ψ be single-valued, finite and continuous therefore lead to restrictions analogous to those for the H atom: the angular momentum is limited to the values $\sqrt{[J(J+1)]}\hbar$, where the rotational quantum number J is an integer (cf. $\sqrt{[l(l+1)]}\hbar$), and the z-component of angular momentum is $M\hbar$, where $M = J, J-1, \cdots, J$. The corresponding restriction on the energy is

$$E_r = \frac{1}{2I} J(J+1)\hbar^2 \tag{2.32a}$$

For practical purposes the rotational energy is expressed in wavenumbers and designated by the letter F:

$$F(J) = \frac{E_r}{100ch} = B_e J(J+1) \tag{2.32b}$$

where the rotational constant B is in cm^{-1} units and is given by

$$B_e = \frac{1}{2Ih\,(100c)}\left(\frac{h}{2\pi}\right)^2 = \frac{h}{8\pi^2\,(100c)I} = \frac{h}{8\pi^2\,(100c)\mu r_0^2} \tag{2.33}$$

Since r_0 is a function of the particular electronic state, so also is B_e. Referring back to the definition of μ (Equation 2.29), it is evident that for two approximately equal nuclei $\mu \approx \frac{1}{2}M$, but if one nucleus is much lighter than the other, $M_A \ll M_B$ say, then $\mu \approx M_A$. A diatomic hydride (HCl, OH, for example) therefore has a particularly large B value, in the range 10–20 cm^{-1}. Most molecules not containing a H atom have B values below 2 cm^{-1}.

Equation 2.32b gives rotational levels at 0, $2B$, $6B$, $12B$, $\cdots$ i.e., the spacing between successive levels increases by $2B$ each time. The first few levels are shown in Fig. 2.26.

Obviously the separation of the nuclei is not really rigidly fixed (vibration would be impossible if this were so), and at high rotational energies one might expect the centrifugal force to increase the separation and hence the effective moment of inertia. This effect is taken account of by writing the rotational energy as

$$F(J) = BJ(J + 1) - DJ^2 (J + 1)^2 \qquad (2.34)$$

where D is a small positive constant. D is of course connected to the 'restoring force' for the vibrational oscillations, and for the Morse potential it is given by $D = 4B^3/\omega^2$. One can see from the relative sizes of ω and B that D is of order $10^{-4} B$, and the correction, which slightly lowers the levels of high J, is very small.

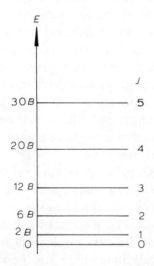

Fig. 2.26 Rotational energy levels. B is the rotational constant defined in the text and J the rotational quantum number.

We have described J as the rotational quantum number, but if the electronic angular momentum is not zero J is actually the resultant of the electronic and rotational angular momenta, the former being itself composed of spin and orbital parts. A given J-level may then be split into several sub-levels characterized by different values of the quantum numbers of the constituent angular momenta, and the expression given above for E_r is only approximately correct. It would take too long here to go into the various ways in which the angular momenta can be coupled, the terminology that describes the possible energy states, and the corresponding energy values. A textbook such as that of Herzberg [8] should be consulted.

2.18 Spectra of diatomic molecules

Because of the very different orders of magnitude of the spacings of the electronic, vibrational and rotational energy levels – each of which is roughly 100 times smaller than the one before – molecular transitions occur in three distinct regions of the spectrum. The pure rotational spectrum, transitions between adjacent rotational levels of the same vibrational and electronic state, is found in the microwave region – say 0.1 mm (100 μm) to 30 cm; the vibration-rotation spectrum, transitions between different vibrational levels of the same electronic state with a fine structure composed of the rotational transitions, falls in the infra-red – say 100 μm to 1 μm; and transitions between different electronic states occur, as for atomic spectra, in the near infra-red, visible and ultra-violet regions. Molecular transitions are traditionally written with the upper state first, which is the opposite way round to atomic transitions. The vibrational and rotational quantum numbers are given a single prime for the upper state and a double prime for the lower state.

Starting with the simplest spectrum, radiative transitions between rotational levels can take place only if the molecule possesses a permanent electric dipole moment. This requirement is easy to understand classically in terms of radiation from a rotating dipole. We shall look at the wave-mechanical representation at the end of this section, indicating there also how the selection rule $\Delta J = \pm 1$ is derived. From the considerations of Section 2.15, homonuclear molecules do not have electric dipole moments and so do not have pure rotational spectra. For heteronuclear molecules the selection rule allows transitions between adjacent J-levels, and it is evident from Fig. 2.26 and Equation 2.32 that their frequencies are $2B$, $4B$, $6B$, $\cdots$ if the centrifugal correction is neglected. The pure rotational spectrum therefore consists of almost equally spaced lines separated by $2B$.

The infra-red spectrum due to transitions between vibrational levels likewise exists only for heteronuclear molecules, the requirement in this case being a dipole moment that varies with the inter-nuclear distance r. Again, the requirement for an oscillating dipole is an obvious one classically, and its wave-mechanical form is indicated below. The selection rule for a harmonic oscillator is $\Delta v = \pm 1$, but for the anharmonic oscillator

$\Delta v = \pm 2, 3, \cdots$ are also allowed as less probable transitions. The infra-red spectrum of a heteronuclear molecule therefore consists of a 'fundamental' at wavenumber ω_e (Fig. 2.24 and Equation 2.30) with weaker 'overtones' close to $2\omega_e$, $3\omega_e \cdots$. The multiples are not exact because of the anharmonicity correction (Equation 2.31). Because of the rotational structure each vibrational transition is in fact a band, or a group of lines spaced at

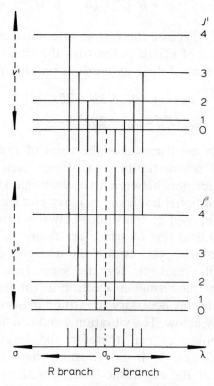

Fig. 2.27 Vibration–rotation band. The diagram is drawn for $B'' \approx B'$ so that the lines are equally spaced at intervals $2B''$. The transition $J' = 0 \leftrightarrow J'' = 0$ (shown dashed) is forbidden.

approximately equal intervals $2B$ on either side of the band origin σ_0. Fig. 2.27 shows the first few rotational levels for each of two vibrational states v' and v'', with the corresponding spectrum underneath. Lines for which $J' = J'' + 1$ fall on the high frequency side of σ_0 and are known as the R branch, while the low frequency side, $J' = J'' - 1$, is the P branch. There is a gap in the position of σ_0, which corresponds to the forbidden transition

$J = 0 \leftrightarrow J = 0$. Again disregarding the D-correction, the wave-numbers of the lines are

$$\sigma = \sigma_0 + B'J'(J' + 1) - B''J''(J'' + 1) \qquad (2.35)$$

which for $J' = J'' + 1$ and $J' = J'' - 1$ respectively becomes

$$\left. \begin{aligned} \sigma_R(J'') &= \sigma_0 + 2B' + (3B' - B'')J'' + (B' - B'')J''^2 \\ \sigma_P(J'') &= \sigma_0 - (B' + B'')J'' + (B' - B'')J''^2 \end{aligned} \right\} \qquad (2.36a)$$

In general $B' \neq B''$ because the average inter-nuclear distance r and hence the moment of inertia varies with the vibrational state, but if the difference is small these expressions simplify to

$$\left. \begin{aligned} \sigma_R(J'') &= \sigma_0 + 2B + 2BJ'' \\ \sigma_P(J'') &= \sigma_0 - 2BJ'' \end{aligned} \right\} \qquad (2.36b)$$

The electronic spectrum takes the form of systems of bands. Fig. 2.28 shows schematically part of one such system, a few bands of the transition between two electronic states A and X. Each pair of vibrational levels (v', v'') may give rise to a band, and each band is composed of transitions between different rotational levels $J' - J''$, only a few of which are shown in the figure. The electronic spectrum appears even if the molecule does *not* have a permanent dipole moment, for the same reason that atomic transitions occur: the change in electron distribution between the two states provides the necessary 'transition moment', as described a little more fully below. The vibrational and rotational energies of homonuclear molecules can therefore be investigated in the electronic transitions. It is not possible here to go into the selection rules for the electronic levels, for these depend on the symmetry properties and the coupling of the angular momenta. We shall simply take A–X as an allowed transition and examine the band structure. Neglecting the anharmonic terms in the vibrational energy, the wavenumbers of the bands are given by Equation 2.30:

$$\begin{aligned} \sigma(v', v'') &= \sigma_e + \omega'(v' + \tfrac{1}{2}) - \omega''(v'' + \tfrac{1}{2}) \\ &= \sigma_{00} + \omega'v' - \omega''v'' \end{aligned} \qquad (2.37)$$

where σ_e is the separation of the potential minima of the A and X states and σ_{00} is the wavenumber of the $(0, 0)$ band. For

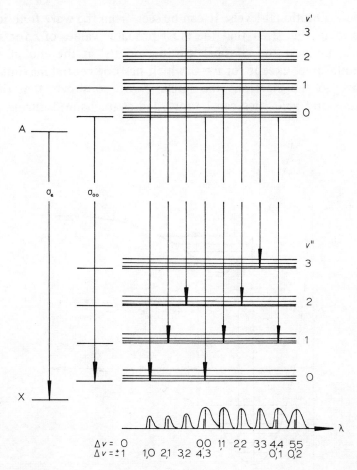

Fig. 2.28 Electronic transition giving rise to a band spectrum. Only the first few rotational levels of each vibrational level are shown, and their spacing is greatly exaggerated. The vertical arrows represent the band origins, corresponding to the vertical lines in the schematic spectrum below. The bands formed by the rotational structure may extend to a considerable distance from the band origins and may overlap one another. The diagram is drawn for

$$\omega' < \omega''.$$

electronic transitions there are no restrictions on Δv, so the entire band system may extend over several units of ω, or a thousand or more Ångströms. If $\omega' \approx \omega''$, bands with the same Δv fall almost on top of one another. The probability of any particular vibrational transition is determined by the Franck Condon principle; this is illustrated in Fig. 2.29, which shows the potential curves for the ground state X and the excited state A with their

first few vibrational levels. It can be seen from the wave functions
sketched in Fig. 2.24 that the most probable values of r for any
given v are near the classical 'turning points' at the end of the
horizontal line, except for $v = 0$ which has one central maximum.
Relative to the electrons the nuclei move so slowly that their
positions and velocities can be regarded as unchanged during the

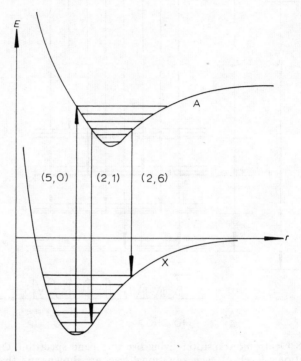

Fig. 2.29 Illustration of Franck-Condon principle. The most probable transitions in emission from $v' = 2$ are to $v'' = 1$ and 6. The most probable
absorption from $v'' = 0$ is to $v' = 5$.

electronic transition, and so the most probable vibrational
transitions are those between levels whose probability maxima
occur at about the same r. These are indicated by the vertical lines
in Fig. 2.29 for the cases of absorption from $v'' = 0$, the (5,0)
band, and emission from $v' = 2$, the (2,1) and (2,6) bands.

Finally, we come to the rotational structure of a single band.
Relative to the band origin σ_0 we have, as before

$$\sigma = \sigma_0 + B'J'(J' + 1) - B''J''(J'' + 1) \qquad (2.35)$$

For electronic transitions, with one exception (when both states have zero resultant electronic orbital angular momentum), the selection rules allow $\Delta J = 0$ as well as $\Delta J = \pm 1$. In addition to the P and R branches given by Equations 2.36 there is now a Q branch, $\Delta J = 0$.

$$\sigma_Q = \sigma_0 + (B' - B'')J'' + (B' - B'')J''^2 \qquad (2.36c)$$

The Q branch overlaps the R branch or the P branch according as $B' > B''$ or $B' < B''$. Another distinction from the infra-red band structure is that the B's of different electronic states are often very different, so that the J^2 term in Equation 2.36a is much more significant. From these equations it can be seen that if $B' > B''$ the terms in J and J^2 have opposite signs for the P branch, which must eventually double back on itself to form a head. If $B' < B''$, the same thing occurs for the R branch instead. In the first case the turning point is at the long wavelength side of the band, which is said to be degraded to the violet; in the second case it is at the other end, and the band is degraded to the red. Because of the sub-division of J-levels when the electronic angular momentum is not zero, each of the P, Q and R branches may itself be sub-divided. A band may therefore consist of several hundred lines, usually very closely packed in the region of the band head and band origin but extending over several hundred cm^{-1}.

As a post-script to this section, the wave-mechanical treatment is outlined below, not in any way rigorously but simply to show where the selection rules come from and to provide a basis for the discussion of molecular oscillator strengths in Section 9.11. The transition probability for electric dipole radiation was shown in Section 2.8 to be proportional to $|R_{if}(er)|^2$ (Equation 2.12). For several electrons

$$R_{if}(er) = \int \psi'(\sum_k er_k)\psi'' \, d\tau$$

Using the Born-Oppenheimer approximation (Section 2.14) we can write

$$\psi = \psi_e(r_k, r)\psi_v(r)\psi_r(\theta, \varphi) \qquad (2.38)$$

where r_k refers to the electron coordinates, r to the inter-nuclear separation, and θ, φ to the angular co-ordinates of the system as a whole. This allows the integral to be separated as

$$R(er) = \int \psi'_e \psi'_v (\sum er_k)\psi''_e \psi''_v \, d\tau \int \psi'_r \psi''_r \, d\tau \qquad (2.39)$$

Let us first consider pure rotational transitions, for which $\psi_e'\psi_v' = \psi_e''\psi_v''$. Then

$$R(er) = \int R_e(r)\psi_v^2(r)\,dr \int \psi_r'(J',M')\psi_r''(J'',M'')\,d\tau \qquad (2.40)$$

where

$$R_e \equiv \int \psi_e^2(\Sigma er_k)\,d\tau$$

The first integral is the average over the vibrational motion of the permanent electronic dipole moment R_e and vanishes if this is zero. The second integral depends on the quantum numbers J', M', J'', M''. Since ψ_r is essentially the same function as the $\Theta_{l,m}\,\Phi_m$ of the central field atom, the same considerations leading there to the selection rule $\Delta l = \pm 1$ (Equation 2.13) lead here to $\Delta J = \pm 1$.

For vibrational transitions Equation (2.40) becomes

$$R(er) = \int \psi_v' R_e(r)\psi_v''\,dr \int \psi_r'\psi_r''\,d\tau \qquad (2.41)$$

The dependence of R_e on r can be expressed by expanding it as a power series:

$$R_e(r) = R_e(r_0) + \frac{\partial R}{\partial r}(r - r_0) + \cdots$$

Substituting this in the first integral of (Equation 2.41) gives

$$R_e(r_0)\int \psi_v'\psi_v''\,dr + \frac{\partial R}{\partial r}\int \psi_v'(r - r_0)\psi_v''\,dr + \cdots$$

The first term vanishes because of the orthogonality properties of the wave functions. The second term vanishes unless R_e varies with the inter-nuclear distance, and if the ψ_v are harmonic wave functions it also vanishes unless $v'' = v' \pm 1$. If the ψ_v are anharmonic functions the integral has a small but non-zero value for other values of Δv. The ψ_r integral is unaffected by the vibrational wave functions, and the selection rule $\Delta J = \pm 1$ holds as before.

Finally, for electronic transitions Equation (2.39) becomes

$$R(er) = \int \psi_v' R_e^{if}(r)\psi_v''\,dr \int \psi_r'\psi_r''\,d\tau \qquad (2.42)$$

where

$$R_e^{if} \equiv \int \psi_e'(\Sigma er_k)\psi_e''\,d\tau$$

R_e^{if} is the electronic transition moment, evaluated between the initial and final electronic states just as in the atomic transitions. Although it is a function of inter-nuclear distance, it does not change rapidly with r as a rule. To a first approximation it may be replaced by an average value $\bar{R}_e^{if}$ and taken outside the r-integral. To the extent that this approximation is valid, one can therefore write R as the product of three integrals, each involving *only* electronic, vibrational or rotational wave functions:

$$R^{if}(er) \approx \bar{R}_e^{if} \int \psi_v' \psi_v'' \, dr \int \psi_r' \psi_r'' \, d\tau \qquad (2.43)$$

The second term is known as the overlap integral and is large only when ψ_v' and ψ_v'' have maxima at approximately the same value of r. This is the wave-mechanical basis of the Franck-Condon principle. The last term is superficially the same as in Equations 2.40 and 2.41, but there is, or may be, one difference. In treating ψ_r as a function of J and M only, no account was taken of any electronic orbital angular momentum. In the ground electronic state, which is where rotation and vibration-rotation spectra are observed, this neglect is usually justified because, as remarked in Section 2.15, only the components of l_k along the inter-nuclear axis are significant, and the resultant Λ of these is frequently zero in the ground state. However, if Λ is not zero it enters into the rotational wave function and modifies the selection rules. If either or both of ψ_r' and ψ_r'' has $\Lambda \neq 0$, the transition $\Delta J = 0$ is allowed.

2.19 Remarks on polyatomic molecules

The structure and spectra of polyatomic molecules are much too complicated to be discussed here, but it is worth pointing out a few features. Although a few polyatomic molecules are of astrophysical importance (e.g., H_2CO, NH_3) the subject is usually regarded as a branch of chemistry rather than of physics. Relatively brief accounts may be found in references [3, 7, 11] and a full account in references [9] and [10].

The complexity of the problem may be understood by considering the number of degrees of freedom of a polyatomic molecule – that is, the number of co-ordinates needed to specify the positions of all the nuclei for a given electronic state. There are three for each nucleus, making a total of $3n$ for a molecule of n atoms. Three of these are required to specify the position of the centre of mass and three more to describe the rotation of the molecule as a whole about each of three axes. The remaining $3n-6$ specify the positions of the nuclei relative to one another and are therefore associated with vibrational motions. (With a linear molecule, as with a diatomic molecule, rotation about the inter-nuclear axis does not change the nuclear coordinates, and so only two degrees of freedom are required for the rotation, leaving $3n-5$ for vibration.) It is possible to choose these $3n-6$ (or $3n-5$) co-ordinates in such a way that each describes a 'normal mode' of vibration, in which all the nuclei vibrate at constant amplitude with the same frequency. The normal mode for the diatomic molecule, for example, is vibration along the inter-nuclear axis, described by the inter-nuclear separation r as the normal co-ordinate. To a first approximation restoring forces are assumed to be proportional to displacements, and each normal mode behaves as a one-dimensional harmonic oscillator, with its energy quantized as for the diatomic case. In simple molecules it is often possible to regard each normal mode as approximately equivalent to either a change in length of the bond between one pair of atoms (stretching vibration) or a change in angle between two bonds (bending vibration). In more complex molecules the normal modes may describe vibrations either of the whole 'molecular skeleton' or of particular groups of atoms such as OH, NH, NO_2, CH, CH_3. Vibrations of the second type are characteristic of the group and almost independent of the particular molecule to which it is attached.

The vibrational frequencies of polyatomic molecules are in the same energy range as those of diatomic molecules – a few hundred to a few thousand cm^{-1}. As with diatomic molecules the vibrational bands have a fine structure of rotational transitions. In general, the larger the molecule the larger its moments of inertia and the smaller the rotational constants B and the spacing of the rotational lines. Moreover, there are often a large number of branches. The band structure is therefore frequently observed as a continuous intensity distribution rather than as a set of resolved lines. The rotational structure is best determined from the pure rotational spectrum in the microwave region in most cases, but if this cannot be done (if, for example, the molecule has no permanent dipole moment) it may be necessary to deduce it from the band structure in the infra-red or Raman spectra. A permanent dipole moment is not necessary for the observation of infra-red absorption in polyatomic molecules because the necessary condition of a *change* in dipole moment is satisfied by at least some of the vibrational modes.

2.20 Raman spectroscopy

Raman spectroscopy is somewhat different, both in principle and in practice, from emission and absorption spectroscopy, but it is such an important tool in molecular analysis that it should be mentioned briefly here. If a gas or a liquid is irradiated with a strong line of arbitrary frequency ν_0 (for example, one of the lines from a mercury lamp), the scattered light is mostly of the same frequency, but one or more weak displaced lines may appear on the long wavelength side at frequencies ν_s, say; these are known as Stokes lines. Even more faintly on the short wavelength side there

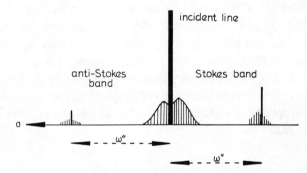

Fig. 2.30 Schematic Raman spectrum. The scattered light contains components at $\pm\omega''$ cm from the incident line, corresponding to vibrational transitions $\Delta v'' = \pm 1$. Rotational structure may appear around both components and the incident line.

may appear the anti-Stokes lines, of frequencies ν_a. The energy difference $\nu_0 - \nu_s$ (or $\nu_a - \nu_0$) is given up to (or taken from) the vibrational or rotational energy of the molecule. This method can be used to investigate molecular vibrational modes that do not lead to infra-red absorption. The condition for Raman spectra to appear is that the molecule must have a polarizability that depends on the vibrational or rotational state. In diatomic molecules this is always the case, and the Raman spectrum consists of a Stokes and an anti-Stokes band at frequencies $\pm\omega''$, each with rotational structure determined by a selection rule that can be shown to be $\Delta J = 0, \pm 2$. The rotational structure also appears about the position of the incident line (Fig. 2.30). Polyatomic molecules also have pure rotational Raman spectra (except for the spherically symmetric class of molecule). Usually some of the vibrational modes are Raman-active and some are not. In certain cases the rules for the appearance of infra-red and Raman spectra are mutually exclusive. Indeed, the type of symmetry in the molecule may be deduced from the mere existence of either or both of Raman and infra-red spectra.

The most severe experimental problem is usually that of inadequate intensity, since the shifted lines are very weak and have to be distinguished from the very intense elastically scattered unshifted line. Until recently, Raman spectroscopy has required spectrographs of very great aperture and has been largely confined to liquid samples, for which the density of scattering centres is of course much greater. In this connection it should be pointed out that free rotation of molecules is not possible in the liquid state, so that no discrete rotational structure appears in Raman spectroscopy of liquids. The same limitation applies equally, of course, to absorption spectra, but for this purpose it is much easier to work with gaseous samples. Raman spectroscopy is, however, one technique for which the laser is the ideal light source, and the experimental problems are nothing like as severe as they were before it became available.

References

Atomic and molecular

1. Dixon, R. N. 'Spectroscopy and Structure', Methuen, 1965
2. Slater, J. C. 'Quantum Theory of Matter', McGraw-Hill, 1968
3. Walker, S. and Straw, H. 'Spectroscopy', Vols. 1 and 2, Chapman and Hall, 1961/2

Atomic

4. Herzberg, G. 'Atomic Spectra and Atomic Structure', Dover, 1944
5. Kuhn, H. G. 'Atomic Spectra', Longmans, 1969
6. Woodgate, G. K. 'Elementary Atomic Structure', McGraw-Hill, 1970

Molecular

7. Banwell, C. N. 'Fundamentals of Molecular Spectroscopy', McGraw-Hill, 1966
8. Herzberg, G. 'Spectra of Diatomic Molecules', Van Nostrand, 1959
9. Herzberg, G. 'Infra-red and Raman Spectra', Van Nostrand, 1947
10. Herzberg, G. 'Electronic Spectra of Polyatomic Molecules', Van Nostrand, 1966
11. Barrow, G. M. 'Introduction to Molecular Spectra', McGraw-Hill, 1962

CHAPTER THREE

Light sources and detectors

Light sources for spectroscopy may be divided into broad-band, or continuum, sources suitable for intensity standards or as backgrounds for absorption spectroscopy and narrow-band, or line, sources suitable for emission spectroscopy. Before discussing individual types of light source it is helpful to deal briefly with some general concepts, such as temperature, brightness, excitation energy and equilibrium.

3.1 Sources of continuous spectra

For a solid, or a gas in thermodynamic equilibrium, the temperature T can be defined thermodynamically. The density of radiation of wavelength between λ and $\lambda + d\lambda$ in the gas or in a cavity in the solid is given by Planck's formula

$$\rho(\lambda, T)\, d\lambda = \frac{8\pi hc}{\lambda^5} \frac{1}{e^{hc/\lambda kT} - 1}\, d\lambda \ \ \mathrm{Jm^{-3}} \qquad (3.1a)$$

where $\rho(\lambda, T)$ is the energy density per unit wavelength interval and k is Boltzmann's constant. Expressed in terms of unit frequency interval, this becomes

$$\rho(\nu, T)\, d\nu = \frac{8\pi h\nu^3}{c^3} \frac{1}{e^{h\nu/kT} - 1}\, d\nu \ \ \mathrm{Jm^{-3}} \qquad (3.1b)$$

The flux of radiation in the waveband $d\lambda$ escaping from a small hole of unit area into unit solid angle defines the luminance or brightness of a black body source:

$$B_0(\lambda, T)\, d\lambda = \frac{\rho(\lambda, T)c}{4\pi}\, d\lambda = \frac{2hc^2}{\lambda^5} \frac{1}{e^{hc/\lambda kT} - 1}\, d\lambda \ \ \mathrm{Wm^{-2}\,ster^{-1}} \quad (3.2a)$$

Expressed as a function of frequency this becomes:

$$B_0(\nu, T)\, d\nu = \frac{2h\nu^3}{c^2} \cdot \frac{1}{e^{h\nu/kT} - 1}\, d\nu \ \text{Wm}^{-2}\text{ster}^{-1} \qquad (3.2b)$$

Fig. 3.1 shows the Planck distribution as a function of wavelength for two different temperatures. With increasing temperature the absolute value of the energy increases and the maximum of the

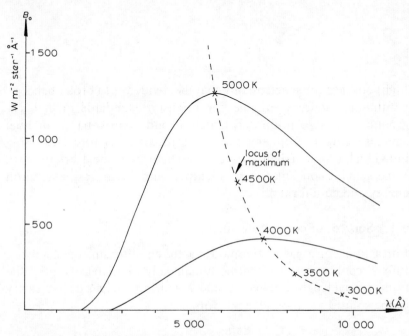

Fig. 3.1 Spectral distribution of black body radiation. Note that $B(\lambda, T)$ is plotted here; curves of $B(\nu, T)$ against ν have a somewhat different shape. The dashed line shows the positions of the maxima at different temperatures. The peak value increases as T^5.

distribution moves towards shorter wavelengths. It can be shown that

$$\lambda_{\max} T = \text{const} = 2.884 \times 10^{-3} \ \text{m K} \qquad (3.3)$$

(Wien's displacement law).

No source can have greater brightness at any wavelength than a black body at the same temperature, a fact that may be expressed by writing

$$B(\lambda, T) = \epsilon_\lambda B_0(\lambda, T) \qquad (3.4)$$

where the emissivity $\epsilon_\lambda \leqslant 1$. ϵ_λ is in general a function of wavelength, but if it is almost constant the source is known as a grey body. An alternative way of describing the brightness at any given wavelength is to allot the source a 'brightness temperature', T_b, which is the temperature of a black body radiating the same energy at that wavelength – i.e.,

$$B(\lambda, T) = B_0(\lambda, T_b) \tag{3.5}$$

Obviously if $\epsilon_\lambda < 1$ the brightness temperature T_b must be less than the true temperature T, and it must vary with λ unless ϵ is constant.

Another concept that is sometimes useful is that of 'colour temperature'. If a source of true temperature T has brightness $B(\lambda_1, T)$ and $B(\lambda_2, T)$ at two different wavelengths, its colour temperature T_c is the temperature of a black body having the same relative brightness, so that

$$\frac{B_0(\lambda_1, T_c)}{B_0(\lambda_2, T_c)} = \frac{B(\lambda_1, T)}{B(\lambda_2, T)} = \frac{\epsilon_1 B_0(\lambda_1, T)}{\epsilon_2 B_0(\lambda_2, T)}$$

by using Equation 3.4. For a grey body ($\epsilon_1 = \epsilon_2$) the colour temperature is identical with the true temperature. Lamp filaments are usually very nearly grey, and this makes them most useful intensity standards in spectroscopy, since it is often relative intensities that are required rather than absolute values. If ϵ increases towards the blue end of the spectrum, the colour temperature is actually higher than the true temperature.

The simplest background source for absorption spectroscopy is an incandescent solid, but any solid necessarily has an upper temperature limit, which is in practice 3655 K, the melting point of tungsten. There are two reasons why higher temperatures may be required. First, with dispersing systems of high resolution or great complexity a very bright source may be necessary to avoid unduly long photographic exposures or photoelectric integration times; the problems get worse towards shorter wavelengths as light losses in the optical system increase and the black body distribution drops rapidly. Secondly, absorption lines appear only if the brightness temperature of the background source in the relevant spectral region is greater than that of the absorbing gas or plasma. It is therefore often necessary to use as a background a hot gas in some form of electric discharge. This is most unlikely to be in thermodynamic equilibrium at a well-defined temperature T, and its radiated energy distribution may bear little resemblance to a Planck function.

3.2   Intensity standards

A primary standard is a source for which the radiated power is known theoretically as a function of wavelength. The only practicable such source is the black body enclosure, or 'hohlraum' (apart from the electron synchrotron at very short wavelengths). A true black body is closely approximated by a furnace with a small observation hole, and the ultimate standard of brightness, or luminance, is such a furnace at a temperature defined by the melting point of gold. It can be seen from Fig. 3.1 that a black body furnace is a feasible laboratory standard in the infra-red, near the peak of the distribution at moderate temperatures, but the sharp drop of intensity on the short wavelength side of the maximum makes it highly inconvenient for the visible and ultra-violet regions. Secondary standards that can be calibrated against the primary standard are therefore essential.

The most useful secondary standards are the tungsten and tungsten-halide strip filament lamps. These lamps are fairly good grey bodies, useful for wavelengths from about 10 000 Å to 2000 Å when fitted with quartz envelopes. They can be run up to a true temperature of about 3000 K, corresponding to a brightness temperature of about 2700 K in the visible. A somewhat higher brightness temperature is produced in the anode crater of a carbon arc (~3800 K at 10 amp), but this is an awkward source, rather overlaid by molecular bands and seldom used except as a measure of desperation in the ultra-violet.

No simple secondary standards exist for the vacuum ultra-violet. Recent work has established the electron synchrotron as a primary standard at very short wavelengths (below say 500 Å), but its intensity falls rapidly with increasing wavelength. Section 3.8 contains a little more information about this source. The alternative approach is to calibrate some arbitrary but reproducible source by means of a detector which either measures absolute intensity or has in its turn been calibrated against a standard source in the visible. The spectral response of the detector must of course be accurately known. A thermopile, for example, measures the energy incident on it, regardless of wavelength, and is suitable for the second role. The first – that of an absolute detector – can be filled by photo-ionization detectors in certain conditions (Section 3.13).

3.3 The excitation of line spectra

The intensity of radiation of frequency ν_{12} emitted by an atom or a molecule as a result of a radiative transition between two discrete states (2) and (1) is determined by the probability of finding the atom or molecule in the initial state (2) as well as by the inherent probability of the particular transition $2 \rightarrow 1$. This last has already been discussed in terms of time-dependent perturbation theory in Section 2.8. Before pursuing it further, it is necessary to investigate the first point – the distribution of atoms and molecules over the various excited states. To avoid repetition, we shall refer in what follows to atoms, but the remarks apply equally to molecules provided the rotational levels are regarded as separate and discrete.

It can be shown by statistical mechanics (see [3] for example) that for a system in equilibrium at temperature T the ratio of the numbers of atoms occupying the two energy states E_2 and E_1 is given by Boltzmann's formula:

$$N_2/N_1 = g_2/g_1 \ e^{-(E_2 - E_1)/kT} \tag{3.6}$$

The statistical weight g of a state is equal to its degeneracy – that is, the number of distinct sub-states having the same energy. The concept of degeneracy was first introduced in Section 2.2 in connection with the hydrogen atom for which, ignoring the very small energy differences due to fine structure, there are $2n^2$ different wave functions having the same value of energy E_n. More generally, the energy of a many-electron atom depends on the electron configuration $n_i l_i$ and on the quantum numbers L, S and J, but not on the quantum number M (Section 2.7). A level of given J therefore has a degeneracy of $2J + 1$ in the absence of an external field – that is, its statistical weight is $2J + 1$. In an external magnetic field each M sub-level has a different energy, giving $2J + 1$ sub-levels each with unit statistical weight. Similarly, each rotational level of the simple vibrator-rotator model of a diatomic molecule has a statistical weight of $2J + 1$ multiplied by the statistical weight of the relevant electronic state (vibrational levels being non-degenerate).

The conditions in which Boltzmann's formula may be expected to hold require some discussion. A source of line spectra cannot be in complete thermodynamic equilibrium, because if the radiating gas is enclosed in a black body cavity the spectral distribution is necessarily continuous, as given by Planck's formula. What, then, is meant by the temperature of a gas in incomplete thermo-dynamic equilibrium? It is shown in Chapter 11 that the Boltzmann distribution still holds provided that the states are populated and depopulated predominantly by collisions rather than by radiative processes. T is then the parameter determining the mean kinetic energy, $\frac{3}{2}kT$, of the colliding particles, assuming the velocities to follow a Maxwellian distribution. Electrons are so much more effective for collisional excitation than neutral atoms that only the electron velocity distribution is relevant if electrons are present in any quantity. The temperature, T_e, must then be taken to refer to the electron kinetic energy; indeed, it is often expressed in electron volts rather than degrees. Since $1\ eV = 1\cdot6 \times 10^{-19}$ joule and $kT = 1\cdot38 \times 10^{-23}\,T$ joule, a useful equivalent for order of magnitude calculations is that $1\ eV$ corresponds to about $10\ 000\ K$ (the exact value is $1\cdot16 \times 10^4\ K$).

The ratio g_2/g_1 in Equation 3.6 is of order unity, so the likelihood of populating an excited state (2) depends on its excitation energy E_2 relative to kT. For spectroscopic purposes k may be usefully expressed in wavenumbers as $0\cdot7\ cm^{-1}$ per degree. At room temperature kT is about $200\ cm^{-1}$, and one would expect the rotational and to some extent the vibrational levels of the ground electronic state of a molecule to be populated. Most excited electronic states of molecules and atoms are some tens of thousands of wavenumbers above the ground state and require electron temperatures of several thousand degrees for appreciable population. For example, in the case of the yellow resonance lines of sodium $E_2 \sim 17\ 000\ cm^{-1}$, and $e^{-E_2/kT}$ is only about $0\cdot1$ even at $10\ 000\ K$.

The excitation of ionic spectra depends on the degree of ionization in the source. In equilibrium – that is, when ionization and the reverse process of recombination are predominantly due to electron collisions – the ratio of ions to neutral atoms is given by Saha's equation (Equation 11.11). This ratio has a certain similarity to the N_2/N_1 ratio of Boltzmann's equation in that it is proportional to $e^{-\chi/kT}$ where χ is the ionization energy; but it is also inversely proportional to the electron density, and there are additional temperature-dependent terms. Explanation and discussion of Saha's equation will be postponed to Chapter 11, but one point should be made here. There may be appreciable ionization even if χ is considerably larger (say 10 times) than kT. For example, at 5000 K and an electron density of 10^{15} cm^{-3} $(10^{21}$ m$^{-3})$, conditions typical of an arc or shock tube, the degree of ionization of calcium $(\chi = 6\cdot1$ eV) is about 70%, while for oxygen $(\chi = 13\cdot6$ eV) it is almost zero. At 10 000 K and the same electron density ionization is virtually complete for Ca and is about 40% for O. Multiple ionization becomes significant at temperatures of this order; about 60% of the calcium would be doubly ionized at the higher temperature in this particular example.

3.4 The Einstein probability coefficients

The intensity of a given spectral line depends not only on the population of the initial level but also on the intrinsic probability of the particular transition, and this last is conveniently defined by the Einstein coefficients. Fig. 3.2 shows two energy levels E_1 and

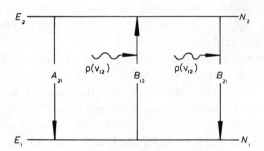

Fig. 3.2 Illustration of the Einstein probability coefficients. N_1 and N_2 are the population densities of the levels E_1 and E_2 respectively. The frequency ν_{12} is given by $h\nu_{12} = E_2 - E_1$.

E_2, populated respectively by N_1 and N_2 atoms cm^{-3}. There are three possible radiative processes connecting the two levels. First, an atom in level (2) may undergo spontaneously a transition to level (1) with emission of energy $h\nu_{12}$; the probability of this occurring in unit time is denoted by the coefficient A_{21}. The number of such transitions per sec per cm^3 is therefore $A_{21}N_2$. Secondly, in the presence of radiation of density $\rho(\nu_{12})$ of the appropriate frequency an atom in level (1) may jump to level (2) with absorption of energy $h\nu_{12}$; the probability of this process is denoted by $B_{12}\rho$, and the number of such transitions is then $B_{12}N_1\rho$ sec^{-1} cm^{-3}. Finally, an atom in state (2) in the presence of radiation $\rho(\nu_{12})$ may undergo an induced transition to level (1) with emission of energy $h\nu_{12}$; the probability for this is denoted by $B_{21}\rho$, and it gives rise to $B_{21}N_2\rho$ additional downward transitions per sec per cm^3. This last process is the least obvious of the three. It is in effect negative absorption, since the emitted radiation has the same direction and phase as the stimulating radiation. Regarding the radiation as a time-dependent perturbation to the atom, it is at least understandable that the perturbation should be as effective for $2 \to 1$ as for $1 \to 2$ transitions.

For electric dipole transitions the Einstein coefficients are proportional to the square of the electric dipole strength, $|R_{if}(er)|^2$, defined in Section 2.8. The proportionality constant will be investigated in Chapter 9 in the course of a more detailed discussion of transition probabilities and their relations to absorption coefficients, oscillator strengths and refractive indices. At this stage we shall simply derive the relationship between the three Einstein coefficients, which is quite general and independent of the particular type of transition – electric dipole, electric quadrupole or magnetic dipole – involved.

For this purpose, suppose the assembly of atoms to be enclosed in a black body cavity at temperature T. Then the population ratio N_2/N_1 and the radiation density $\rho(\nu_{12})$ are given by Boltzmann's formula (Equation 3.6) and Planck's formula (Equation 3.1b) respectively. In equilibrium the rate at which atoms enter level (2) must be equal to the rate at which they leave it. By the principle of detailed balance this equality must hold separately for collisional and radiative processes and for transitions to and from all other levels individually. In particular, the rate of radiative transitions upward into (2) from (1) must equal the rate downward from (2) to (1).

Therefore

$$B_{12}N_1\rho(\nu_{12}) = A_{21}N_2 + B_{21}N_2\rho(\nu_{12}) \qquad (3.7)$$

Re-arranging this:

$$\rho(\nu_{12}) = \frac{A_{21}N_2}{B_{12}N_1 - B_{21}N_2} = \frac{A_{21}}{B_{12}(N_1/N_2) - B_{21}}$$

By Boltzmann's equation

$$N_1/N_2 = g_1/g_2 \; e^{h\nu/kT}$$

where $E_2 - E_1$ has been set equal to $h\nu$ and the subscript on the ν dropped for convenience.

$$\therefore \quad \rho(\nu) = \frac{A_{21}}{B_{12}(g_1/g_2)e^{h\nu/kT} - B_{21}}$$

This is identical with the Planck expression

$$\rho(\nu) = \frac{8\pi h\nu^3/c^3}{e^{h\nu/kT} - 1}$$

for arbitrary T if and only if

$$\left. \begin{array}{c} g_1 B_{12} = g_2 B_{21} \\[2mm] A_{21} = \dfrac{8\pi h\nu^3}{c^3} B_{21} \end{array} \right\} \qquad (3.8)$$

The Einstein coefficients are intrinsic atomic properties (it was seen in Section 2.8 that they can in principle be calculated from the wave functions) and cannot be changed by moving the atoms out of the black body enclosure. Therefore, although the relations (3.8) were derived for thermodynamic equilibrium, they must still hold in any other conditions. It can be seen that induced emission is necessary for the balancing process: B_{21} cannot vanish without taking A_{21} with it. It can also be seen by comparing the two expressions for ρ that the induced emission corresponds to the -1 term in the denominator of the Planck equation. This term can obviously be neglected if $h\nu/kT \gg 1$; this is the condition of validity for Wien's approximation to the black body function: $\rho(\nu) = 8\pi h\nu^3/c^3 \; e^{-h\nu/kT}$. Since $g_1 \sim g_2$ the induced emission or 'negative absorption' can similarly be neglected by comparison with the 'positive absorption' in normal conditions if $h\nu \gg kT$. This inequality holds good throughout the visible and ultra-violet regions except at extremely high temperatures. For $\lambda = 1\mu m$ and

$T = 5000$ K, $e^{h\nu/kT} = e^{2\cdot9} \approx 17$, and for shorter wavelengths correspondingly higher temperatures give the same result. As a useful guide, Wien's approximation is accurate to 1% for $\lambda T \leqslant 0\cdot3$ cm K. However, abnormal conditions, in which induced emission cannot be neglected, can be created by inflating the population of level (2) well above its thermal equilibrium value; this is just what is done in a laser, as will be seen later. In the infra-red and microwave regions $h\nu$ is comparable to or less than kT. At room temperature $kT \approx 200$ cm^{-1}, which corresponds to a wavelength of 50 μ. Induced emission is then obviously important. In other words, if levels (1) and (2) are almost equally populated, the radiation will induce approximately equal numbers of upward and downward transitions. Moreover, since $B_{21}/A_{21} \propto \lambda^3$, at long wavelengths the induced emission can become greater than the spontaneous emission at sufficiently high radiation density. This is the principal reason that masers were invented before lasers.

Before leaving the coefficients, it should be pointed out that some authors define the B-coefficients in terms of the specific intensity of radiation instead of the radiation density. If the radiation is isotropic, the flux per steradian crossing unit area is $I(\nu) = \rho(\nu)c/4\pi$. Since the two definitions must give the same actual transition probability, $B_{21}^{I}I(\nu) = B_{21}^{\rho}\rho(\nu)$, leading to $B_{21}^{I} = (4\pi/c)B_{21}^{\rho}$ and $A_{21} = (2h\nu^3/c^2)B_{21}^{I}$. If ρ is defined in terms of unit *wavelength* interval,

$$\frac{\rho(\lambda)}{\rho(\nu)} = \frac{d\lambda}{d\nu} = \frac{c}{\lambda^2},$$

and $B^{\rho,\lambda} = (\lambda^2/c)B^{\rho,\nu}$.

3.5 The different types of light source

Apart from the temperature, or degree of excitation, of a light source, there are a number of other characteristics that may be important, such as the width of the spectral lines emitted, the extent to which thermal equilibrium is attained, the spectral region in which the source is useful, whether it be pulsed or continuously running, and so on.

Line width forms the subject of Chapter 8. It will be seen that the two most important line-broadening processes are Doppler broadening, due to the thermal motion of the emitting atoms, and pressure broadening. At high densities, particularly at high electron and ion densities, pressure broadening may be large, but in low pressure laboratory sources Doppler broadening is dominant. In a non-equilibrium source such as a low pressure glow discharge the gas kinetic temperature, which determines the Doppler width, may be close to the temperature of the surroundings (room temperature, or below if the source is cooled) when the electron temperature, which determines the excitation, is in the region of 10 000 K. In an arc at atmospheric pressure, on the other hand, the two temperatures are usually similar (a few thousand degrees), and Doppler and pressure broadening may both be quite large.

Sources suitable for the measurement of transition probabilities, which have to satisfy particular requirements as to thermal equilibrium, will be further discussed in Chapter 10. The next few sections are concerned with the traditional spectroscopic emission sources and some of their recently developed counterparts, including sources suitable for the infra-red and the far ultra-violet and a short discussion of masers and lasers. Since many of the new types of source are not mentioned in standard textbooks on experimental spectroscopy, a fairly extensive list of references is given at the end of the chapter.

3.6 Traditional sources: flames, arcs, sparks and glow discharges

3.6.1 Flames

Most flames have temperatures of the order of 2000 K, although certain stoicheiometric flames can reach about 4000 K. Because of the low excitation energy ($\sim\frac{1}{4}$ eV) the most important atomic lines excited are the resonance lines of the alkalis and alkaline earths and these are often used to measure the temperature of the flame by the reversal method (Section 11.16). The principal use of the flame is in the study of the molecules and radicals formed in the combustion process and in atomic absorption spectroscopy and spectrochemical analysis [12].

3.6.2 Arcs

The traditional type of free-burning arc runs typically at a few amps and a few tens of volts, producing temperatures of the order of 5000 K; this is high enough to excite neutral spectra of atoms and molecules and a number of ionic lines. The arc is struck between metal or carbon electrodes, usually in air at atmospheric pressure, and has a column about 1 cm long. Where practical, the arc can be struck between pole pieces made of the metal whose spectrum is required, as in the case of the iron and copper arcs used widely through the visible and quartz ultra-violet for reference wavelengths. Other elements may be introduced as salts in an arc run between carbon electrodes, or as trace impurities in copper electrodes. Arcs may also be run in other gases and under reduced pressure.

Various special arcs have been developed to reach higher temperatures. High current arcs (up to 500 amps) with water-cooled electrodes can produce axial temperatures as high as 30 000 K. The arc column is restricted to a few mm diameter, either by a helical swirl of gas or water (vortex arc) or by a series of water-cooled diaphragms with axial holes (wall-stabilized arc). It is possible to introduce powdered solids as well as vapours into these arcs. Wall-stabilized arcs are particularly important in the measurement of transition probabilities, as described in Chapter 10, when high stability and close approach to thermal equilibrium are essential. Another development of the high current arc has led to the plasma jet: gas sucked in to the constricted arc column is heated and spewed out through a hole in the anode. With a current of 200 amps (current density 2×10^9 amp m^{-2}) temperatures of around 20 000 K are produced in the jet. ([13], [14], [15]).

3.6.3 Sparks

Sparks require high voltage (10–20 kV) from a transformer or induction coil to break down an insulating gap. The transformer may be connected directly across the spark gap, but it is usually used to charge a capacitor, and the stored energy, a few joules, is dissipated across the gap every half cycle. Several degrees of ionization can be reached in atoms of the metal forming the spark electrodes or of the gas through which it passes. The old names 'arc and spark spectra' for the spectra of neutral atoms and ions respectively come from these traditional methods of exciting them, although high degrees of ionization are also attained in a high current arc.

In the vacuum ultra-violet, moderate degrees of ionization can be attained in a spark through a gas at low pressure confined in a capillary tube. The degree of excitation can conveniently be controlled by varying the gas pressure and the circuit inductance. Below 500 Å, and for very high ionization, a vacuum spark is necessary. The highest possible excitation is attained with an open spark at a voltage approaching 100 kV, when the electron temperature may be as high as 4 keV. In spite of the high electron density ($\sim 10^{20}$ cm^{-3}) the spark is too fast for full ionization equilibrium to be reached. Although 20 stages of ionization have been achieved, something around 10 is much more common. Rather less violent excitation is achieved with a sliding spark, in which the voltage break-down takes place across the surface of an insulator. The spectrum contains lines of electrode and insulator materials and of any low-pressure gas introduced into the gap [10], [11].

3.6.4 Glow discharges

Gases and vapours are often conveniently excited in the positive column of a glow discharge, or in an electrodeless discharge, operating at pressures of a few torr. Molecular, atomic, and ionic spectra may all be excited in this way. Excitation is by electron bombardment throughout the length of the positive column, which can easily be made to extend for a metre. A discharge of this type has to be run off a high voltage source (a kilovolt upwards) with a ballast resistor to stabilize it and limit the current to a couple of amps or less.

Another form of glow discharge, the hollow cathode, makes use of the cathode glow. At suitable pressures, the cathode glow will jump into and fill a hollow cathode a few mm in diameter, producing an intense and fairly compact source. Material from the walls of the cathode is sputtered into the discharge and there excited, so that metals and salts as well as gases can be studied. For high resolution work this source has great advantages: the electric field in the cathode glow is very small, and the whole cathode can readily be cooled to liquid air temperatures (provided the current is limited to 20 milliamps or so), thereby reducing both Stark and Doppler broadening (see Chapter 8). The spectra of ions, as well as neutral atoms, can be excited in this source [16].

Electrodeless discharges, run from radiofrequency or microwave sources, are particularly useful for producing high purity spectra.

The gas, or solid with appropriate carrier gas, is sealed into a pyrex or quartz tube placed in a coil or resonant cavity, and there is no contamination from electrode material.

3.7   Recently developed sources

3.7.1   Shock tubes

These were originally developed for aerodynamics, but have become a most useful spectroscopic source. A shock wave is produced (in the case of a pressure driven shock tube) by the bursting of a diaphragm between a region of high pressure light gas and one of low pressure heavier gas. As the shock wave travels down the tube, through the low-pressure gas, it leaves behind it a region of almost uniform hot gas at a few thousand degrees and at a pressure of a few atmospheres lasting, at any given observation point, for perhaps a few milliseconds. The temperature and pressure of the shocked gas is comparable to that of an arc (or, in the case of a low temperature shock, to a flame), but the source has the advantage of a relatively large homogeneous region in which both temperature and pressure can be measured and independently controlled. The transit time is sufficiently long for easy time resolution. The spectra of the atoms or molecules of the shocked gas, or of atoms introduced in the form of powdered solids, can be studied either in emission or, using a suitable flash as background, in absorption. The heating effect may be increased by a factor of two by reflecting the shock wave from the end of the tube [17, [18], [21], [24].

3.7.2   Pinch discharges

The pinch discharge is another instance of a device developed for an entirely different purpose – in this case, controlled thermo-nuclear reactions – that has found useful applications in spectro-scopy. It works on the principle that the magnetic field associated with a pulse of high current can be used to squeeze a column of ionized gas rapidly in towards the axis. As the collapse is completed, the ordered radial velocities are converted by collisions to random or thermal velocities, giving a narrow column of very hot gas lasting, typically, for a few microseconds before it expands and cools. The high degree of ionization attained in the pinch and its short duration makes it comparable to a spark, but because the plasma has dimensions measured in centimetres rather than fractions of a millimetre and characteristics that are to a large extent controllable and reproducible the source can also be used for investigating atomic processes such as line broadening. The operation of a pinch discharge may be illustrated by the linear or

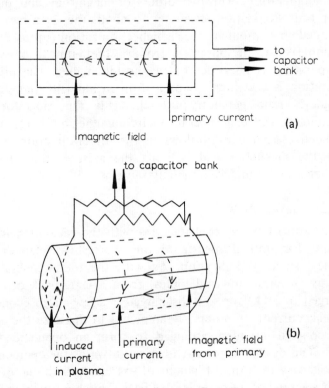

Fig. 3.3 Current and magnetic field in (a) *Z*-pinch and (b) a θ-pinch discharge. In both figures the direction of the current is indicated by dashed lines and the direction of the magnetic field by solid lines.

Z-pinch. A capacitor bank consisting of a large number of capacitors in parallel is charged to several kilovolts, and the stored energy, some hundreds or thousands of joules, is discharged through a gas at low pressure (a few millitorr to one torr) between electrodes at either end of a cylindrical discharge tube, which may be a metre or so in length. The inductance of the whole circuit is kept very low, so that the discharge takes only a few microseconds and the peak current is of order 10^5 amp. The lines of force of the associated magnetic field are circles concentric with the axis of the discharge tube (see Fig. 3.3a) and the charged particles are therefore driven radially in towards the axis. Typical values for the electron temperature and density at maximum constriction are 10^5 K and 10^{23} m^{-3} (10^{17} cm^{-3}) respectively, but with suitable choices of circuit parameters and gas pressure it is possible to vary the temperature over two orders of magnitude and the density over four. Ionization stages up to about the sixth may easily be attained in a Z-pinch. Much higher ionization (up to Fe XV, for example) can be obtained from a theta-pinch, in which the temperature is an order of magnitude higher. The principle of operation is the same as for a Z-pinch but with the current and magnetic field geometry interchanged: the capacitor bank is discharged through a coil encircling the discharge tube, the induced plasma current flows in the 'theta direction' (Fig. 3.3b) and the magnetic field is along the axis, so that the charged particles are again driven radially inwards [19]-[24].

3.7.3 Plasma focus

This form of pinch discharge vies with the vacuum spark for the record for optical spectra of the shortest wavelength and the highest stage of ionization. As with the theta-pinch the stored energy (several thousand joules) from a large bank of capacitors charged to 30 kV or so is discharged in a few microseconds, giving a peak current of order 10^6 amp, but in this case the electrodes are co-axial cylinders arranged to form an open-ended 'plasma gun'. The discharge current is driven towards the open end, where it collapses to form a filament of extremely high energy density, with a temperature of order 1000eV. This blob of high temperature plasma has a volume of about 0·01 cm^3 and lasts for only $1/10\,\mu$sec, but the electron density is great enough ($\sim 10^{19}$ cm^{-3}) to produce a high degree of ionization in this time. The spectra come from both the electrode material and the filling gas. For example, lines of the helium- and hydrogen-like spectra of 16 and 17 times ionized argon, between 3 and 4 Å, have been photographed with this device [22]-[24].

3.7.4 Laser-produced plasmas

A plasma in the same temperature range as the theta pinch (50–500 eV) but with a very much higher electron density,

10^{20} cm^{-3} or more, may be produced by focusing a Q-spoiled giant pulse ruby or neodynium laser on a solid target. With a peak laser power of order 100 megawatts, stages of ionization in the range 12 to 15 are attained despite the short lifetime of the plasma (only a few nanoseconds) because of the high electron density. Such a plasma is particularly useful for study of the high ionization stages of metals used as the target material, in contrast to the pinch discharges in which the spectra are mainly those of ions of the gas in the discharge tube [24].

3.7.5 Beam foil source

Beam foil spectroscopy uses a source of an entirely different nature. Its most important applications are to the measurement of radiative lifetimes, as described in Chapter 10. A beam of ions accelerated to around 1 MeV in a van der Graaf accelerator is fired through a thin foil and emerges, by some process which is not properly understood, as a mixture of ions of various degrees of ionization and in various states of excitation, but with its beam character unimpaired and its velocity essentially unaltered. A large number of ionic (and in certain cases a few neutral atomic) lines are emitted in the region beyond the foil as the ions return to their ground states. The degree of ionization reached is up to 5–7, and because of the non-thermal character of the excitation the excited levels may differ considerably from those obtained in other sources. There is usually considerable difficulty in identifying the excited levels, partly because the large Doppler width (see Section 10.7) results in the blending of a number of lines and partly because the radiated energy is rather low owing to the small particle density in the beam ($\sim 10^5$ cm^{-3}). The intensity of the beam depends rather critically on the foil as to material, thickness and ageing. The material must be light to reduce scattering and energy loss, and carbon is usually used in thicknesses of about 10 microgram cm^{-2} (say about 500 Å). As an aid to identification of the spectra, ions of different charge may be separated spatially by means of electric fields, but even so most experiments yield a large residue of unidentified lines. The source has, however, proved useful in the analysis of a number of ionic spectra, particularly in following along iso-electronic spectra at an intermediate stage between the low ionization produced by an arc and the very much higher ionization produced by plasma sources [25].

3.8   Absorption spectroscopy

3.8.1   General

Absorption spectroscopy requires a background source of radiation that is continuous over the relevant spectral region, and an absorption vessel. Except for the permanent gases, the latter is normally a furnace, running usually in a reducing or inert atmosphere. The temperature attainable is limited by the melting or softening point of the furnace materials, which means in practice some 3000 K. However, various flash heating techniques have been developed for studying the absorption of short-lived molecular radicals (flash photolysis) and of very refractory solids (flash pyrolysis). Moreover, it is possible to use most emission sources in absorption provided there is available a background source of effectively higher brightness temperature, such as a flash tube. Windows tend to present a problem in absorption spectroscopy: below 1100 Å solid materials are not transparent, and in the longer wavelength regions the windows are apt to have material condensed or sputtered on to them and to react with certain hot vapours. It is frequently necessary to use a buffer gas system to protect them. In the window-less region differential pumping is often used between the absorption vessel and the background source on one side and the spectrograph slit on the other.

The simplest form of background continuum is the tungsten filament lamp, but this cannot be used above a brightness temperature of about 3000 K and is limited to the transmission region of quartz (2000–30 000 Å in round numbers). The carbon arc mentioned in Section 3.2 provides a slightly hotter source, corresponding very closely to black body radiation at about 3800 K. Comparable brightness temperatures in the visible and quartz ultra-violet can be obtained from high pressure gas arcs (mercury and the inert gases). Flash tubes, which are most important sources of continuous radiation in the far ultra-violet, can also be used at longer wavelengths. Flash tubes constructed with very low inductance so as to minimize the time constant can attain very high brightness temperatures (of order 50 000 K), so that lines from an arc or shock tube can be seen in absorption. The very short period of the flash (a few microseconds) makes this

source a highly suitable background for transient absorption, such
as that in shock tube and photolysis work.

3.8.2 The far infra-red region

The most convenient source of continuous radiation for most of
this spectral region is the Nernst glower, which is an electrically
heated filament made of a mixture of rare earth oxides operating
at about 2000 K. At wavelengths above 10 μm, more energy may
be obtained from a silicon-carbide strip or rod (Globar). Another
possible source is a high pressure mercury arc lamp contained in a
quartz envelope. Above about 2·7 μm the quartz is opaque and
emits black body radiation, and at longer wavelengths still – above
about 150 μm – the envelope again transmits the arc radiation.

At still longer wavelengths (1 mm) we reach the microwave
region, where the entire technique changes. Instead of using a
disperser to scan a continuum over a non-selective detector, one
has available a narrow-band radiator and tuned detector which are
together scanned over the required spectral region.

3.8.3 The far ultra-violet region

Continuous emission over the region from the quartz ultra-violet
down to the resonance line of helium at 584 Å can be obtained
from discharges in hydrogen and the inert gases, although below
1100 Å these must be run in window-less tubes. The molecular
hydrogen continuum, excited in a positive column discharge of up
to an amp or so through a capillary tube, has long been used from
the quartz ultra-violet to 1600 Å, when it ceases to be a continuum; and the so-called Hopfield continuum of helium, excited
in a condensed discharge at about 10 kV, is a well-established
background source from about 1000 Å to 600 Å. The gap between
these two continua has more recently been filled with the other
rare gases: a discharge run in any one of them (either a microwave
or a condensed discharge) at a fairly high pressure (a few hundred
torr) emits a continuum over a few hundred Ångströms extending
to the long wavelength side of the resonance lines. The regions
covered by the different inert gases overlap nicely. Much fuller
descriptions of these sources and the ones mentioned below are
given in [10].

The one source that covers the whole ultra-violet region is the
flash-tube, already mentioned as a source of high brightness for

the visible region. In the original version devised by Lyman the flash was produced by discharging a capacitor through a capillary tube containing low pressure gas. This had the disadvantages that the capillary was eroded rather rapidly and the continuous spectrum was overlaid by emission and absorption lines from the material scoured off the walls. In the Garton-type flash-tube these disadvantages have been largely eliminated by use of a wider-bore tube; to achieve the necessary current density, the duration of the

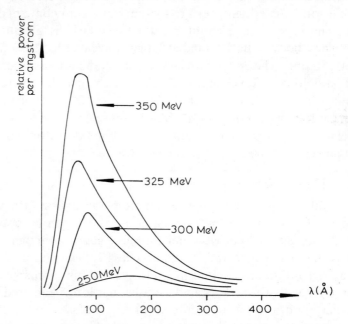

Fig. 3.4 Spectral distribution of synchrotron radiation for different electron energies. For a synchrotron of given radius, the peak power varies with the electron energy E according to $P \propto E^7$. The peak wavelength varies as E^{-3}.

flash is reduced to a microsecond or so by keeping the inductance of the circuit very low. A discharge of a few microfarads at 10 kV produces a continuum usable down to about 300 Å.

At still shorter wavelengths the flash tube continuum becomes progressively more of a line spectrum, and there have been several attempts to devise an alternative source. The most successful to date, developed by Balloffet, Romand and Vodar and called after them the BVR source, is a condensed spark in vacuum with a uranium pin acting as anode. The continuous radiation is emitted from a very small blob of hot gas just off the tip of the anode. Under typical operating conditions the flash from a $0·1$ μF capacitor charged to 29 kV lasts about a microsecond and produces a continuum down to a few Ångströms.

A source of an entirely different type, available to relatively few experimenters, is the bremsstrahlung radiation emitted by the electrons in a synchrotron accelerator. The electrons are accelerated towards the centre of the circular orbit, and, according to classical electromagnetic theory, at low energies they should emit radiation in all directions perpendicular to the orbital radius. At high energies, when relativistic effects become important, the radiation is concentrated into a narrow cone in the direction of instantaneous motion of the electron. The greater the energy, the smaller the angle of this cone. The radiation is therefore virtually confined to the orbital plane, and furthermore it is plane polarized with the electric vector in this plane. The spectral distribution of the radiation depends on the electron energy: the peak of the distribution shifts to shorter wavelengths as the energy is increased, as shown in Fig. 3.4. The power radiated at any wavelength can be calculated (it depends only on the radius of the orbit and the instantaneous energy of the electrons), and the synchrotron radiation can therefore be used as an intensity standard in the far ultra-violet – albeit hardly as a convenient laboratory standard.

3.9 Masers and lasers

The action of masers and lasers depends on induced emission, as represented by the Einstein coefficient B_{21}. *M*aser stands for *microwave* amplification by stimulated emission of radiation and *l*aser for *light* ditto. Masers were developed some time before lasers, mainly because the energy losses from the competing process of spontaneous emission are much less severe at long wavelengths ($A_{21}/B_{21} \propto \nu^3$ – equation 3.8).

Consider a collection of atoms or molecules with energy levels (1) and (2) as in Fig. 3.2, irradiated with light of frequency ν_{12}. Every atom in (1) has a probability per unit time $B_{12}\rho(\nu_{12})$ of capturing a photon and thus attenuating the light, whereas every atom in (2) has a probability $B_{21}\rho$ of adding a photon and thus amplifying the light. The characteristics of laser radiation arise entirely from the fact that the emitted radiation has exactly the same phase and direction as the stimulating radiation. This means that many atoms can be made to radiate coherently if the amplification exceeds the attenuation. In the normal way there are more atoms in (1) than in (2), so that absorption over-rides stimulated emission – by a factor $e^{h\nu/kT}$ in thermal equilibrium. To obtain a net amplification the population must somehow be inverted to make $N_2 > N_1$. This in itself is not enough, at any rate in the visible region, because a large number of the excited atoms decay spontaneously – i.e., incoherently – and are wasted from the point of view of coherent amplification. To offset this loss, the coherent amplification is increased by enclosing the radiation in a resonant cavity tuned to the frequency ν_{12}. In the visible and infra-red regions such a cavity is formed by a pair of highly reflecting parallel mirrors, one at each end of the material in which the population has been inverted. The standing waves due to the multiple reflections between the ends of this cavity have a much greater amplitude than the travelling wave representing a single passage through the material. Since the cavity is tuned only for light travelling exactly perpendicular to the mirrors, the beam that emerges through the end (one of the mirrors has a small transmissivity) is highly directional – the spread can be as little as 10 cm per km. Moreover, because the radiation behaves as if it originated in a single undamped oscillator rather than in a large number of unrelated oscillators there is no spread of frequency from Doppler or pressure shifts, or even from the natural lifetime (see Chapter 8). The wavelength band can therefore be extremely narrow and is limited essentially by the thermal noise or the stability of the resonant cavity. For a gas laser $\Delta\nu/\nu$ may be of order 10^{-14}, which is the size of the thermal and mechanical fluctuations of the mirrors. This may be compared with the normal Doppler limit of about 10^{-6}. It is this concentration of energy, both spatially and spectrally, that accounts for the great intensity of masers and lasers. In most cases the energy is emitted in pulses lasting a few microseconds or less, so the concentration is also temporal. The peak power radiated by a ruby laser of a few megawatts is something like 10^{12} times as high as the power radiated by the sun in the same narrow solid angle and frequency band, but the pulse lasts for only about 10^{-7} sec.

The various ways in which population inversion can be achieved will not be discussed here, since a large number of textbooks are available. The general principle is to choose a level (2) that is metastable and to populate it by decay from some higher level (3) that can be reached by absorption of shorter wavelength radiation from the ground state, a process known as optical pumping. In the ruby laser, for example, the 6943 Å red light is produced by the decay of the chromium impurity atoms from a metastable level (2) to the ground state (1), as shown in Fig. 3.5a. Level (2) is a sharp level rather than the band to be expected in the solid state because the partly filled $3d$ shell of the chromium ion is shielded by the

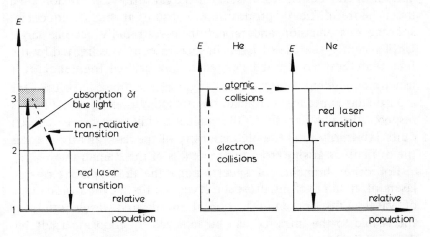

Fig. 3.5 Energy level diagrams for (a) a ruby laser and (b) a He–Ne laser. The length of the horizontal line for each energy level indicates its relative population.

outer electrons from the crystal fields. This level is populated by non-radiative transitions from the broad band (3), the energy being given up to the crystal lattice. A flash of continuous radiation serves to excite atoms to level (3). The laser operates in pulses because of the flash excitation and the need for dissipation of the lattice energy. The multiple reflections are obtained by polishing and silvering the end faces of the ruby itself. A different method of optical pumping is illustrated by the helium-neon gas laser, as shown in Fig. 3.5b. Since all the levels are sharp, energy can be absorbed only in very narrow frequency bands, and excitation by a flash of continuous radiati n is very inefficient. Instead, level (2) in neon is populated by collisions with metastable helium atoms of the same energy, produced in a continuously running discharge. The (2) → (1) laser transition does not end on the ground state, so that (1) has a relatively small population, and population inversion can be achieved without (2) being metastable. Both (1) and (2) are in fact split into several sub-levels, and several infra-red and red laser lines are produced by this process.

A wide variety of masers and lasers has now been developed, both gaseous and solid, covering the entire spectral region from the microwave to the near ultra-violet – although generally only at certain fixed frequencies. Lasers for the far ultra-violet region are a much more difficult proposition, because at high frequencies spontaneous emission and reflection losses rapidly become large. Until recently the use of lasers in spectroscopy was limited by the fact that they could not be tuned to any desired frequency; it is usually coincidental that a fixed frequency source can be used for a particular experiment. Raman spectroscopy is exceptional in this respect: the wavelength of the incident radiation can be chosen quite arbitrarily, and the high intensity of the laser greatly reduces the difficulties arising from the weakness of the Raman lines.

For other branches of spectroscopy the development of dye lasers offers an exciting prospect. Dyes fluoresce over a wide range of wavelengths, and amplification at any desired wavelength from the visible to the infra-red can be achieved by choosing a suitable dye and picking out the required waveband with a diffraction grating or interferometer. By means of frequency-doubling techniques the wavelength range can be extended into the ultra-violet. If a tunable dye laser is used as the background source for absorption spectroscopy, a conventional spectrograph can be dispensed with. Moreover, the intensity of a dye laser can easily be made great enough to saturate the population of any suitable excited state that can be reached by a radiative transition from the ground state. By this is meant that the populations of the ground state and the excited state can be made effectively equal. From Equations 3.7 and 3.8 one finds in the limit of very large ρ: $N_2/N_1 = B_{12}/B_{21} = g_2/g_1 \sim 1$. It then becomes possible to investigate a variety of effects in the excited state, such as absorption to a yet higher state, collisional and radiative decay processes, and transfer of excitation energy to neighbouring states.

It should be added that lasers are an extremely useful asset in a spectroscopy laboratory for such humble but essential tasks as lining up spectrographs, adjusting interferometers, and obtaining instrumental line profiles. They are also widely used in plasma spectroscopy as diagnostic tools for the direct measurement of temperature and electron density, a brief account of which is given at the end of Chapter 11.

No particular references for further reading on masers and lasers have been cited at the end of this chapter. So many new books and monographs on all aspects of the subject are constantly appearing that the choice is wide and still increasing.

3.10 General remarks on detectors from 1 μm to the far ultraviolet

Apart from the human eye, with its very limited range of sensitivity (roughly 4000–7000 Å), the detector used most widely until fairly recent years was the photographic plate. The wavelength range for normal photographic emulsions is about 2300 Å–7000 Å. Special infra-red sensitive emulsions have been pushed up beyond 12 000 Å, but have very poor sensitivity. In the ultra-violet plates with emulsions almost free of gelatine are sensitive down to the X-ray region, but the contrast is low compared to the visible region.

Great advances have been made with photo-electric detectors in the last 30 years or so. Photomultipliers are now used over much the same wavelength range as photographic plates. Below the transmission limit of quartz, say 2000 Å, two adaptations are possible: most commonly, the window of a conventional photomultiplier is coated with a phosphor (usually sodium salicylate) which converts the incident light to sufficiently long wavelength to get through the window; the efficiency of this process drops as the wavelength is reduced to about 1000 Å, but then remains almost constant down to 300 Å. Alternatively, open photomultipliers having metal cathodes and dynodes that are not poisoned on exposure to air may be used, either in the spectrograph vacuum or in a separately pumped chamber; the high work function of the cathode prevents these multipliers from responding to wavelengths above 2000/3000 Å (they are sometimes called 'solar-blind' multipliers for this reason), so that scattered light of longer wavelength is not detected.

Since photographic and photoelectric recording can be used over much the same wavelength range, the choice between them depends on the particular experiment. From the mechanical point of view, high resolution spectrographs are easier to construct than high resolution spectrometers, but the main advantages of photography are the long time integration possible for very weak spectra and the so-called multiplex advantage – that is, the simultaneous recording of a large number of spectral elements. Photo-electrically, the latter can only be done by Fourier transform spectroscopy (see Chapter 6), and even this requires a source constant over an appreciable time interval or reproducible from flash to flash. On the other hand, photoelectric detectors are greatly superior when time resolution is needed, as when following the emission from a pulsed discharge or a shock tube or when making lifetime measurements. In addition, their linear response gives them a big advantage in intensity measurements and line profile work and their efficiency is higher, as will be seen below.

3.11 The photographic plate

The response of a photographic emulsion to light is usually represented in the form shown in Fig. 3.6, where the density d is plotted against the logarithm of the illumination E for constant exposure time. d is defined as $\log_{10} 1/T$, where T is the trans-

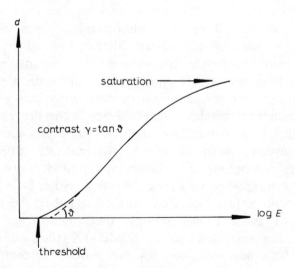

Fig. 3.6 Response curve of photographic emulsion.

missivity of the blackened emulsion – i.e., the ratio of transmitted to incident intensity. It is T that is measured when the plate is scanned with a microdensitometer. A curve rather similar to that of Fig. 3.6 is obtained if d is plotted against the logarithm of the exposure time t at constant E. If d were a function of the product Et, the two curves would have exactly the same shape, but because of the nature of the multi-stage processes by which a single grain is blackened the reciprocity law between illumination and exposure time does not usually hold strictly. Simultaneous doubling of the illumination and halving of the time does not result in the same plate blackening. This is often expressed by taking d to be a

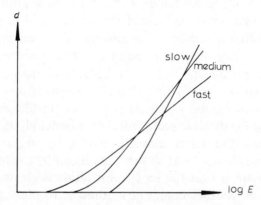

Fig. 3.7 Response curves for emulsions of different speeds.

function of $E t^q$, where q, the Schwarzchild factor, is a little smaller than one.

At both ends, near threshold and near saturation, the curve of Fig. 3.6 flattens out in a way that makes it obviously unsuitable for measuring relative values of E. Intensity measurements should be made as far as possible on the linear part of the curve. The slope of this defines the contrast γ of the plate, $\gamma = \tan \theta$. γ for spectrographic emulsions may vary from about 0·6 to 6, but for medium speeds it is usually fairly close to one. It depends on the wavelength, the conditions of development, the strength of the exposure – whether short and sharp or long and weak – and even the particular batch of plates as well as the type of emulsion. Typical response curves for slow (insensitive), medium and fast (sensitive) emulsions are shown in Fig. 3.7. A fast emulsion means a low threshold, but as the contrast is usually low the plate is not necessarily fast for large exposures.

Another important consideration is the grain size. Fast, low contrast plates have large grains and hence low resolution. The resolution is also affected by the scattering of the light in the emulsion. While the physical size of the grains is usually in the range 1 to 5 μm, the resolution limit is worse by a factor of about 5. It is reasonable to take about 10 μm for a fine emulsion and 25 μm for a coarse emulsion when matching instrumental to plate resolution limits.

So many factors influence the response of a plate that if accurate intensity measurements are to be made photographically a response curve must be plotted for each individual plate. This is done by making a number of exposures of known intensity ratio with the same exposure time and at the same wavelength. It is *not* safe to keep E constant and vary t because of the possible failure of the reciprocity law. With a stigmatic spectrograph it is usually simplest to photograph a neutral step filter (in the visible region) or a rotating logarithmic sector placed immediately in front of the slit. Although the latter actually works by varying the total exposure time, it has been shown that it can be treated as varying intensity provided that the speed of rotation is greater than about 10 cycles/sec. Another common method is to photograph a continuous source on a spectrograph of fairly large dispersion, either using a stepped slit or taking a succession of exposures with different slit widths. In the ultra-violet grids are sometimes used, their transmission being determined in the visible.

Once the plate has been developed, it must be run through a microdensitometer to record the transmissivity at every point of interest. The calibration exposures are used to construct a response curve. The relative values of E for any desired points on the microdensitometer trace can then be read off from the curve. It should be quite clear that accurate photographic photometry is a very tedious process. It is also inherently a rather inaccurate one, however careful the calibration, because the sensitivity is not necessarily uniform over the whole plate. Furthermore, the 'dynamic range' is rather limited, in that the moderately straight part of the response curve covers only about two orders of magnitude in intensity.

3.12   The photomultiplier

A photomultiplier consists of a photocathode, from which an electron is ejected by an incident photon, followed by an assembly of dynodes, each of which accelerates the electrons and emits several secondary electrons for every one that hits it. The total gain of the multiplier part depends on the number of stages and the applied voltage, but is usually in the range 10^6 to 10^8 for about 11 stages. The overall response depends in addition on the quantum efficiency of the cathode – that is, the number of electrons ejected per incident photon. This can be as high as 10–20%, which may be compared with the figure of between 100 and 1000 photons required to blacken one photographic plate grain.

The rough equivalent of 'threshold' in a photomultiplier is dark current, which is the current due to thermal emission from the cathode when there are no incident photons. At room temperature this is of order 10^{-15} to 10^{-14} amp cm^{-2}. A typical cathode has an area of about $0 \cdot 1$ cm^2, so that at the output end the dark current is a few thousandths of a microamp. This may be reduced by cooling the multiplier. At low light levels it is usual to discriminate against the dark current by chopping the light signal and using a.c. amplification. At the other end of the scale, the maximum intensity is limited by the saturation current of the cathode, usually about 10^{-7} to 10^{-6} amp. This means that the photo-multiplier can be used over an intensity range of about 10^9, which is an improvement of some 7 orders of magnitude over that of the photographic plate. With the gain suitably adjusted to avoid saturation, the photomultiplier has a virtually linear response over this range. It obviously offers a much easier and more reliable way of making intensity measurements provided that the source is continuously running or reproducible. Non-reproducible flashes may have to be handled photographically, although it is sometimes possible to allow for fluctuations by comparing the output from two or more multipliers.

Apart from the linearity of response and the dynamic range, the main properties of interest in a detector are its space and time resolution and its limiting sensitivity. Space resolution is not relevant to a photomultiplier, nor time resolution to a photo-graphic plate. In either case, auxiliary apparatus is required – a slit for the photomultiplier and a shutter or rotating mirror for the plate. The time resolution of a photomultiplier, as determined by its response time, is a few nanoseconds, while the space resolution of a plate, as we have seen, is of order 10μm. The ultimate sensitivity in both detectors is limited by the noise level (more familiarly described as fog in the photographic case). Comparison is a little difficult because the multiplier measures energy per unit time and the plate measures energy per unit area. Moreover, the noise of the former depends on whether the fluctuations are in the signal current or in the dark current. It can, however, be shown [26] that the ultimate sensitivity is comparable to within about an order of magnitude for photographic times of the order of seconds. If the exposure time can be increased indefinitely the plate must inevitably win in the end.

3.13   Detectors for the far ultra-violet

In addition to the special photographic plates and photo-multipliers that can be used through the ultra-violet, detection by photo-ionization becomes possible below about 1300 Å where the photon energy becomes high enough (> 9 eV) to ionize the permanent gases. For continuous (d.c.) detection an ionization chamber is used, working on the plateau or saturation region of the curve of ion current versus voltage, where the ion current is virtually independent of the chamber voltage and is proportional to the incident intensity. The efficiency, in terms of ion pairs per photon, may easily be 100% – indeed, if the photon energy is high enough for double ionization it may be greater than this. The pulsed version of the detector is essentially a Geiger counter: the original photo-electron from the incident photon is accelerated to make several collisions with gas molecules so that an avalanche takes place, using gas amplification. Both types are difficult to use in the wavelength region 1040–300 Å because of the lack of any window material. This is particularly true of the Geiger or photon counter, owing to the relatively high gas pressure required (~ 100 torr). Below 300 Å cellulose and thin metal films start to transmit, and the detector can be used from there right on to the X-ray region. The filler gas at the longer wavelengths is usually nitric oxide or a similar molecule, but the inert gases are preferable at short wavelengths because of their higher ionization potential. With a clever choice of window material and filler gas it is possible to arrange the long and short wavelength cut-offs to leave a fairly narrow band (as little as 50 Å) of sensitivity. Selective detectors of this type replaced the spectrometer in some of the early experiments with rockets.

Photo-ionization detectors have also been used to measure absolute intensities and to calibrate sources as intensity standards in the vacuum ultra-violet, as mentioned in Section 3.2. If every absorbed photon yields one photo-electron, the output current of an ion chamber is equal to the number of absorbed photons. The rare gases fulfil this requirement, and moreover their absorption coefficients are so high that a low gas pressure is enough for complete absorption. The ion chamber can be used in this way with the rare gases in decreasing order of atomic weight from 1022 Å, the ionization limit of xenon, to 250 Å where the ejected photoelectrons have sufficient energy to cause secondary ionization in helium and the 1:1 correspondence breaks down. The photon counter, however, can take over at this point since it records the pulse produced by each absorbed photon rather than the number of electrons.

A full account of all these detectors is to be found in Samson's book [10].

3.14 Detectors for the infra-red

The infra-red region presents a more difficult problem in detection, because at wavelengths above about 13 000 Å the photons do not have enough energy to eject an electron from a cathode or activate an emulsion, and the light sources are also weaker. At a wavelength of about 1 mm the crystal detectors of microwave techniques become usable, so we are concerned here with the span 1 μm–1000 μm. Infra-red detectors for this region may be divided into two classes, thermal detectors and photo-conductive devices. In both classes the response is proportional to the absorbed power W, but, whereas thermal detectors measure the rate at which energy is absorbed independently of its wavelength, photo-conductive detectors, like photomultipliers, measure the rate at which photons are absorbed, $dn/dt = W/h\nu$, so that ideally their response increases linearly with wavelength for constant power over the sensitive range. The two classes also differ in intrinsic response time – that is, the speed with which they respond to changes of signal. As a rough guide, this is measured in milliseconds for thermal and microseconds for photo-conductive detectors, as compared with nanoseconds for photomultipliers. Moreover by comparison with photomultipliers at shorter wavelengths infra-red detectors are noisy as well as slow. Since the power radiated by infra-red sources is relatively low, the noise level of the detector is a particularly important characteristic and will be discussed at the end of this section.

Thermal detectors are sensitive throughout the whole infra-red region. Thermocouples and thermopiles measure the rise in temperature from absorbed radiation as a thermoelectric emf, and bolometers measure it as a change in resistance. Bolometers are the more commonly used and are often operated at liquid helium temperature to reduce the thermal noise and increase the sensitivity. A rather different type of detector, the Golay cell, works by the expansion of an inert gas heated by the incident radiation. The gas is contained in a small cell, one wall of which is a flexible membrane. This distorts under pressure, and the signal is obtained from the deflection of a light beam reflected off it. The Golay cell is an order of magnitude slower and less sensitive than a cooled bolometer, but the minimum signal detectable is not very different, and the Golay cell has the advantage of operating at room temperature.

Photo-conductive cells are semi-conductors whose electrical resistance decreases when exposed to light. The change in resistance is proportional to the rate at which photons are absorbed and can be detected as a change in voltage across a load resistor in series with the photo-conductor. The mechanism may be described as an internal photo-electric effect: the photons have insufficient energy to throw an electron right out of the surface, as in a photo-cell or the photo-cathode of a photomultiplier, but below a certain cut-off wavelength they do have enough energy to free an electron from the crystal lattice and so increase the number of free electrons and/or holes that act as current carriers. The effect is enhanced by cooling the semi-conductor to reduce the number of thermally excited electrons. Until recently photo-conductors could be used only in the near infra-red, the cut-off wavelength for such crystals as lead sulphide being a few micrometres, but new types of 'doped' semi-conductor – that is, crystals containing small amounts of selected impurities – can operate (at liquid helium temperature) up to about 100 μm. Indeed, it has now been found that In-Sb detectors can be used right into the millimetre wave region because the mobility of the *free* electrons is increased by the absorption of photon energy and the free electrons can exist at a temperature higher than that of the crystal. For this reason such detectors are known as 'hot electron' detectors [6].

The ultimate useful sensitivity of a detector is limited by the noise. This is usually expressed by setting the minimum detectable signal power equal to the 'noise equivalent power' W_n, which is to say that a signal is just detectable when the signal/noise ratio is unity. Noise may have three origins: fluctuations in the background radiation and in the light source itself, electrical and thermal fluctuations in the detector, and fluctuations in the amplifying circuit. The last of these can usually be kept below the other two. The importance of background radiation stems from the fact that the peak of the black-body curve at room temperature is at 10 μm. To eliminate the large constant background and to make a.c. amplification possible it is general practice to chop the light signal at some frequency f and use a phase-sensitive amplifier of band width Δf behind the detector. f is limited by the requirement that $1/f$ must be large compared to the response time of the detector, and f may range from about 10 cycles per second for slow thermal detectors to a few kilocycles for photoconductors. Although chopping reduces background noise, it does not eliminate it entirely, because random fluctuations in the background level have components in the frequency range Δf. The detector is sometimes enclosed in a cooled container to reduce the background fluctuations as far as possible, but of course a certain amount of background can always get in through the aperture that admits the signal. This type of noise, resulting from the irregular arrival of photons, is known as photon noise. At shorter wavelengths signal fluctuations are likely to contribute more than background fluctuations to photon noise. By signal fluctuations in this context is meant statistical departures from a mean, not a drift or a change in mean level of signal that can be compensated with a double beam system such as is commonly used in infra-red absorption spectroscopy.

While the minimum power detectable is usually limited by photon noise in the visible, the limit in the infra-red is more likely to be set by detector noise, in most cases thermal noise for thermal detectors and current noise for photo-conductors. These different types of noise are discussed in the references given for the infra-red region [5, 7-9]. The analysis is too long to be reproduced here, but shows that for both types of noise the mean square power fluctuations are proportional to Δf and to the sensitive area – i.e., $W_n \propto A^{1/2} (\Delta f)^{1/2}$. For thermal noise $W_n \propto T^{5/2}$, and for current noise $W_n \propto f^{-1/2}$. In practice for a given wavelength, area and bandwidth it seems to be generally possible to find some photo-conducting detector with lower W_n than a thermal detector.

A final remark may be made about bandwidth. W_n can seemingly be reduced indefinitely simply by reducing Δf. However, a small bandwidth is associated with a long response time for the system as a whole (as distinct from the intrinsic response time of the detector) since the two quantities are related by $t \approx 1/\Delta f$. Limiting the band width therefore increases the time necessary to scan a spectrum and also limits the time resolution if the signal is varying with time. In fact, the effect of limiting the band width to Δf is simply to integrate the signal over time t, and the relation $W_n \propto 1/t^{1/2}$ is analogous to the well-known result from error theory that the error in the mean of a set of observations is inversely proportional to the square root of the number of observations.

References

General (including black-body, Einstein coefficients, etc.)
1. Harrison, G. R., Lord, R. C. and Loofbourow, J. R. 'Practical Spectroscopy', Prentice-Hall, 1948
2. Sawyer, R. A. 'Experimental Spectroscopy', Dover, 1961
3. Cowley, C. R. 'The Theory of Stellar Spectra', Gordon and Breach, 1970
4. Ditchburn, R. W. 'Light', Blackie, 1963

Infra-red region
5. Martin, A. E. 'Infra-red Instrumentation and Techniques', Elsevier, 1966
6. Martin, D. H., ed., 'Spectroscopic Techniques for far Infra-red, Sub-millimetre and Millimetre Waves', North Holland, 1967
7. Kruse, P. W., McGlauchlin, L. D. and McQuistan, R. B. 'Elements of Infra-red Technology', Wiley, 1962
8. Smith, R. A., Jones, F. E. and Chasmar, R. P. 'Detection and Measurement of Infra-red Detection', Oxford, 1968

9. Conn, G. K. T. and Avery, D. G. 'Infrared Methods', Academic Press, New York and London, 1960

Ultra-violet region

10. Samson, J. A. R. 'Techniques of Vacuum Ultra-violet Spectroscopy', Wiley, 1967
11. Garton, W. R. S. Spectroscopy in the Vacuum Ultra-violet, in 'Advances in Atomic and Molecular Physics', 2, Academic Press, New York and London, 1966

Special sources

12. (Flames) Gaydon, A. G. and Wolfhard, H. G. 'Flames', Chapman and Hall, 1960
13. (Arcs) Wiese, W. L. Electric Arcs in 'Methods of Experimental Physics', 7B (ed. Marton), Academic Press, New York and London, 1968
14. (Arcs) Lochte-Holtgreven, W. Production and Measurement of High Temperatures, *Rep. Prog. Phys.* 21, 312, 1958
15. (Arcs) Foster, E. W. Measurement of Oscillator Strengths, *Rep. Prog. Phys.* 27, 469, 1964
16. (Hollow cathodes) Tolansky, S. 'High Resolution Spectroscopy', Methuen, 1947
17. (Shocktubes) Kolb, A. C. and Griem, H. R. High Temperature Shock Waves, in 'Atomic and Molecular Processes' (ed. Bates), Academic Press, New York and London, 1962
18. (Shocktubes) Pain, H. J. and Rogers, E. W. E. Shock Waves in Gases, *Rep. Prog. Phys.* 25, 287, 1962
19. (Pinches, etc.) Green, T. S. 'Thermonuclear Power', Newnes, 1963
20. (Pinches, etc.) Glasstone, S. and Lovberg, R. H. 'Controlled Thermonuclear Reactions', Van Nostrand, 1960
21. (Pinches, etc.) Gross, R. A. and Miller, B. Plasma Heating by strong Shock Waves, in 'Methods of Exptl. Phys.,' 9A (ed. Griem and Lovberg), Academic Press, New York and London, 1970
22. (Pinches, etc.) Mather, J. W. Dense Plasma Focus, in ditto, 9B
23. (Pinches, etc.) Niblett, G. B. F. Production and Containment of high density Plasmas, in 'Physics of Hot Plasmas' (ed. Rye and Taylor), Oliver and Boyd, 1970
24. (Pinches, etc.) Burgess, D. D. Spectroscopy of Laboratory Plasmas, *Space Science Reviews*, 13, 493, 1972
25. (Beam foil) Bashkin, S. Beam Foil Spectroscopy, *Applied Optics*, 7, 2341, 1968
26. Lochte-Holtgreven, W. and Richter, J. in 'Plasma Diagnostics', (ed. Lochte-Holtgreven), North Holland, 1968

CHAPTER FOUR

Dispersion and resolving power. Prism spectrographs

4.1 General remarks

Dispersion can be achieved by refraction, diffraction, or inter-ference – that is, by prisms, gratings or interferometers. In a way this distinction is rather artificial, because all these instruments work by superposing a number of rays of varying phase, the number being infinite for the prism, in the range 10^5 to 10^3 for the grating, and from 30 down to 2 for the interferometer. The important properties of a disperser are its spectral range, its resolving power and dispersion, and its light grasp, to which perhaps one should add the ease of interpretation of the output.

On spectral range the grating wins easily, as it can be used from the X-ray region to the microwave region – and indeed beyond, although it is not needed where tunable sources and detectors are available. Prisms and interferometers have to be made of some transparent solid, and in the case of prisms this limits the range to between $2000\,\text{Å}$ and $40\,\mu\text{m}$. Fluorite prism instruments are possible in principle below $2000\,\text{Å}$ as far as $1400\,\text{Å}$, but are scarcely ever used in practice. Interferometers can be taken further into the infra-red than prisms, since the solid can take the form of a thin film; they have in fact been used right through to the sub-millimetre region. As regards maximum resolving power, the interferometer leads by an order of magnitude over the grating, which in turn has at least an order of magnitude over the prism. Light grasp is strongly inter-related to resolving power, but if we compare the light grasp for the same resolving power the order of merit is again found to be interferometer–grating–prism. However, in ease of interpretation the order is reversed: only the prism gives directly an unambiguous spectrum; in the grating overlapping orders may have to be sorted out, and in the Fabry–Perot interferometer this problem is much worse. With the Michelson interferometer used in Fourier transform spectroscopy the output cannot be interpreted as a spectrum at all until it has been subjected to a Fourier transform.

We shall revert to the comparison of interferometers with diffraction grating instruments at the end of Chapter 6, after both types have been separately described. In this chapter a general discussion of resolving power applicable to all three types of spectrometer is followed by some remarks on slit width and illumination relevant to those instruments that use slits – prism and grating spectrometers. The description of prism instruments that follows has been kept very brief, because the limited spectral range and resolving power of these instruments renders them relatively unimportant for most of the line shape and intensity studies with which the last part of this book is concerned.

4.2 Dispersion and resolving power

Leaving aside for the moment Fourier transform spectroscopy, all dispersers, including conventional multiple beam interferometers, separate different wavelengths by spreading them out spatially. Dispersion is the measure of this spreading. If rays of wavelengths λ and $\lambda + \delta\lambda$ emerge from the disperser at angles θ and $\theta + \delta\theta$ respectively, the angular dispersion is defined as $\delta\theta/\delta\lambda$, or $d\theta/d\lambda$. The linear dispersion is the linear separation of the two images in the focal plane of the camera lens or concave grating, so that for focal length f the linear dispersion $dl/d\lambda$ is equal to $fd\theta/d\lambda$. In practice more frequent use is made of the reciprocal dispersion $d\lambda/dl$. The correct S.I. unit for reciprocal dispersion is nm mm^{-1}, but the traditional form Å/mm makes the meaning considerably clearer. Explicit expressions for the dispersion of prisms, gratings and Fabry-Perot interferometers will be derived when the individual instruments are described.

If two close lines are to be separated, one has to consider their widths as well as the separation of their centres. A spectrograph illuminated by an ideal source of monochromatic light does not form an infinitely narrow spectral line. Even if there are no geometrical aberrations or instrumental imperfections, diffraction effects cannot be eliminated. The intensity distribution in the image of a monochromatic line is known as the slit function, or instrument function. It is this that determines the resolving power, or the ability to separate two lines very close in wavelength. If two such lines of wavelength λ and $\lambda + \delta\lambda$ are just resolved, then the resolving power is defined by

$$\mathscr{R} = \lambda/\delta\lambda = \nu/\delta\nu = \sigma/\delta\sigma \qquad (4.1)$$

The question then is what we mean by 'just resolved'. This must depend on the form of the instrument function, which we now examine.

In an ideal, aberration-free, prism or grating instrument the instrument function is determined by diffraction due to the finite aperture of the lens-disperser system. The limiting aperture is normally the rectangular cross-section $a \times b$ of the prism or grating (see Fig. 4.1). Taking the slit to be vertical, the spectrum is

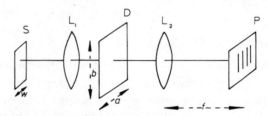

Fig. 4.1 Fraunhofer diffraction at a rectangular aperture. S is the slit, width w, in the focal plane of L_1, D is the aperture $a \times b$ of the dispersing element, and P is the photographic plate in the focal plane of L_2.

dispersed horizontally, and we are concerned only with the intensity distribution in the horizontal direction. For an infinitely narrow slit w (the effect of finite slit width will be considered later), the intensity distribution is the Fraunhofer diffraction pattern for an aperture of width a (see, for example, [4], [5]). In terms of the phase difference α between the centre of the aperture and one edge of it for a beam emerging at angle φ, this is given by

$$I = I_0 \, (\sin \alpha/\alpha)^2 \qquad (4.2)$$

where $\alpha = (2\pi/\lambda)\tfrac{1}{2}Q \sin \varphi = \pi a\varphi/\lambda$ for small φ (see Fig. 4.2). Fig. 4.3 shows this distribution. The central maximum corresponds to

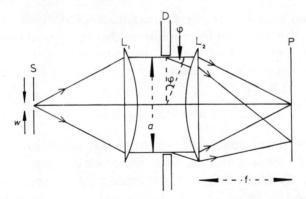

Fig. 4.2 Ray diagram for Fraunhofer diffraction at a rectangular aperture. This figure is a top view of Fig. 4.1, showing the aperture of width a. The width scale is much exaggerated.

$\alpha = \varphi = 0$ and the first minimum on either side to $\alpha = \pm\pi$, or $\varphi_m = \pm\lambda/a$. φ_m therefore measures the angular width of the line.

Suppose we now add to the picture the diffraction pattern for a second line of wavelength $\lambda + \delta\lambda$. The central maxima of the two lines will be separated by an angle $\delta\theta$ that depends on the dispersion according to $\delta\theta = d\theta/d\lambda \; \delta\lambda$. If $\delta\theta$ is small compared to

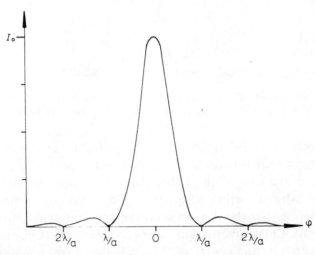

Fig. 4.3 Fraunhofer diffraction pattern: intensity distribution. The intensity is given by

$$I = I_0 \left(\frac{\sin \alpha}{\alpha} \right)^2 \qquad \text{where} \qquad \alpha = \pi a \varphi/\lambda.$$

φ_m, the two lines will not be distinguishable. Rayleigh's criterion is taken as the condition for resolution: that the central maximum of one line should fall on the first minimum of the other, so that the central maxima are separated by φ_m. Equating $\delta\theta$ to φ_m, we have $d\theta/d\lambda \ \delta\lambda = \lambda/a$

$$\therefore \qquad \mathscr{R} = \lambda/\delta\lambda = a \ d\theta/d\lambda \qquad\qquad (4.3)$$

For ideal prism and grating instruments, which have diffraction-limited instrument functions, the resolving power is given directly by Rayleigh's criterion and Equation 4.3. Explicit expressions for $\mathscr{R}$ will be derived when these instruments are discussed.

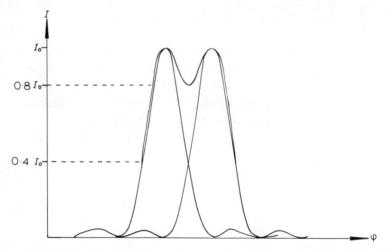

Fig. 4.4 Illustration of Rayleigh criterion for resolution of two spectral lines. The heavy curve is the superposition of the two light curves.

Interferometers cannot be treated in quite the same way because their instrument functions have a rather different intensity distribution. In the case of the Fabry-Perot interferometer this is given by the Airy function (Equation 6.6), and the criterion for resolution is that the dip between the two line peaks must be the same as for the diffraction line shape. Numerically, this means that the intensity at the centre of the dip between two equal lines that are just resolved is 80% of the peak intensity of either, since the diffraction pattern of a single line falls to 0·4 of its peak intensity half way between the centre and the first minimum (Fig. 4.4). The line shape produced by Fourier transform spectroscopy is, in general, different again. However, in this case the line shape can be altered by numerical weighting in the transform process, which is frequently arranged to produce a diffraction line shape.

The term 'resolution limit' is sometimes used instead of resolving power. This is the interval of wavelength, $\delta\lambda$, frequency, $\delta\nu$, or wavenumber, $\delta\sigma$, between two lines that are just resolved by Rayleigh's criterion. The resolution limit is expressed in Å, $\sec^{-1}$, or cm^{-1}, whereas resolving power is dimensionless.

It is useful to look at resolution from a slightly different point of view that is applicable to both diffraction and interference devices. If the disperser imposes a path difference Δ between the extreme rays passing through it (Fig. 4.5), the emerging wave-front

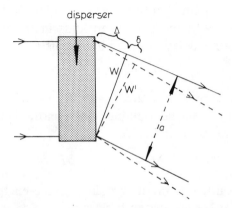

Fig. 4.5 Path difference imposed by disperser. The disposer imposes path difference Δ between the extreme rays of a wavefront W of wavelength λ. The path difference imposed on W' of wavelength $\lambda + \delta\lambda$ is $\Delta + \delta$.

W for wavelength λ is indistinguishable from W' for $\lambda + \delta\lambda$ if the maximum phase difference between them is less than about 2π – that is, their maximum separation δ is less than about a wavelength. Adopting a phase difference of 2π as the criterion that $\delta\lambda$ should be just resolved by the path difference Δ gives

$$\frac{2\pi\Delta}{\lambda} - \frac{2\pi\Delta}{\lambda + \delta\lambda} = 2\pi \qquad \text{or} \qquad \Delta . \delta\lambda = \lambda^2$$

since $\delta\lambda \ll \lambda$

$$\therefore \qquad \mathscr{R} = \lambda/\delta\lambda = \Delta/\lambda \qquad (4.4)$$

i.e., the resolving power is the number of wavelengths in the path difference between extreme rays. It is easily shown that for diffracted wave fronts the criterion of a phase difference of 2π, corresponding to $\delta = \lambda$, is equivalent to Rayleigh's criterion: the latter requires W and W' to diverge at an angle $\varphi_m = \lambda/a$. But $\delta = a\varphi_m$

$$\therefore \quad \delta = \lambda$$

Equation 4.4 leads to a particularly convenient expression for the resolution limit in wavenumbers;

$$\delta\sigma = \delta\lambda/\lambda^2$$

$$\therefore \quad \delta\sigma = 1/\Delta \tag{4.5}$$

The resolution limit in cm^{-1} is just the reciprocal of the extreme path difference in cm. Multiplying both sides by c gives a similar relationship between the resolution limit in frequency units and the maximum time delay imposed by the instrument:

$$\delta\nu = c/\Delta = 1/\delta t$$

Both these expressions are really forms of the uncertainty principle, which can be obtained in a more familiar form by multiplying through by h to give

$$h\delta\nu \sim h/\delta t \qquad \text{or} \qquad \delta E.\,\delta t \sim h$$

δE may be regarded as the uncertainty in the energy of the wave packet and δt as the associated time spread.

In practice, any resolution criterion is somewhat arbitrary. Lines considerably closer than the resolution limit may be distinguished if their intensity is comparable and the signal/noise ratio is good. On the other hand, a weak satellite at the theoretical resolution limit will not necessarily be resolved from its strong neighbour, particularly if the signal/noise ratio is bad.

4.3   Slit width and illumination

The remarks on slit width in this section apply only to prism and grating instruments. Fabry-Perot and Michelson interferometers produce fringes with circular symmetry, and the restrictions on the width of the input beam are much less severe, as will be seen in Chapter 6. In prism and grating spectrographs, where the spectral line is the image of the slit in light of the appropriate wavelength, a wide slit will produce a wide line. On the other hand, narrowing the slit will not decrease the line width beyond the diffraction limit. As seen in the last section, the angle between the central maximum and the first minimum for monochromatic light of wavelength λ diffracted at an aperture of width a is λ/a (Figs. 4.2 and 4.3). The least possible width of the image of an infinitely narrow slit measured from minimum to minimum is therefore $2f\lambda/a$, where f is the focal length of the lens L_2. As the slit is

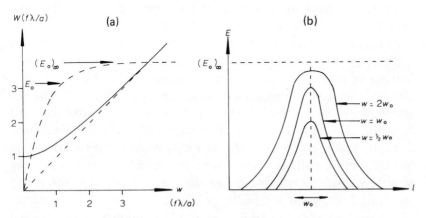

Fig. 4.6 Effect of slit width on (a) peak illumination and image width and (b) intensity distribution in image plane. In (a) the solid curve W is half the distance between the diffraction minima – i.e. the width of the image at 0·4 peak intensity – and w is the slit width. Both W and w are in units of $f\lambda/a$. The broken curve shows the variation of peak illumination E_0 with w. In (b) the image line shapes are shown for different values of w. w_0 is the 'optimum slit width' $\lambda f/a$.

widened, the pattern changes little until the width of the geometric slit image becomes comparable with the diffraction width – i.e., $w \sim \lambda f/a$, where w is the slit width and L_1 and L_2 are assumed to have the same focal length f, giving unit magnification. As w is increased beyond this point, the width of the image increases at the same rate. This is shown in Fig. 4.6a, where the width W of the image, measured at half the peak intensity, is plotted against w, both in units of $\lambda f/a$.

The slit width determines the total light flux that gets into the spectrograph as well as the line width. The broken curve of Fig. 4.6a shows how the illumination E_0 at the peak of the central maximum varies with w. Evidently the optimum slit width is about $\lambda f/a$. This can also be seen in Fig. 4.6b, which shows the illumination in the focal plane of L_2 for three different slit widths, $\frac{1}{2}\lambda f/a$, $\lambda f/a$, and $2\lambda f/a$. All these curves apply to the usual case of non-coherent illumination of the slit, which implies that the source itself or an image of it is fairly close to the slit. The effects are a little different for coherent illumination, which requires either a distant source or collimated light. A fuller discussion can be found in [2, 3].

In most applications f/a is in the range 10 to 25, and the optimum slit width from the near infra-red to the near ultra-violet is of order $10\ \mu$m. In the far ultra-violet the theoretical optimum width may drop to less than $1\ \mu$m, but as the resolution is usually limited by plate grain or mechanical alignment in this region it is rare to use a slit appreciably narrower than $10\ \mu$m. In the far infra-red the optimum slit width may be about a millimetre, but even this may have to be exceeded to provide sufficient light flux.

In the visible region the slit may easily be adjusted by looking at the diffraction pattern formed near the prism or grating when the source is distant enough for the slit itself to act as a diffracting aperture illuminated with coherent light. The diffraction pattern, as imaged by the collimating lens of focal length f', has a central maximum of width $2f'\lambda/w$, taken from minimum to minimum. When this just over-fills the field of view, which is of width a, we have $a \approx \lambda f'/w$. This agrees with the condition above if $f = f'$.

It is evident from Fig. 4.6 that, for monochromatic light, increasing the slit width beyond the optimum does not increase E_0 appreciably and cannot therefore increase the blackening at the centre of the line on a photographic plate. The extra flux entering the spectrograph is merely spread over a larger area of plate. However, for a continuous spectrum, or a spectral line whose true width exceeds $\lambda f/a$, the blackening of the photographic plate *does* increase as the slit is opened, because effectively one is superposing on the centre of one 'line' light from the edges of a lot of other 'lines'. The exposure time in absorption spectroscopy may therefore be reduced by opening up the spectrograph slit, but it must be remembered that this is likely to mask a weak and narrow absorption line whose hole in the continuum will be partly filled by the overlap from the continuum on each side. Opening the slit can also increase the signal from a photomultiplier, bolometer, photo-conductor or similar detector, even if the light is strictly monochromatic, because these devices integrate the energy over the whole illuminated area. In practice, this is restricted by an exit slit in the focal plane of L_2 set to the same width as the entrance slit. Obviously both slits must be widened together if the gain in signal is to be realized, and any such gain is made at the expense of resolution.

The energy reaching the detector depends on how the slit is illuminated as well as on its width. For maximum throughput of

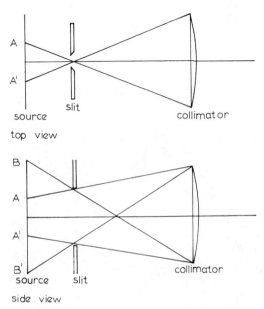

Fig. 4.7 Illumination of collimator lens. For maximum illumination the source should have a width AA' and height BB'.

energy the light must enter the slit from a solid angle at least as large as the solid angle that can be accepted by the collimator or equivalent aperture. In other words, the source must subtend at the slit a solid angle at least as large as that of the collimator. It can be seen from Fig. 4.7 that a source of width AA' satisfies this requirement; but because the slit is much longer than it is wide, the source should extend over a larger distance BB' in the vertical direction. While every point between A and A' sends light through the whole collimator, the regions AB and A'B' illuminate only part of the collimator but nevertheless contribute to the total through-put. Maximum illumination is achieved if the source is large enough or can be brought close enough to cover the whole height BB'. If this condition is satisfied, the light flux reaching the detector cannot be increased by interposing a condensing lens between source and slit. This statement should perhaps be

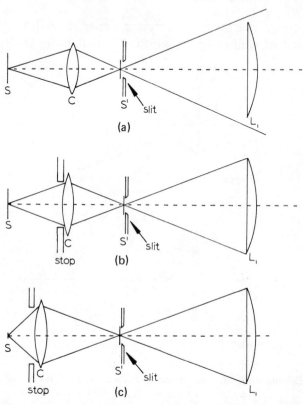

Fig. 4.8 Matching of condensing system to spectrograph aperture: (a) collimator over-filled; (b) correct matching; (c) system for small source. In each case S' is the image of the source S formed on the slit by the condensing lens C, and L_1 is the collimating lens of the spectrograph.

elaborated, since at first sight it appears that if the source is larger than the slit one ought to gain by focusing a reduced image of the source on the slit. Certainly the illumination is increased by reducing the size of the image, but there is a general theorem of optics to the effect that the *brightness* of an image (the flux per unit area *per unit solid angle*) can never exceed the brightness of the object. The flux accepted by the spectrograph is the product of the slit area, the solid angle subtended by the collimator, and the brightness of the source, and all of these quantities are independent of the condensing system. As shown in Fig. 4.8a, although more of the source area can be used by matching it to the slit, the flux is spread over a larger solid angle, and the extra light misses the collimator and is wasted. Actually, because of the scattered light it is much better not to overfill the collimator in this way. The ideal condensing system is shown in Fig. 4.8b, with the solid angles just matched by stopping down the condensing lens. Fig. 4.8c shows the correct condensing system for a small source (a capillary tube, for example): the image formed by the condenser is sufficiently large that the whole slit height can be used, and the aperture of the condenser is again chosen to match the solid angle accepted by the spectrograph. The alignment of source and condenser is very critical in a properly matched system of this kind. The source need be only $2°$ off the optic axis of an $f/25$ spectrograph for the entire beam to miss the collimator. It should also be remarked that it is sometimes undesirable to focus the source on the slit – when, for example, uniform slit illumination is required.

4.4 Prism spectrographs

A conventional prism spectrograph is shown in Fig. 4.9. Light from the slit S is collimated by the lens L_1, refracted by the prism D, and focused by the camera lens L_2 on the photographic plate P. The spectral lines on the plate are images of the slit in light of different wavelengths. For visual observation with a spectroscope, L_2 is replaced by the telescope lens and the eyepiece is focused on the plane P. With photo-electric detection, the spectrum is usually scanned by tracking the exit slit and photomultiplier along the plane P. This type of mounting is not at all suitable for use as a monochromator because the direction of the emergent beam varies with the wavelength.

4.5 Deviation and dispersion

Fig. 4.10 shows the path of a ray through a principal section of a prism of angle α. It can be seen from the geometry that the deviation θ is given by $\theta = d_1 + d_2$ where $d_1 = i_1 - r_1$ and

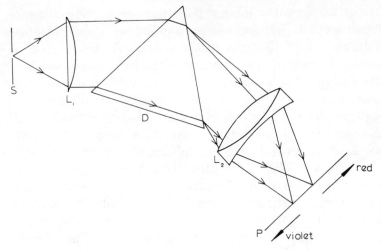

Fig. 4.9 Components of a conventional prism spectrograph. S is the slit, L_1 and L_2 the collimating and camera lenses respectively, D the prism and P the focal plane of L_2.

$d_2 = i_2 - r_2$. But $\alpha = r_1 + r_2$, so $\theta = i_1 + i_2 - \alpha$. Prism spectrographs are normally used near the position of minimum deviation, for which it can be shown that the ray must traverse the prism symmetrically. i_1 is then equal to i_2, and the ray passes through the prism parallel to the base, so that $\theta = 2i - \alpha$ and $\alpha = 2r$. Combining these relations with Snell's law of refraction $\sin i = n \sin r$:

$$\sin i = \sin \frac{\theta + \alpha}{2} = n \sin \frac{\alpha}{2} \qquad (4.6)$$

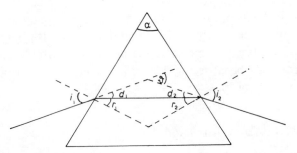

Fig. 4.10 Path of ray through prism. The deviation θ is the sum of the deviations d_1 and d_2 at the two faces. i_1, r_1 and i_2, r_2 are the angles of incidence and refraction, respectively, at the two faces.

The change of deviation θ with wavelength determines the angular dispersion of the prism. Writing

$$\frac{d\theta}{d\lambda} = \frac{d\theta}{dn}\frac{dn}{d\lambda},$$

it can be seen from Equation 4.6 that the angular dispersion depends on the prism angle α and on the prism material by way of the refractive index n and its variation with wavelength $dn/d\lambda$ (i.e., the dispersion of the prism material). Differentiating Equation 4.6,

$$\frac{d\theta}{dn} = \frac{2\sin\alpha/2}{\cos\frac{1}{2}(\theta + \alpha)} = \frac{2\sin\alpha/2}{\cdot\sqrt{(1 - \sin^2\frac{1}{2}(\theta + \alpha))}} = \frac{2\sin\alpha/2}{\sqrt{(1 - n^2\sin^2\alpha/2)}}$$

$$\therefore \qquad \frac{d\theta}{d\lambda} = \frac{2\sin\alpha/2}{\sqrt{(1 - n^2\sin^2\alpha/2)}}\frac{dn}{d\lambda} \qquad (4.7)$$

The dispersion obviously increases with increasing α, but so too does the amount of material required for a prism of given aperture. An angle of 60° is nearly always chosen as the best compromise. Putting $\sin\alpha/2 = \frac{1}{2}$, equation 4.7 becomes

$$\frac{d\theta}{d\lambda} = \frac{1}{\sqrt{(1 - (n/2)^2)}}\frac{dn}{d\lambda}$$

It so happens that for values of n in the range 1·4 to 1·6 – which includes most suitable crystals and glasses – the first term on the right hand side is equal to n to within 4% (the relation is exact for $n = \sqrt{2}$). To a good approximation, therefore,

$$\frac{d\theta}{d\lambda} \simeq n\frac{dn}{d\lambda}$$

The dispersion of any material is a function of wavelength and increases sharply near a region of absorption. Prisms therefore disperse most efficiently near their absorption limits. A few figures for refractive index and dispersion in the visible and ultra-violet are given in Table 4.1, together with the angular dispersion of a 60° prism of the material. The figures for glass are only approximate, because they vary greatly from one type to another. Any glass prism gives considerably higher dispersion than quartz down to its transmission limit, which is usually between 3700 Å for high dispersion (flint) glass and 3300 Å for low dispersion (crown) glass. The last two columns of Table 4.1 give the linear dispersion and reciprocal dispersion obtained with a camera lens of focal length 50 cm, typical of a medium sized instrument. For a large instrument $f \sim 150$ cm, and the reciprocal dispersion is accordingly about three times smaller. The figures for linear dispersion assume the lenses to be achromatic, so that f is the same for all wavelengths. It is quite common, however, to use simple lenses, particularly in the ultra-violet owing to the expense of quartz-fluorite achromats, and compensate for the variation of focal

Table 4.1

| | | | | | For $f = 50$ cm | |
λ in Å	Material	n	$dn/d\lambda$ in Å^{-1} ($\times 10^{-5}$)	$d\theta/d\lambda$ in radians per Å ($\times 10^{-5}$)	$dl/d\lambda$ in mm/Å ($\times 10^{-2}$)	$d\lambda/dl$ in Å/mm
2000	Fused silica	1·55	13	21	10	10
	Crystal quartz	1·65	16	28	14	7
3000	Fused silica	1·49	2·8	4·3	2·1	46
	Crystal quartz	1·58	3·2	5·3	2·6	38
4000	Fused silica	1·47	1·1	1·60	0·8	125
	Crown glass	1·53	1·3	2·03	1·0	100
	Flint glass	1·76	5·1	10·6	5·3	19
6000	Crown glass	1·52	0·36	0·54	0·3	370
	Flint glass	1·72	1·0	2·0	1·0	100

The figures for crystal quartz refer to the ordinary ray. The figures for crown glass and flint glass refer to the particular types UBK 7 and SF 1 respectively.

length with wavelength by tilting the plate, as shown in Fig. 4.11. A wavelength range $\Delta\lambda$ occupying Δl mm as measured perpendicular to the optic axis is then spread over a length $\Delta l/\cos\varphi$ of plate. The actual linear dispersion is $\sec\varphi \, dl/d\lambda$. As φ is usually 60°–70°, this represents an increase by a factor of between 2 and 3.

The angular dispersion of a quartz prism at 2000 Å is about the same as that of a diffraction grating with 12 000 lines per cm used in the second order. The angular dispersion of a grating, however, changes only slowly with wavelength if the same order is used, and it is apparent from the table that the performance of the prism must fall below that of the grating very rapidly as the wavelength is increased.

Crystalline quartz is a doubly refracting material. Prisms are cut with the optic axis parallel to the base of the prism so that the ordinary and extraordinary rays coincide for the position of minimum deviation. Two very slightly displaced images can still be formed, however, because quartz is also an optically active material. This means that the refractive index is slightly different

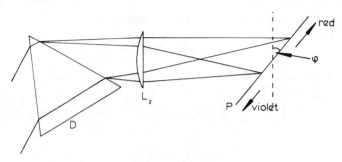

Fig. 4.11 Plate tilt in a prism spectrograph. D is the prism and L_2 a simple camera lens. The plate P is tilted through angle φ from the normal to the axis.

for right- and left-circularly polarized light. Unpolarized light incident on the prism emerges as two slightly divergent beams, circularly polarized in opposite senses. There are, however, two varieties of quartz, right- and left-handed, in which this effect is equal and opposite, and the image doubling can be avoided by making a 60° prism out of two 30° prisms, one of each variety, as in the Cornu prism. In the Littrow mounting (Section 4.9) the effect is self-cancelling because the beam traverses the prism once in each direction.

4.6 Resolving power

From Equation 4.3 the resolving power of a prism is $a \, d\theta/d\lambda$ where a is the effective aperture. If the prism is filled – that is, if its face is illuminated all over – the effective aperture of the system is determined by the prism face of length l. It can be seen from Fig. 4.12 that at minimum deviation $a = l \cos i$. In terms of the prism base t,

$$t = 2l \sin \alpha/2$$

so we have

$$a = \frac{t \cos i}{2 \sin \alpha/2}$$

and

$$\frac{\lambda}{\delta\lambda} = a \frac{d\theta}{d\lambda} = t \cos i \frac{dn}{d\lambda} \frac{1}{\sqrt{(1 - n^2 \sin^2 \alpha/2)}}$$

But at minimum deviation $n \sin \alpha/2 = \sin (\theta + \alpha)/2 = \sin i$ from Equation 4.6, so the resolving power simplifies to

$$\mathscr{R} = \lambda/\delta\lambda = t \, dn/d\lambda \qquad (4.8)$$

The full resolving power is, of course, realized only if the full aperture of the prism is used and if the spectrograph slit is narrow enough. From Table 4.1 it can be seen that the dispersion $dn/d\lambda$ in

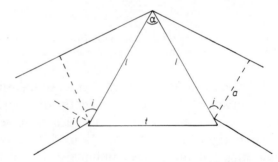

Fig. 4.12 Geometry of ray passing through prism at minimum deviation. t is the prism base and a the width of the beam.

the visible/ultra-violet is of order 10^3 per cm. Taking 10 cm as an upper limit for the prism base, one arrives at 10^4 as a reasonable order of magnitude for the resolving power of a prism.

4.7   Prism materials for the far ultra-violet and the infra-red

The high dispersion of quartz near its transmission limit gives prism spectrographs a high performance at the edge of the vacuum ultra-violet region, about 2000 Å. Traditionally quartz is usable to about 1850 Å, but specially pure varieties of fused silica now available transmit to 1700 Å. Below this limit, fluorite or lithium fluoride prisms may be used down to about 1400 Å, but the loss of light is considerable in this region, and the crystals are available only in small sizes. Prism spectrographs below 1700 Å are in fact so inefficient that they are scarcely ever used.

Going now to the infra-red region, glass prisms can be used from the visible up to about 2·5 μm and quartz (crystalline or fused) to about 3·5 μm. It would appear from the trend evident in Table 4.1 that the dispersion might be extremely small at such long wavelengths, but in fact the dispersion of any transparent material increases at *both* ends of its region of transparency. On approaching the long wavelength cut-off at a few μm, therefore, $dn/d\lambda$ increases, while n continues to decrease. In the case of quartz, $dn/d\lambda$ goes through a minimum at about 1·5 μm, and the 6000 Å value is regained at about 3 μm.

The most common material for the region 3– 16 μm is rock salt, but it has the disadvantage of being very hygroscopic. Some of the other alkali halides are useful over limited bandwidths in the region up to 16 μm. The dispersion of all these materials is low compared to that of glass and quartz at shorter wavelengths – typically about 0·005 per μm, or 5×10^{-7} per Å. In the last 20 years or so the range has been extended to considerably longer wavelengths by the development of new crystal-growing techniques. Prism spectroscopy up to 40 or 50 μm is feasible with materials such as thallium bromoiodide and cesium iodide. References [7, 8] give useful data for transmission, refractive index, and dispersion as a function of wavelength for all the materials used in the infra-red.

4.8 Image defects

4.8.1 Line curvature

With the slit at the focus of the collimator lens, all rays from the centre of the slit pass through the prism parallel to a principal section of it. Rays from the ends of a slit of finite length, however, pass through the prism at a small inclination to the principal section. The prism angle is effectively larger for these rays, and they are deviated more. The slit image, instead of being a straight line, is an arc whose ends are curved towards shorter wavelengths. The radius of curvature is of the same order of magnitude as the camera focal length; it depends on the refractive index and so increases to shorter wavelengths. Curvature does not affect the resolution of a spectrograph, but it does significantly reduce that of a photoelectric spectrometer using straight entrance and exit slits.

4.8.2   Lens aberrations

Line curvature is an aberration introduced by the prism, but the slit image is also affected by all the normal lens aberrations from the collimator and camera lenses. Of these, chromatic aberration is hardly a defect in this context, since it may be compensated by tilting the photographic plate as already described. The gain in dispersion from plate tilt is often a positive advantage and may be considerable in the quartz ultra-violet.

Spherical aberration may be contributed by both collimator and camera lenses, but is rarely significant because of the small apertures. The off-axis rays from the ends of the slit produce coma, which is an important defect because of its asymmetry. It can be reduced by 'bending' the lenses – that is, by adjusting the curvature of the two faces while keeping the focal length constant. In the camera lens astigmatism must also be taken into account, because if the spectrum is to cover an appreciable length of photographic plate one has to use rays at a considerable angle to the optic axis. Astigmatism means that there is no longer a point to point correspondence between the slit and its image. Stigmatism is a desirable property of prism spectrographs for such purposes as space resolution of a source and intensity calibration, so it should be corrected as far as possible. One is usually left with curvature of the field as the outstanding aberration, and this can be allowed for by using a plate-holder curved to match it.

4.9   Types of spectrograph

This section is intended only as a very cursory survey, as an enormous number of prism mountings and associated spectrographs have been devised. Much fuller information may be found in the books of Sawyer and Harrison, Lord and Loofbrough.

In small or medium sized instruments not required as mono-chromators, the straight through design of Fig. 4.9 is the most usual. The collimator and camera lenses of medium spectrographs usually have the same focal length, giving unit magnification, and there is a single 60° prism. Small spectrographs may have a shorter focal length camera lens. This type of instrument is an awkward shape for focal lengths of a metre or more, and the Littrow mounting generally replaces it. This uses a 30° prism with a silvered back or a 60° prism with a mirror immediately behind it, so that the light retraces its path through the instrument. The lens then serves as both collimator and camera lens, saving some expense as well as space. The light is usually brought in from the

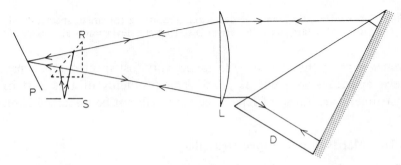

Fig. 4.13 Littrow mounting for prism spectrograph. D is a 30° prism and L acts as both collimator and camera lens. The slit S and reflecting prism R are above the plane of the rest of the diagram.

side with the help of a totally reflecting prism, as shown in Fig. 4.13. A disadvantage of this type of mounting is the scattered light from unwanted reflections at the lens surfaces.

Constant deviation spectrographs are useful in many appli-cations, especially as monochromators when it is highly desirable to have a fixed direction for the emergent beam. The simplest form is the Wadsworth mounting, in which a 60° prism is combined with a plane mirror as shown in Fig. 4.14. Another method, at the cost of reduced dispersion, is to combine three 60° prisms of two different glasses. This is the form usually adopted for small direct-vision spectroscopes. A spectrograph with a con-stant deviation of 90° similar to Fig. 4.14 can be constructed by using a four-sided prism, replacing the mirror of the Wadsworth mount by a total internal reflection.

When intensity is of paramount importance, as, for example, in stellar spectroscopy, it becomes essential to use an instrument of very wide aperture. Whereas most conventional instruments are

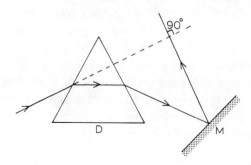

Fig. 4.14 Wadsworth (constant deviation) mounting for prism spectrograph. The mirror M rotates with the prism D so that the deviation is always 90°.

somewhere in the $f/10$ to $f/25$ range, very fast spectrographs may have apertures as great as $f/0.5$. The problems of this sort of instrument are those of lens design and will not be discussed here.

4.10   Merits of prism spectrographs

The principal advantages and disadvantages of prism instruments may be summarized quite briefly. On the credit side, small prism spectrographs are relatively cheap and simple to construct and simple to adjust, although this is not true of scanning spectrometers. Over the concave grating (but not over the plane grating) prisms have the advantage of astigmatism. This makes them suitable for cross-dispersion with the Fabry-Perot interferometer, an application for which high dispersion is usually unnecessary. In the quartz ultra-violet region, where prism instruments are at their most efficient, their light losses are likely to be less than those of grating instruments. Finally, the spectrum is quite unambiguous, with no problem of overlapping of orders. The simplicity of interpretation of the spectrograms makes the prism spectrograph the most generally useful instrument for preliminary survey work.

The most important disadvantages are the rather low dispersion and resolving power as compared with gratings and interferometers. The light flux that can be accepted is also low by comparison with an interferometer of comparable resolving power. Prisms cannot be used in the far infra-red or the far ultra-violet. Finally the non-linear wavelength scale (resulting from the change of dispersion with wavelength) may be a disadvantage in some applications.

References

1. Sawyer, R. A. 'Experimental Spectroscopy', Dover, 1961
2. Harrison, G. R., Lord, R. C. and Loofbourow, J. R. 'Practical Spectroscopy', Prentice-Hall, 1948
3. Bousquet, P. (a) 'Spectroscopie Instrumentale', Dunod, 1969; (b) (Transl. Greenland. P.) 'Spectroscopy and its Instrumentation', Hilger, 1971
4. Ditchburn, R. W. 'Light', Blackie, 1963
5. Born, M. and Wolf, E. 'Principles of Optics', Pergamon, 1965
6. Samson, J. A. R. 'Techniques of Vacuum Ultra-violet Spectroscopy', Wiley, 1967
7. Smith, R. A., Jones, F. E. and Chasmar, R. P. 'Detection and Measurement of Infra-red Radiation', Oxford, 1968
8. Conn, G. K. T. and Avery, D. G. 'Infrared Methods', Academic Press, New York and London, 1960

CHAPTER FIVE

Diffraction gratings

A diffraction grating consists of a large number of very close, equally spaced grooves ruled on a plane or concave surface. In the case of a plane grating, light from a slit set parallel to the rulings is collimated and then either transmitted by or reflected from the periodic structure formed by the rulings. These behave as a large number of coherent sources, and for any given wavelength the secondary wavelets from them interfere constructively at certain angles and destructively at others. The interfering wavefronts are brought to a focus by the camera lens, and images of the slit appear in the focal plane at positions corresponding to the angles for the interference maxima. The concave grating differs only in that the grooves are ruled on a concave mirror blank instead of on a flat, and the grating acts as its own collimator and camera.

5.1 Theory of plane grating: condition for maxima

Figs. 5.1 and 5.2 show a plane wave incident at angle i on a transmission and a reflection grating respectively. The path difference between contributions from adjacent rulings to a wave diffracted at angle θ is $\Delta = d\,(\sin i + \sin \theta)$, where θ is taken to be positive if it is on the same side of the normal as i and negative if it is on the opposite side. For light of wavelength λ the condition that contributions from all rulings reinforce one another at angle θ is

$$\Delta = d(\sin i + \sin \theta) = n\lambda \qquad (5.1)$$

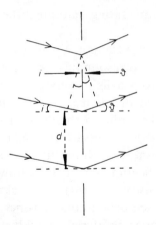

Fig. 5.1 Angles of incidence and diffraction for transmission grating. A plane wave is incident at angle *i* on the grating (spacing *d*) and is diffracted at angle θ (shown here as positive).

where n is an integer, the order number. Zero order, $\theta = -i$, corresponds to straight-through transmission and specular reflection respectively. If θ is negative and $|\theta| > i$, the order number is negative, as shown in Fig. 5.2c for the reflection grating. In practice, transmission gratings are scarcely ever used outside teaching laboratories, and gratings in this chapter are to be

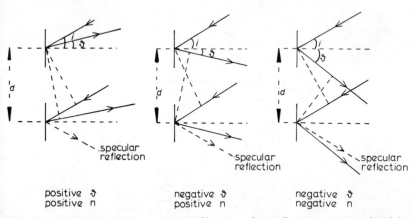

specular reflection

specular reflection

specular reflection

positive θ
positive n

negative θ
positive n

negative θ
negative n

Fig. 5.2 Angles of incidence and diffraction for reflection grating (positive and negative orders). In (a) the diffracted and incident rays are on the same side of the normal, so θ is positive and so is the order n. In (b) and (c) they are on opposite sides, so θ is negative. In (b) $|\theta| < i$, so n is positive. In (c) $|\theta| > i$, so n is negative.

understood as reflection gratings, although the theory is applicable to both.

Evidently a given wavelength λ can have several maxima at different values of θ corresponding to different orders n. The complete spectrum from a light source is repeated several times, and spectra of different orders may well overlap; for example, the second order of 3000 Å coincides with the first order of 6000 Å.

Equation 5.1 gives the necessary information for measurement of wavelengths, and one can obtain the dispersion of the grating by differentiating it (Section 5.3). It tells, however, nothing about the line shape or intensity. In order to assess the resolving power and general performance of a grating it is necessary to investigate how the intensity varies in the immediate neighbourhood of a maximum and how it is distributed between orders.

5.2   Intensity distribution from ideal diffraction grating

The diffraction grating is a periodic structure of period d (the ruling spacing) and width $W = Nd$, where N is the total number of rulings. Whatever form the rulings may take, the grating should behave for large N like an aperture whose width is that of the emerging beam, $W \cos \theta$. This would give the line a diffraction shape with a width determined by $\lambda/W \cos \theta$ (Section 4.2). The analysis below shows that this is indeed the case. The particular

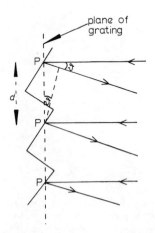

Fig. 5.3 Diffraction of plane wave at grating. The points P are corresponding points of the periodic structure, of spacing d. Light is incident normally.

form of the rulings determines the overall distribution of energy between different orders, or the intensity envelope. This is easily evaluated for the conventional bar-and-space model of the grating, but is much harder to treat in a real grating.

To find the line shape, let us take a plane monochromatic wave incident normally on an arbitrary periodic structure (Fig. 5.3). If the angle of incidence is not zero, the diffraction pattern is simply shifted as a whole, and the line shape is not affected. The path difference between rays from corresponding points P on adjacent rulings is $d \sin \theta$, and the corresponding phase difference is

$$\beta = \frac{2\pi d \sin \theta}{\lambda} \tag{5.2}$$

If the complex amplitude from the element P of each ruling is da'_0, the contribution from the rth point P is $da'_0 e^{-ir\beta}$, and the sum over all rulings is

$$da = da'_0 \sum_{r=0}^{n} e^{-ir\beta} = da'_0 \frac{1 - e^{-iN\beta}}{1 - e^{-i\beta}}$$

This has now to be integrated over all elements of one groove. Writing $\int_{\text{groove}} da'_0 = A'_0$, we have

$$A = A'_0 \frac{1 - e^{-iN\beta}}{1 - e^{-i\beta}}$$

for the amplitude from the entire grating in the direction θ. The intensity is obtained by multiplying the complex amplitude by its complex conjugate

$$A^\star \equiv A'_0{}^\star \frac{1 - e^{+iN\beta}}{1 - e^{+i\beta}}$$

$$\therefore \quad I = AA^\star = I'_0 \frac{1 - \cos N\beta}{1 - \cos \beta} = I'_0 \frac{\sin^2 N\beta/2}{\sin^2 \beta/2} = I'_0 \frac{\sin^2 N\delta}{\sin^2 \delta} \tag{5.3}$$

where $I'_0 = A'_0 A'_0{}^\star$ and

$$\delta = \frac{\pi d \sin \theta}{\lambda} \tag{5.4}$$

I'_0 is the intensity envelope from a single groove and may be expected to vary only slowly with θ, as will be seen shortly. Let us look first at the behaviour of the term $\sin^2 N\delta / \sin^2 \delta$. This has principal maxima of value N^2 whenever $\delta = n\pi$ so that $\sin \delta = \sin N\delta = 0$, where n is a positive or negative integer. Using Equation 5.4, the condition for principal maxima becomes $n\lambda = d \sin \theta$ in agreement with Equation 5.1. Between these principal maxima will be secondary maxima and minima, as shown schematically in Fig. 5.4, corresponding to the N times more rapid fluctuations of $\sin N\delta$. The minima occur for $\delta = (m/N)\pi$, where m is another integer, except that whenever m/N is itself an integer we

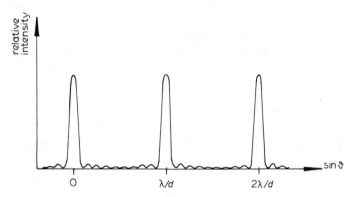

Fig. 5.4 Grating diffraction pattern shown schematically. The figure is drawn for 10 rulings (N = 10), giving 9 subsidiary maxima between the principal maxima. In practice $N \gg 10$, and the subsidiary maxima are much more numerous and much weaker.

get back to a principal maximum. There are $N - 1$ of these minima between each pair of principal maxima at positions given by $d \sin \theta = (m/N)\lambda$. If N is large, so that $\sin \delta$ changes slowly compared to $\sin N\delta$, the $N - 2$ secondary maxima fall half-way between the minima. Their intensity is given by $I = I'_0 / \sin^2 \delta$, and so as $\sin \delta$ increases towards its maximum value of unity half way between orders the intensities of the secondary maxima fall rapidly towards the fraction $1/N^2$ of that of the principal maxima. With N of order $10^4 - 10^5$, we can safely forget about all secondary maxima except those close to the principal maxima.

The principal maxima are of course the 'spectral lines'. It can easily be shown that these lines have the anticipated diffraction

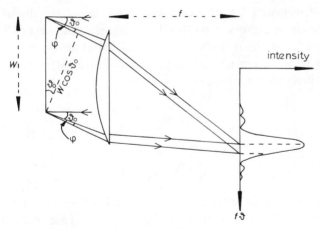

Fig. 5.5 Intensity distribution of a principal maximum of a diffraction grating. Light is incident normally on grating of width W. θ_0 satisfies the condition $n\lambda = d \sin \theta_0$ for a principal maximum and φ is a very small angle.

shape if θ is replaced by $\theta_0 + \varphi$, where θ_0 defines the centre of the line according to Equation 5.1, $n\lambda = d \sin \theta_0$, and φ is small (Fig. 5.5). Then

$$\sin (\theta_0 + \varphi) = \sin \theta_0 \cos \varphi + \cos \theta_0 \sin \varphi \approx \sin \theta_0 + \cos \theta_0 . \varphi$$

and from Equation 5.4

$$\delta = n\pi + \pi d/\lambda \cos \theta_0 \, \varphi$$

An integral number of πs has no effect on the value of $\sin^2 \delta$ or $\sin^2 N\delta$, so for small φ Equation 5.3 becomes

$$I = I_0' \frac{\sin^2 N\delta}{\delta^2} = I_0 \frac{\sin^2 N\delta}{(N\delta)^2} \tag{5.5}$$

where

$$I_0 = N^2 I_0' \text{ and } N\delta = \frac{\pi}{\lambda} Nd \cos \theta_0 \varphi = \frac{\pi}{\lambda} W \cos \theta_0 \varphi.$$

Equations 5.5 and 4.2 are identical if $W \cos \theta_0$ replaces a. The intensity distribution for the central maximum and the neighbouring secondary maxima therefore conforms to the Fraunhofer diffraction pattern for an aperture of width $W \cos \theta_0$.

There remains to be considered the shape of the intensity envelope. In the much simplified bar-and-space model of Fig. 5.6, where each element reflects (or transmits) uniformly over its width b, I_0' in Equation 5.3 is the Fraunhofer diffraction pattern from a single slit of width b:

$$I_0' = \text{const.} \frac{\sin^2 \beta}{\beta^2} \tag{5.6}$$

where

$$\beta = \pi b \sin \theta / \lambda$$

β is the phase difference between the centre and edge of one grating element. This distribution has its first minimum at $\sin \theta = \pm \lambda / b$. Since $b \sim \frac{1}{2}d$, the central maximum has an angular spread somewhat larger than the distance between orders. In real

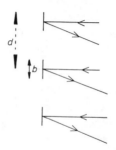

Fig. 5.6 'Bar and space model' of grating. Light is reflected uniformly from the 'bars' of width b and spacing d.

gratings I_0' may depart considerably from Equation 5.6, depending on the form of the ruling. Modern gratings are nearly always 'blazed' to direct a high proportion of the intensity at a particular diffraction angle (Section 5.7). It remains true, however, that the envelope is a relatively slowly varying function of θ.

5.3   Dispersion

The dispersion is found by differentiating Equation 5.1,

$$n\lambda = d (\sin i + \sin \theta)$$

to give

$$d\theta / d\lambda = n / d \cos \theta \tag{5.7}$$

Near the grating normal, where $\cos\theta \approx 1$, the dispersion is n/d. Gratings for the visible and ultra-violet usually have 600 or 1200 lines/mm. The first of these, used in the first order, has $d\theta/d\lambda = 6 \times 10^{-5}$ rad/Å, which gives a reciprocal dispersion of 16 Å/mm in a one-metre spectrograph. The angular dispersion of a 60° glass prism at 6000 Å is almost an order of magnitude smaller than this, although that of a quartz prism just above 2000 Å is about the same (see Table 4.1). As an example of the much greater dispersion easily available, the 1200 line/mm ruling used in the second order in a three-metre spectrograph gives a reciprocal dispersion of about 1·3 Å/mm.

Since $\cos\theta$ changes only slowly near $\theta = 0$, the dispersion is nearly constant near the normal to the grating, giving an almost linear wavelength scale, in contrast to the non-linear scale of the prism spectrograph. Another point of contrast is that the grating deviates shortest wavelengths least, while the prism – except in the infra-red – deviates them most.

The ambiguity of grating spectra, arising from the overlapping of orders, has already been mentioned. For example, the third order of a wavelength λ, the second order of $3/2\,\lambda$ and the first order of 3λ will all coincide when low orders are used. The unwanted orders can generally be eliminated by using suitable filters or detectors of limited spectral sensitivity. Prisms or small additional gratings are sometimes introduced in front of the slit to act as 'order-sorters' or pre-dispersers. Overlapping orders sometimes serve a useful purpose in acting as reference lines in spectral regions where good standards do not exist. But in the far ultra-violet and infra-red, where detectors of limited band width and filters are often not available, they can be a very great nuisance. It should be remembered that the entire visible region covers only one octave (doubling of wavelength), whereas the ultra-violet, even for $\lambda > 100$ Å, covers 5 octaves and the infra-red 10 octaves. An additional problem in the infra-red is the relatively greater intensity of the higher orders of shorter wavelengths.

5.4 Resolving power

The resolving power of the diffraction grating can be approached in two slightly different ways. The conventional treatment is to use the general results for a diffraction line shape, which we have shown to be applicable to the grating, and simply insert the appropriate values into Equation 4.3: $\mathscr{R} = a\,d\theta/d\lambda$. If W is the width of the grating (Fig. 5.7), the effective aperture for the diffracted beam is $a = W\cos\theta$. Taking $d\theta/d\lambda$ from Equation 5.7,

$$\mathscr{R} = nW/d = nN \qquad (5.8)$$

since $N = W/d$.

The fundamental importance of the grating width is brought out more clearly by using the second approach of Section 4.2 – namely, that the resolving power is equal to the number of wavelengths in the extreme path difference, Δ/λ (Equation 4.4). From Fig. 5.7,

$$\Delta = \Delta_1 + \Delta_2 = W(\sin i + \sin\theta)$$

$$\therefore \quad \mathscr{R} = W(\sin i + \sin\theta)/\lambda \qquad (5.9a)$$

(This equation can of course be obtained from Equation 5.8 by substituting for n from Equation 5.1). The significance of

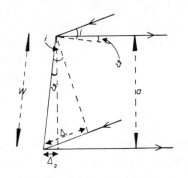

Fig. 5.7 Effective aperture of grating. For light diffracted at angle θ, the effective aperture of a grating of width W is $a = W\cos\theta$. The extreme path difference between incident and diffracted light is $\Delta_1 + \Delta_2$.

Equation 5.9a is that $\mathscr{R}$ is seen to depend on the grating width and the angles of incidence and refraction. For a given wavelength, $(\sin i + \sin\theta)$ depends only on n/d and is the same for a fine ruling in a low order as for a coarse ruling in a high order, an important principle which is put into practice in the echelle grating. For the commonly used auto-collimating mounting,

$$i = \theta \quad \text{and} \quad \mathscr{R} = 2W\sin i/\lambda \qquad (5.9b)$$

showing that the maximum attainable resolving power approaches $2W$ as the grating is turned towards grazing incidence. The resolution limit, from Equation 4.5, is

$$\delta\sigma = \frac{1}{\Delta} = \frac{1}{W(\sin i + \sin \theta)} \geqslant \frac{1}{2W} \qquad (5.10)$$

The approximate relationship $\delta\sigma \sim 1/W$ is a particularly easy way to remember the resolution limit of a grating. W must of course be in cm to put $\delta\sigma$ in cm^{-1}.

There are practical limits to the width over which a grating can be ruled with sufficient accuracy, and it sometimes pays actually to decrease the effective width by blanking off bad parts of surface. A conventional grating for the visible or ultra-violet has 600 or 1200 lines per mm, with a ruled area up to 20 cm wide. A 1200 line/mm grating in the second order could have a theoretical resolving power of 480 000. The theoretical resolving power is seldom realized in practice for a grating as large as this, for reasons that will be discussed briefly below, but it remains true that for this type of grating resolving powers of a few hundred thousand are quite feasible. This is a couple of orders of magnitude beyond the reach of a prism spectrograph. In the far ultra-violet, however, and usually in the far infra-red, the resolving power is limited by the slit width rather than the grating.

Echelle gratings make use of the coarse ruling – high order combination, having $n \sim 100$, $N \sim 10^4$ and $W \sim 25$ cm. The resolving power therefore approaches 10^6. In echelles (as also in modern conventional gratings) the groove is specially shaped to concentrate the diffracted light at a particular angle, so as to obtain a high intensity at these large order numbers (Section 5.7).

5.5   The concave grating

It was first shown by Rowland in 1882 that if a grating is ruled on a concave mirror instead of on a plane surface, it can act as its own collimator and camera lens. The spectrograph is then reduced to the three essential elements of slit, grating and detector, and so it can be used in the far ultra-violet where lenses do not exist and mirror reflectivities are very low. Rowland showed that if the slit and grating both lie on a circle whose diameter is equal to the radius of curvature of the grating and if the grating is tangent to this circle, then for small apertures (the equivalent of paraxial rays in a lens system) the spectral lines are in focus on the circle. The positions of the lines are given, as for the plane grating, by

$$d(\sin i + \sin \theta) = n\lambda \qquad (5.1)$$

where d is the distance between rulings as measured along a chord across the grating arc.

The Rowland condition can be derived quite simply for rays in a plane perpendicular to the grating if the grating is short enough

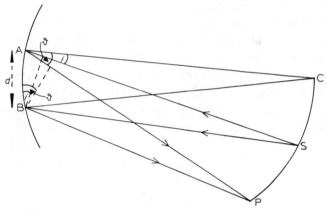

Fig. 5.8 Angles of incidence and diffraction for concave grating. C is the centre of curvature of the grating, and A, B are adjacent rulings. S is the slit and P its image. The distance AB is much exaggerated.

to be considered coincident with the circle instead of tangent to it. Fig. 5.8 shows two adjacent rulings A and B of a grating whose centre of curvature is at C. Rays from the slit S are incident at angle i, are diffracted at angle θ, and intersect at P. To a first approximation the path difference between the two incident rays SA and SB is $d \sin i$ and that between the two diffracted rays AP and BP is $d \sin \theta$. The total path difference for the two rulings is therefore $d(\sin i + \sin \theta)$, and it follows that Equation 5.1 represents the condition for a principal maximum at angle θ. If P is to be a sharp image of S, it is necessary that diffracted rays from the whole width of the grating that satisfy Equation 5.1 should

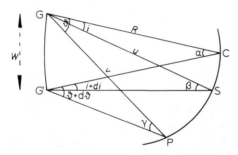

Fig. 5.9 Ray diagram for concave grating. GG′ is the grating, of width W, subtending angles α, β, γ at C, S and P respectively. R is the radius of curvature of the grating, and u, v are the object and image distances. The grating size is much exaggerated.

intersect at P. Fig. 5.9 shows the grating GG′ of width W and radius of curvature R with S at distance u and P at distance v. If the angles of incidence and diffraction are i and θ at G, they can be written as i + di and θ + dθ at G′ if W is small compared to u and v. The relation between di and dθ is obtained by differentiating Equation 5.1:

$$\cos i \, di + \cos \theta \, d\theta = 0$$

From the geometry of the figure, if GG′ subtends angles α, β and γ at C, S and P respectively,

$$i + \alpha = i + di + \beta \qquad \text{and} \qquad \theta + \alpha = \theta + d\theta + \gamma$$

$$\therefore \qquad di = \alpha - \beta \qquad \text{and} \qquad d\theta = \alpha - \gamma$$

In terms of the distances R, u and v,

$$\alpha = W/R, \quad \beta = W \cos i/u, \quad \gamma = W \cos \theta/v$$

The condition that P be a sharp spectral line is therefore

$$\cos i \left(\frac{1}{R} - \frac{\cos i}{u} \right) + \cos \theta \left(\frac{1}{R} - \frac{\cos \theta}{v} \right) = 0 \qquad (5.11)$$

In mountings based on the Rowland circle this is satisfied by making both the expressions in parentheses identically zero – i.e., $u = r \cos i$ and $v = R \cos \theta$. It can be seen from Fig. 5.10 that S and P must then lie on a circle of diameter R, which is the Rowland circle. This is the simplest and most aberration-free way

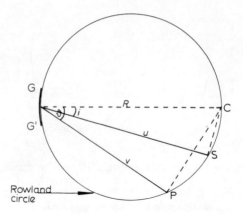

Fig. 5.10 Geometry of Rowland circle. The letters have the same meanings as
in Figs. 5.8 and 5.9.

of mounting the concave grating, but Equation 5.11 can be
satisfied by other types of mounting, such as the Wadsworth and
Seya.

Since the basic grating equation holds for both plane and
concave gratings, the expressions for dispersion and resolving power
derived for the former are valid for the latter. The angular
dispersion is

$$\frac{d\theta}{d\lambda} = \frac{n}{d \cos \theta}$$

and the linear dispersion is

$$R \frac{d\theta}{d\lambda} = \frac{nR}{d \cos \theta}.$$

Most concave gratings have a radius of curvature from one to three
metres, but values as large as 10 metres have been used. Above this
the aperture becomes impracticably low because, owing to the
difficulties of ruling on a concave surface, the width of a concave
grating is even more severely limited than that of a plane grating.

5.6   Astigmatism of the concave grating

Since the concave grating is essentially a mirror with either or both
of the incident and emergent rays at a large angle to the optic axis,
the image suffers from astigmatism. If the slit is accurately parallel

to the grating rulings, each point of the slit is imaged as a short vertical line at the optimum horizontal focus, and there is no loss of definition. In general, however, the intensity of illumination is reduced, and there is no longer a point-to-point correspondence between the slit and its image. This means that the source cannot be resolved spatially, nor can the plate be calibrated for intensity with a step filter or diaphragm in the usual position immediately in front of the slit (Section 3.11), although it is sometimes

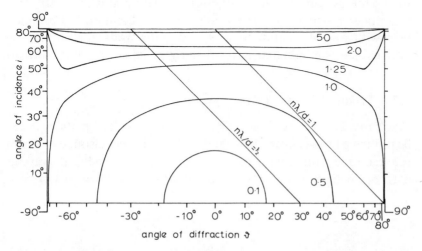

Fig. 5.11 Astigmatism of concave grating. The figure by each curve gives the image length of a point source on the slit in units of the grating ruling length. The scales along the axes are linear in $\sin \theta$ and $\sin i$ respectively, so that a given value of ($\sin i + \sin \theta = $ const) represents a straight line of unit gradient. The lines $n\lambda/d = 1$ and $n\lambda/d = \frac{1}{2}$, for example, correspond to 8333 Å and 4166 Å respectively in the first order of a grating with 1200 line mm^{-1}.

possible to put the diaphragm at the Sirks focus outside the slit, the position at which a point object is imaged as a sharp horizontal line on the Rowland circle. The length z of the line image from a point on the slit is proportional to the vertical aperture – that is, to the length l of the rulings. In auto-collimation (Eagle-type mounting), for which the astigmatism is least, $z = 2\,l\,\sin^2\theta$, which is about $\frac{1}{2}l$ for the second order of a 1200 line/mm grating in the visible. By superposing the line images from successive points of the slit, one can see that the illumination of the complete image is reduced at the ends but not at the centre, provided that a sufficient length of slit is illuminated. Ideally, the effective slit length should be approximately equal to z. This is usually impracticable, partly because it requires such a large source and partly because of spectral line curvature, an aberration arising from the different paths taken by rays from different points of the slit. The combination of curvature with astigmatism effectively broadens the spectral line and limits the usable length of the rulings.

Astigmatism is discussed in detail in [1] and [2]. Fig. 5.11 shows the astigmatism as a function of angle and wavelength. It is sometimes possible to compensate for a small amount of astigmatism with a weak cylindrical lens, and in the Wadsworth mounting astigmatism is eliminated altogether by means of an extra mirror.

5.7 Production and errors of gratings

The early gratings were usually ruled on speculum metal. At the beginning of the century Rowland and, later, Michelson, experts in the required combination of art and science, produced gratings with resolving powers of a few hundred thousand which were not

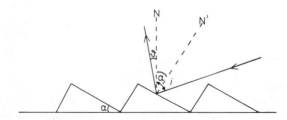

Fig. 5.12 Geometry of blazed grating. For explanation, see text.

improved upon for some 40 years. Modern gratings are ruled on glass blanks covered with a thin evaporated film of aluminium, and highly sophisticated servo-systems have been developed to control the accuracy of the ruling. Most gratings now in use are in fact replicas: a master grating is ruled, and an epoxy resin casting of the surface is formed as a sandwich between the master and a supporting glass blank. The master is then separated and the replica aluminized. Both plane and concave gratings can be produced in this way. The quality of the replicas may actually be better than that of the master because the well burnished parts of the ruling are transferred from the bottoms to the tops of the grooves, with an appreciable gain in efficiency.

Modern gratings are usually blazed – that is, the shape of the groove is controlled so as to concentrate a large fraction of the incident intensity into diffraction at a particular angle. This concept was introduced by Wood for fairly coarsely ruled gratings suitable for the infra-red region, known as echelette gratings, and improvements in ruling technique have made it feasible for finely ruled gratings as well. The principle is illustrated in Fig. 5.12: the preferred direction is that corresponding to specular reflection from the 'step' of each groove. If the step is inclined at angle α to the surface of the grating and N and N′ are the normals to the grating and the step respectively, then

$$i - \alpha = -\theta + \alpha$$

(θ is negative because it is on the opposite side of N to i)

$$\therefore \quad \alpha = \frac{i + \theta}{2}$$

Combining this with the grating equation $n\lambda = d(\sin i + \sin \theta)$,

$$(n\lambda)_{\text{blaze}} = 2d \sin \alpha \cos (i - \alpha)$$

Gratings are frequently used in auto-collimation, in which case the angles of incidence and diffraction are equal. Then $i = \theta = \alpha$, the steps are perpendicular to the axis of the spectrograph, and the blaze wavelength is given by

$$(n\lambda)_{\text{blaze}} = 2d \sin \alpha \qquad (5.12)$$

The blaze wavelength is therefore fixed for a given grating used in auto-collimation, although of course a grating blazed for, say, the first order of 6000 Å is also blazed for the second order of 3000 Å. However, the blaze wavelength changes only slowly with the angles of incidence and diffraction, so the efficiency is maintained over an appreciable angle on each side of the optimum. It is possible to get more than 80% of the incident intensity of light of the relevant wavelength at the blaze angle.

The principal of blazing is carried yet further with the echelle grating. This has very widely-spaced grooves, forming right-angled steps. The light is incident at an angle greater than 45° to the

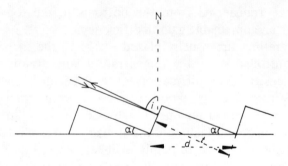

Fig. 5.13 Geometry of echelle grating. For explanation, see text.

grating and is reflected normally from the narrow side of the step
(Fig. 5.13), so that $i = \theta = 90° - \alpha$. The grating equation then gives
for the blaze wavelength

$$(n\lambda)_{\text{blaze}} = 2d \sin i = 2d \cos \alpha = 2t$$

The large value of d means that the order number is very high
($n \sim 100$), and as it is possible to control this relatively coarse
ruling accurately over a wide area (say 25 cm at 80 lines/mm) the
number of rulings N may be of order 10 000, giving a resolving
power of about 10^6. Because of the large order number, successive
orders recur at small wavelength intervals, and the overlap problem
is very severe. Differentiating the grating equation,

$$\Delta\lambda/\lambda = -\Delta n/n$$

so that for

$$\Delta n = 1, \qquad \Delta\lambda = \lambda/n$$

$\Delta\lambda$ is the free spectral range; for $n \sim 100$ it is only a few tens of
Ångströms in the visible and quartz ultra-violet. It is useful to
remember that the free spectral range in wavenumbers is given by

$$\Delta\sigma = \Delta\lambda/\lambda^2 = 1/\lambda n = 1/2t \qquad (5.13)$$

which is constant for a given grating. This is identical with the free
spectral range of a Fabry Perot interferometer of plate separation t
(Section 6.2). The overlapping problem is usually remedied by
cross dispersion. A subsidiary prism or grating of small dispersion,
orientated at right angles to the echelle, displaces successive orders
vertically with respect to one another. The resulting spectrum is an
odd-looking square array crossed by sloping bands of lines.

Errors or imperfections in the ruling of any type of diffraction grating result in light leaving the grating at angles other than those defined by Equation 5.1. Any sort of periodic error produces false lines, or 'ghosts'. With modern ruling techniques these can be kept below one part in 10^5 of the intensity of the parent line and so seldom matter much. Variations in spacing and defects of the figure of the blank produce wings to the lines, over and above the diffraction wings. These may or may not be symmetric and are primarily responsible for the difference between actual and theoretical resolving power. Random errors of any kind result in a general background of scattered light, known as noise or 'grass'. Sometimes the defects originate in a particular part of the grating, and it then pays to mask off this area.

The resolving power of the spectrograph as a whole is also affected by aberrations of the optical system, some of which are considered briefly in the next section, and by vibrations, stray light and temperature changes. When the linear dispersion is small (short focal length) the resolution of a spectrograph may be 'plate-limited': the resolution limit of the grating, $\delta\lambda$ (see Section 4.2) cannot be attained if it corresponds to a distance on the plate, $\delta\lambda \cdot dl/d\lambda$, smaller than the grain size of the emulsion – that is, 10–25 μm (Section 3.11).

More detailed information on the ruling of diffraction gratings may be found in a recent review article [5].

5.8 Grating mountings

Many different types of mounting have been designed for diffraction gratings, some of them for special problems. This section describes only the types in most common use.

Plane gratings are generally used in one of two mountings. The Littrow mounting, Fig. 5.14, is essentially the same as that already described for the prism spectrograph (Section 4.9). The grating replaces the 'half-prism' with its reflecting back surface, and a single lens acts as both collimator and camera lens. In the visible this is usually a glass achromat, but in the ultra-violet it is often a simple quartz lens that can be moved along the axis of the spectrograph to focus different spectral regions. To change the spectral region it is necessary to rotate the grating, re-focus the lens and change the tilt of the plate.

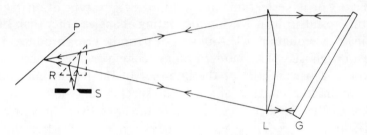

Fig. 5.14 Littrow mounting of plane grating. The slit S and reflective prism R are both above (or below) the plane of the rest of the diagram.

The second common mounting for the plane grating is the Ebert type shown in Fig. 5.15. The concave mirror M both collimates and focuses the light, and, since there are no lenses, the mounting is achromatic and particularly suitable for the infra-red. The spectral region is changed simply by rotating the grating. The arrangement shown in Fig. 5.15 is the out-of-plane, or 'up and over' mounting, which has the advantage of low scattered light. Higher resolution, especially for photo-electric detection, is achieved by the Fastie, or in-plane, arrangement with the entrance

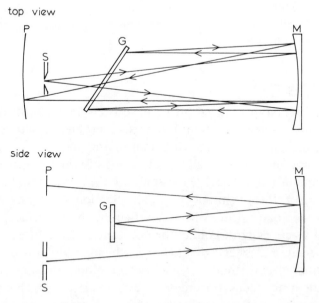

Fig. 5.15 Ebert mounting of plane grating. This diagram shows the 'up and over' mounting, with the slit S vertically below the plate holder P.

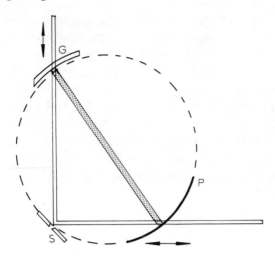

Fig. 5.16 Rowland mounting of concave grating. The dotted circle is the Rowland circle. The arrows show the motions of the plate-holder P and grating G for scanning the spectrum.

and exit slits side by side; the effects of spectral line curvature can then be overcome by using curved slits. The variant of the Ebert mounting in which the single mirror is replaced by two adjacent mirrors is known as the Czerny-Turner mounting. All Ebert-type mountings show very little coma and almost negligible astigmatism.

All common mountings of the concave grating except for the Wadsworth and the Seya use the Rowland circle in some form.

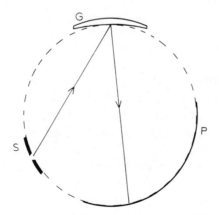

Fig. 5.17 Paschen–Runge mounting of concave grating. The Rowland circle is again shown dotted. The slit S, grating G and plate holder P are all fixed.

Rowland's original mounting has the simplest geometry but is scarcely ever used now because of its awkward shape and the difficulty of making it sufficiently rigid. Fig. 5.16 shows the principle: the plate and grating are joined by a rigid bar sliding on perpendicular ways with the slit mounted at their intersection. The Paschen–Runge mounting, Fig. 5.17, is somewhat similar, but in this case the slit and grating are both fixed on the Rowland circle while the plate holder extends round a long enough arc of the circle to cover the whole spectrum. Two different fixed slits may be used to allow different angles of incidence.

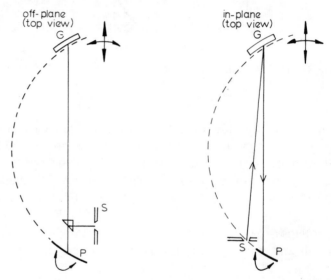

Fig. 5.18 Eagle mounting of concave grating. For explanation see text. The arrows show the motions of the grating G and plate holder P when scanning the spectrum. The dotted arc is part of the Rowland circle.

The most compact arrangement for the concave grating is the Eagle mounting, which is a kind of modified Littrow, using the grating in auto-collimation. The axis of the spectrograph forms a chord of the Rowland circle with the grating and plate holder at its ends, as shown in Fig. 5.18. In instruments intended for the visible and near infra-red the slit is usually mounted slightly above or below the plate holder and in the side of the instrument, a small quartz prism being used to reflect the light in as with the Littrow. This is known as off-plane mounting. For the vacuum ultra-violet the slit and plateholder are usually displaced laterally with respect to one another (in-plane mounting); this is also the form usually adopted for monochromators. In either case the slit is fixed, and the spectral region is changed by moving the grating along the spectrograph axis as well as rotating both grating and plate holder. In this mounting $i = \theta$, so the grating equation becomes

$$n\lambda = 2d \sin \theta \qquad (5.14)$$

Except at very short wavelengths it is not possible to work near the grating normal, and the dispersion is therefore not quite linear. On the other hand, the wavelength range is greater than in the other Rowland circle mountings and the astigmatism is less, making illumination easier.

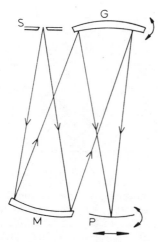

Fig. 5.19 Wadsworth mounting of concave grating. For explanation, see text. The arrows again show the motions of grating and plate holder when scanning.

The Wadsworth mounting, Fig. 5.19, gets away from the Rowland circle and eliminates astigmatism by using an auxiliary concave mirror as a collimator. Since the grating is now illuminated by parallel light, the spectrum is focused at a distance $\frac{1}{2}R$ instead of R, thus halving the dispersion. An additional disadvantage is that the focal surface is parabolic so that the plate holder has to have adjustable curvature as well as adjustable tilt. The principal advantage is that the spectrum near the normal to the grating is stigmatic, and spherical aberration and coma are small provided that the slit is kept close to the grating.

In a monochromator ease of scanning is more important than very high resolution, and the cumbersome linkages required in Rowland mountings when the spectrum is scanned by rotating the grating make these types of mounting rather unsuitable. Various simpler mountings have been developed for monochromators, at the expense of some loss of definition. The best known of these is the Seya, in which the exit and entrance slits are fixed at an optimum angle, about 70°, and the spectrum is scanned by rotating the grating about an axis through its surface. With a

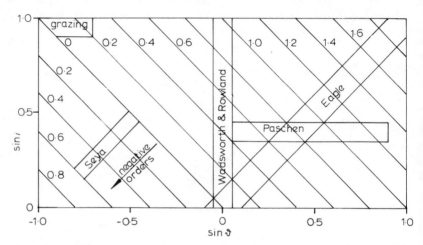

Fig. 5.20 Angular range covered by various mountings of concave grating. As in Fig. 5.11 the scales along the axes are linear in sin θ and sin i respectively so that the diagonals are lines of constant λ/d; for example, the line labelled 0·6 corresponds to $\lambda = 5000$ Å for a 1200 line mm^{-1} grating used in the first order.

one-metre grating of 1200 lines/mm the focus is within 0·1 mm of the exit slit over the whole visible and near ultra-violet.

The angular ranges covered by these different mountings are conveniently summarized by Fig. 5.20. The grazing incidence mounting is described in Section 5.10.

5.9 Gratings in the infra-red

Although concave gratings may be used in the near infra-red, plane gratings of the echelette type nearly always replace them for most of the infra-red region, using the Ebert-Fastie or Czerny-Turner mountings already described. The gain in intensity from a stigmatic mounting is important in the infra-red, where the available radiant energy is nearly always low, and the extra reflections in these mountings can be tolerated because of the high reflectivities in the infra-red. The echelette is blazed to give the maximum intensity for a particular wavelength λ_b and the efficiency is maintained above 60–70% of the maximum from $2/3\,\lambda_b$ to $3/2\,\lambda_b$ – that is, over about an octave, or doubling of the wavelength. Some 5 or 6 gratings are needed to cover the whole infra-red. The distance d between rulings should be of the same order of magnitude as the wavelength. The ruling is therefore much coarser than for gratings used in the visible, varying from a few lines/mm to a few hundred lines/mm. The resolution limit of most gratings is of the order of a few tenths of a wavenumber (corresponding to a resolving power dropping from about 10^4 to 10^2 as one goes further into the infra-red), but the actual resolving power of the spectrograph is usually limited by the slit width rather than the grating. Slit-limited resolving power can be evaluated as follows: if the slit has width w, two wavelengths $\delta\lambda$ apart cannot be resolved unless they diverge at an angle $\delta\theta \geqslant w/f$ where f is the focal length of the system. $\delta\theta$ is obtained from the angular dispersion (Equation 5.7) as

$$\delta\theta = \frac{d\theta}{d\lambda}\,\delta\lambda = \frac{n}{d\cos\theta}\,\delta\lambda.$$

For an auto-collimating mounting $n\lambda = 2d\sin\theta$ (Equation 5.14), so

$$\delta\theta = \frac{2\tan\theta}{\lambda}\,\delta\lambda \sim \frac{\delta\lambda}{\lambda} \qquad \text{for} \qquad \theta \sim 30°.$$

The resolving power is then given by $\lambda/\delta\lambda \sim f/w$. The focal length cannot be made very long, because this would reduce the aperture (the width of the grating being limited to, say, 30 cm in the far infra-red). At $f/10$ slit widths of a few mm give resolving powers of several hundred, and there is clearly no advantage in using gratings of much higher resolving power than this.

The elimination of unwanted orders is a particular problem of the infra-red. For a thermal source in this region the energy is proportional to the square of the frequency, so that higher orders of shorter wavelengths come up with high intensity. Prisms may be used as pre-dispersers up to their limit of transmission ($\sim 40\,\mu m$), but above this some sort of filtering becomes necessary.

'Reststrahlen' crystals and Christiansen filters have been used for many years: the first of these relies on selective reflection exhibited by different crystals over different wavebands, and the second on selective scattering, which occurs when a powder is suspended in a medium having a different refractive index except at one particular wavelength (which then forms the centre of the pass band). Recently developed filters have used various crystal powders incorporated in thin sheets of black polythene. Details of these various filters may be found in the references to infra-red work given at the end of the chapter.

While pure oxygen and nitrogen do not absorb in the infra-red, atmospheric air does, owing to the presence of water and carbon dioxide. Infra-red spectrographs are either evacuated or operated in a dry gas.

5.10 Gratings in the vacuum ultra-violet

Spectroscopic techniques undergo a change for wavelengths below about 2000 Å for two reasons, which have already been mentioned: oxygen begins to absorb, and, for practical purposes, so does quartz (although special forms of fused silica can be used in small thicknesses to as low as 1650 Å). Absorption by the atmospheric gases continues down to the soft X-ray region – about 4 Å – so that it is necessary to use vacuum spectrographs all through the far ultra-violet. (It is, however, occasionally possible to exploit a 'window' in the oxygen absorption centred on the Lyman α line of hydrogen at 1216 Å to do away with a spectrograph altogether at this one wavelength). The quartz cut-off means using either mirror optics or concave gratings, since lenses of fluorite, lithium fluoride, and suchlike materials cannot be made large enough for an instrument of reasonably high resolution. The vacuum ultra-violet is, in fact, the region where the concave grating reigns supreme, for it requires only one reflection – that from the grating itself – and this property becomes increasingly important as one goes to shorter wavelengths because of the decrease in reflectivity of all materials. Aluminium, for example, has a percentage reflectivity in the nineties in the visible and near ultra-violet, but this is down to about 70% at 1500 Å, even when the film is protected from oxidation with a flash coating of magnesium oxide, and it continues to fall rapidly at shorter wavelengths. Below 1000 Å gold and platinum appear to be the best reflectors, but their reflectivities are down in the 10–20% region. The loss of intensity from the two additional mirrors that would be required with a plane grating would be prohibitive. The concave grating is most often used in an Eagle-type mounting, which is the most compact and convenient shape for a vacuum tank. The grating equation has the auto-collimation form

$$n\lambda = 2d \sin \theta \qquad (5.14)$$

where for maximum intensity θ should be as near as possible to the blaze angle for the grating. The high quality gratings used in the visible, with 600 or 1200 lines/mm, are also used in the vacuum ultra-violet (with an appropriate blaze angle), although even finer rulings (up to 3600 lines/mm) exist. λ/d and hence θ are smaller in this region, and consequently the astigmatism is much smaller than in the visible.

Below about 300 Å reflectivities are so low that it becomes necessary to use the grating at grazing incidence. Light incident at a grazing angle φ ($= 90° - i$) smaller than the critical angle φ_c for that particular wavelength is totally reflected. The critical angle is theoretically related to the wavelength by the expression

$$\sin \varphi_c = \lambda \left(\frac{e^2 N}{(4\pi\epsilon_0)\pi m c^2} \right)^{\frac{1}{2}}$$

where e and m are the charge and mass of the electron and N is the number of electrons per cubic metre in the surface material. Looked at the other way round, for a given angle of incidence, this expression gives the minimum wavelength for which there is total reflection. Putting in the constants we have

$$\lambda_{min} (\text{Å}) = 3\cdot3 \times 10^{17} \, N^{-\frac{1}{2}} \sin \varphi.$$

For $\varphi = 1°$, this gives $\lambda_{min} \sim 6\cdot6$ Å for aluminium or glass and $\lambda_{min} \sim 2\cdot5$ Å for a heavy metal such as platinum. To get down to 1 Å requires a grazing angle of about 20′. Wavelengths of this order have actually been photographed, but it is obvious that surface imperfections on the grating face must be extremely important; much work is going into the improvement of gratings (production of lamellar gratings, for example), for the soft X-ray region in view of its present astrophysical importance.

The grazing incidence mounting is shown schematically in Fig. 5.21, with the slit, grating and plateholder all lying on the Rowland circle. Since only a small segment of the circle is used, a grazing incidence spectrograph takes up much less room than a normal incidence mounting, and a two-metre grating is quite a compact instrument. On the other hand, the mounting and its adjustment have to be more accurate because of the very fine angles involved, and the astigmatism is enormous. The resolving power is always either slit- or plate-limited; the optimum slit width in this region is in any case impracticably small.

At wavelengths of a few Ångströms optical and X-ray spectroscopy overlap, and the methods of X-ray spectroscopy may also be

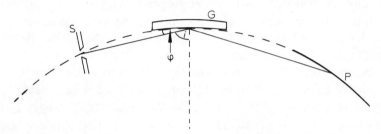

Fig. 5.21 Grazing incidence mounting of concave grating. The dotted arc is part of the Rowland circle. The grazing angle is the complement of the angle of incidence *i*.

used – in particular, the Bragg crystal spectrometer. However, whereas a two-dimensional ruled grating will produce a maximum of a given wavelength somewhere for *any* angle of incidence (within the critical angle), a three-dimensional crystal grating will do so only for an angle of incidence satisfying the Bragg condition $n\lambda = 2t \sin \varphi$ for the distance t between the planes. The latter instrument must therefore be scanned through different angles of incidence.

Two further problems of the far ultra-violet should be briefly mentioned. One is scattered light of longer wavelength, which is preferentially reflected at all surfaces. Selective detectors such as 'solar-blind' photomultipliers can sometimes be used to reduce this, and in the far ultra-violet it is often possible to use thin metallic films as filters (for example, aluminium below 700 Å). The second problem is that of overlapping orders, which becomes acute at short wavelengths because the actual wavelength gap between successive orders becomes very small.

References

1. Sawyer, R. A. 'Experimental Spectroscopy', Dover, 1961
2. Harrison, G. R., Lord, R. C. and Loofbourow, J. R. 'Practical Spectroscopy', Prentice-Hall, 1948
3. Ditchburn, R. W. 'Light', Blackie, 1963
4. Born, M. and Wolf, E. 'Principles of Optics', Pergamon, 1965
5. Loewen, E. G. Diffraction Gratings for Spectroscopy, *J. Phys. E.* **3**, 953, 1970
6. Stroke, G. W. Diffraction Gratings, in 'Encyclopaedia of Physics' **29**, 426, Springer, 1967
7. Bousquet, P. (a) 'Spectroscopie Instrumentale', Dunod, 1969; (b) (Transl. Greenland, P.) 'Spectroscopy and its Instrumentation', Hilger, 1971
8. Martin, A. E. 'Infra-red Instrumentation and Techniques', Elsevier, 1966
9. Samson, J. A. R. 'Techniques of Vacuum Ultra-violet Spectroscopy', Wiley, 1967

CHAPTER SIX

Interferometers

Different forms of interferometer are now used for a large number of very different purposes – for example, high precision metrology, measurements of refractive index and density, wavefront analysis, study of surfaces, and microscopy. For high resolution spectroscopy, however, only two types of interferometer are important: the Fabry–Perot and the Michelson. The Fabry–Perot is a multiple beam instrument, capable of extremely high resolution from the near infra-red to the quartz ultra-violet. The Michelson is a two-beam interferometer whose present importance comes from its use as a scanning instrument to give the Fourier transform of a spectrum, and its particular advantages become most important in the far infra-red. Many other types of interferometer in current use are variations of the Michelson, and a brief description of one of these, the Mach–Zehnder, is given at the end of this chapter because of its applications in plasma spectroscopy and in the measurement of oscillator strengths.

THE FABRY–PEROT INTERFEROMETER

6.1 Basic ideas

The Fabry–Perot interferometer consists of two glass, or, more probably, fused silica, plates, usually between 2 cm and 15 cm in diameter, held accurately parallel to one another at a fixed distance t. The spacers are normally made of quartz or invar (for which the coefficient of thermal expansion is small), and the plates are kept in optical contact with them at three points by some form of spring-loaded mounting. A range of spacers from under 1 mm to several cm may be needed. An interferometer set up with one such spacer is often known as an etalon.

The facing surfaces of the plates are flat to better than 1/50 of a wavelength and are coated with metal or dielectric films of high reflectivity (usually about 90%). A light ray incident at angle θ undergoes multiple reflections, as shown in Fig. 6.1. Successively

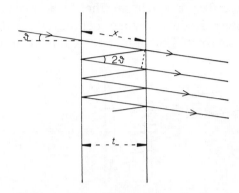

Fig. 6.1 Path of light ray through Fabry–Perot interferometer. The thickness of the interferometer plates is ignored because all rays have identical paths within it.

reflected rays emerge parallel with a constant path difference between them, Δ. It can be seen from the figure that $\Delta = x(1 + \cos 2\theta)$ where $x = t/\cos \theta$.

$$\therefore \quad \Delta = t \frac{2 \cos^2 \theta}{\cos \theta} = 2t \cos \theta \qquad (6.1)$$

A camera lens brings all these rays to a focus at a point in its focal plane. There will be constructive interference at this point if $\Delta = n\lambda$ where n is a positive integer. Allowing for the refractive index μ of the medium between the plates, the condition for a maximum of intensity is

$$2\mu t \cos \theta = n\lambda \qquad (6.2)$$

(μ is used in this chapter to avoid confusion ·ith order number n.) Strictly speaking, the angle θ in the medium is related to the angle of incidence on the interferometer, θ', by $\mu \sin \theta = \mu_{air} \sin \theta'$; but, as the medium between the plates is normally a gas at about atmospheric pressure, the angles differ by only about $1:10^3$ or less, and it is not necessary to distinguish between them.

The interference pattern formed by the Fabry–Perot consists of fringes of equal inclination. If an extended source is used, as shown in Fig. 6.2, all rays incident at angle θ, from whatever part of the source, are superposed by the camera lens L and, if Equation 6.2 is satisfied, form a bright point at P. The figure has cylindrical symmetry about the axis $0'0$, and if it is rotated about this axis P describes a circle in the focal plane of L. The interference pattern produced by monochromatic light is therefore a set of concentric rings. The radii of the intensity maxima are $f\,\theta$,

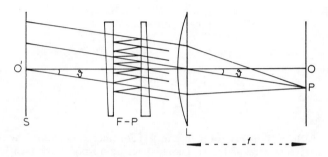

Fig. 6.2 Optical arrangement for Fabry–Perot interferometer. The figure shows the paths of those rays from all points of the source that are inclined at angle θ to the axis. Only a few of the multiple reflections are shown. The wedge angles of the Fabry–Perot plates (see text) are much exaggerated.

where θ satisfies Equation 6.2 with consecutive values of n, and the centre of the pattern is 0, on the normal to the Fabry–Perot through the centre of L. Since $\cos\theta$ decreases as θ increases, the order of interference n is highest for the innermost fringe. t may be anything from a fraction of a millimetre to several centimetres, so n is in the range 10^3 to 10^5.

In practice the source is usually in the focal plane of a second lens L', as shown in Fig. 6.3. Rays incident at the angle θ on the interferometer then all come from the same point P' of the source, and there is a point-to-point correspondence between P and P'. Although the lens L' plays no part in forming the fringes, it serves to image the source in the same plane as the fringes, thereby increasing the intensity of illumination. The fringes are of course visible only over the area of the source image in this case. For

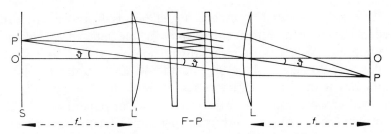

Fig. 6.3 Optical arrangement with source imaged in same plane as fringes. Lens L′ has been added to Fig. 6.2. The only rays incident on the interferometer at angle θ now *all* come from the point P′.

t = 1 cm, λ = 5000 Å and $f = f'$ = 50 cm, eight rings will be visible if the source is 1 cm in diameter.

The Fabry–Perot plates are made with small wedge angles. shown much exaggerated in Figs. 6.2 and 6.3, to throw the 'ghosts' from reflections off the back surfaces well clear of the true ring pattern.

6.2 Dispersion and free spectral range

The dispersion of the Fabry–Perot is found by differentiating Equation 6.2:

$$n \frac{d\lambda}{d\theta} = 2\mu t \sin \theta$$

μ can be set equal to 1, and, since in practice only very small values of θ are used, $n \approx 2t/\lambda$ and $\sin \theta \approx \theta$,

$$\therefore \quad \frac{d\theta}{d\lambda} = \frac{1}{\lambda\theta}$$

The angular dispersion is independent of t and is actually infinite at the centre of the ring system. To illustrate orders of magnitude, let us find the reciprocal linear dispersion 1 mm from the centre for a wavelength of 5000 Å. With f in cm, $\theta = 0\cdot1/f$.

$$\therefore \quad \frac{d\lambda}{dl} = \frac{1}{f} \cdot \frac{d\lambda}{d\theta} = \frac{\lambda\theta}{f} = \frac{0.1\,\lambda}{f^2}$$

For f = 50 cm, this is 0·02 Å/mm, an order of magnitude better than can be achieved with any grating except a large echelle, which is a far bulkier instrument.

The order number n in a Fabry–Perot is from one to three orders of magnitude higher than in an echelle, and the problem of overlapping orders is correspondingly more severe. For $t = 1$ cm and $\lambda = 5000$ Å, $n = 4 \times 10^4$, and the 40 000th order of 5000 Å coincides with the 40 001st order of 4999·88 Å - i.e., the free spectral range is only 0·12 Å. It is more convenient to work in wavenumbers, for which purpose Equation 6.2 can be re-written as

$$n = 2t\sigma \cos \theta \approx 2t\sigma \qquad (6.3)$$

Then $\Delta n = 2t\Delta\sigma$. Setting $\Delta n = 1$ gives for the free spectral range

$$\Delta\sigma = 1/2t \qquad (6.4)$$

which is identical with Equation 5.13. For $t = 1$ cm as before, the range between orders is 0·5 cm^{-1}. Except in very special cases it is essential to use some form of auxiliary dispersion with a Fabry–Perot interferometer, as described in Section 6.5.

6.3   Intensity distribution in the interference pattern

As one scans along a diameter of the ring pattern of a Fabry–Perot interferometer illuminated with monochromatic light, the intensity passes through a maximum whenever Equation 6.2 is satisfied and a minimum whenever $(n + \frac{1}{2})\lambda = 2t \cos \theta$. The intensity distribution in the interference fringes is not described by the Fraunhofer diffraction function discussed in Sections 4.2 and 5.2, and the intensity in the minima is not zero. One cannot therefore obtain the resolving power by applying Rayleigh's criterion directly, nor is it possible to obtain it from the maximum path difference across the beam (Equation 4.4) unless one knows how many of the multiply reflected rays emerging from the interferometer (Fig. 6.1) are effectively contributing to the pattern. It was pointed out in Section 4.2 that the modified form of Rayleigh's criterion, applicable to any 'line shape', states that two lines of equal intensity are just resolved if the intensity half way between them is 0·8 of the peak intensity of either (Fig. 4.4). This 20% dip is the criterion adopted to define two lines just resolved by the Fabry–Perot. Before we can get any further, therefore, we have to look into the line profile produced by the Fabry–Perot.

Suppose fractions s and r of the incident amplitude A_0 are transmitted and reflected at each partially reflecting surface. (s has been used to avoid confusion with the plate separation t). The

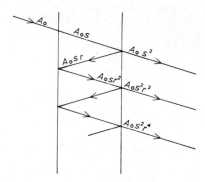

Fig. 6.4 Amplitudes of successively reflected rays in Fabry–Perot interfero-meter. s and r are the fractions of the incident amplitude transmitted and reflected at each partially reflecting surface.

amplitudes of the emergent rays in Fig. 6.4 are $A_0 s^2$, $A_0 s^2 r^2$, $A_0 s^2 r^4$, etc., and the phase difference between successive rays is δ, where from Equation 6.1

$$\delta = 2\pi \frac{\Delta}{\lambda} = \frac{4\pi t \cos\theta}{\lambda} = 4\pi\sigma t \cos\theta \qquad (6.5)$$

When all these rays are superposed in the focal plane of the lens, the resultant complex amplitude is

$$A = A_0 s^2 (1 + r^2 e^{i\delta} + r^4 e^{2i\delta} + r^6 e^{3i\delta} + \cdots)$$

which for a large number of rays becomes the sum of an infinite geometric series:

$$A = \frac{A_0 s^2}{1 - r^2 e^{i\delta}}$$

To get from complex amplitude to intensity, A must be multiplied by its complex conjugate (see, for example, Born and Wolf, chapter 7 [5]).

$$I = AA^\star = \frac{s^4}{(1 - r^2 e^{i\delta})(1 - r^2 e^{-i\delta})} A_0 A_0^\star$$

Introducing the transmissivity $T = s^2$, the reflectivity $R = r^2$ and the incident intensity $I_0 = A_0 A_0^\star$, we have

$$I = \frac{T^2}{1 - 2R \cos \delta + R^2} I_0 = \frac{T^2}{(1 - R)^2 + 2R(1 - \cos \delta)} I_0$$

$$= \frac{T^2}{(1 - R)^2 + 4R \sin^2 \delta/2} I_0$$

$$\therefore \quad I = I_0 \left(\frac{T}{1 - R}\right)^2 \frac{1}{1 + (4R/(1 - R)^2) \sin^2 \delta/2} \quad (6.6)$$

Equation 6.6 is known as the Airy distribution. The maxima are given by $\delta/2 = n\pi$, or $2\sigma t \cos \theta = n$, which is identical with Equation 6.2, and have intensity

$$I_{max} = I_0 \left(\frac{T}{1 - R}\right)^2 \quad (6.7)$$

In the ideal case of no absorption, $T + R = 1$ and the peaks are transmitted without loss. The minima occur at values of δ half way between the maxima, when $\delta/2 = (n + \frac{1}{2})\pi$, and have intensity

$$I_{min} = I_0 \left(\frac{T}{1 + R}\right)^2 \quad (6.8)$$

The contrast of the fringes, defined by $C = I_{max}/I_{min}$, is given by

$$C = \left(\frac{1 + R}{1 - R}\right)^2 \quad (6.9)$$

For $R \sim 90\%$, C is a few hundred.

The Airy distribution can conveniently be written in terms of I_{max} as

$$I = \frac{I_{max}}{1 + f \sin^2 \delta/2} = \frac{I_{max}}{1 + f \sin^2 \pi\sigma\Delta} \quad (6.10a)$$

where

$$f \equiv \frac{4R}{(1 - R)^2}$$

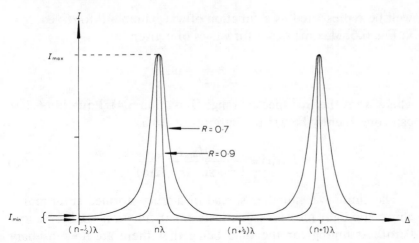

Fig. 6.5 Airy distribution as a function of path difference. The intensity distribution is shown for two different reflectivities, $R = 0.7$ and $R = 0.9$.

and $\delta/2 = \pi\sigma(2t \cos \theta) = \pi\sigma\Delta$ by Equations 6.5 and 6.1. Fig. 6.5 shows the distribution plotted as a function of Δ for two different reflectivities, $R = 0.7$ and $R = 0.9$. The sharpness of the fringes obviously depends critically on R. The distribution can equally

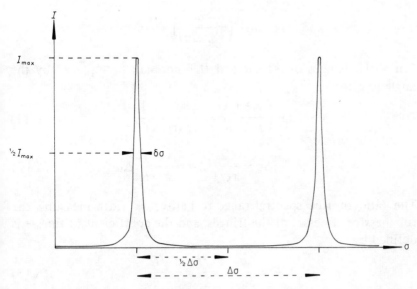

Fig. 6.6 Airy distribution as a function of wave number. $\Delta\sigma$ is the free spectral range. The intensity distribution is shown for $R = 0.9$, and the half-value width $\delta\sigma$ in this case is 1/28 order.

well be represented as a function of wavenumber σ for fixed Δ, as in Fig. 6.6. Maxima occur for values of σ given by

$$\sigma = \frac{n}{\Delta} \approx \frac{n}{2t} = n\,\Delta\sigma$$

where $\Delta\sigma$ is the free spectral range (Equation 6.4). Equation 6.10a can conveniently be written in terms of $\Delta\sigma$ as

$$I(\sigma) = \frac{I_{max}}{1 + f\sin^2\,\pi\sigma/(\Delta\sigma)} \tag{6.10b}$$

The finite wavenumber spread of a fringe formed from mono-chromatic light measures the 'instrument function' of the Fabry-Perot. Assuming for the time being that there are no subsidiary causes of broadening, an expression for the half-value width $\delta\sigma$ (the width at half peak intensity) can be obtained quite easily from Equation 6.10b. Moving out a distance $\frac{1}{2}\delta\sigma$ from the maximum will drop I by a factor 2 if

$$1 + f\sin^2\left(\pi\,\frac{\delta\sigma}{2(\Delta\sigma)}\right) = 2$$

i.e. $\sin^2\left(\pi\,\dfrac{\delta\sigma}{2(\Delta\sigma)}\right) = 1/f$

For sharp fringes $\delta\sigma \ll \Delta\sigma$ and the sine can be replaced by the angle to give

$$\sin\left(\frac{\pi\delta\sigma}{2(\Delta\sigma)}\right) \approx \frac{\pi\delta\sigma}{2(\Delta\sigma)} = \frac{1}{\sqrt{f}} \tag{6.11}$$

$$\therefore \quad \delta\sigma = \frac{2}{\pi\sqrt{f}}\,\Delta\sigma = \frac{1-R}{\pi\sqrt{R}}\,\Delta\sigma \tag{6.12}$$

The ratio of free spectral range to half-value width measures the fine-ness or 'finesse' of the fringes, and the coefficient of finesse is defined by

$$F = \frac{\Delta\sigma}{\delta\sigma} = \frac{\pi\sqrt{f}}{2} = \frac{\pi\sqrt{R}}{1-R} \tag{6.13}$$

Two lines cannot be resolved unless their wave numbers differ by at least $\delta\sigma$, so the Fabry-Perot should have a limit of resolution of about $\delta\sigma$ and a resolving power proportional to F. It will be seen in the next section that this is indeed the case.

Table 6.1 shows how the finesse increases rapidly with R. F is, however, usually limited to about 30 for two possible reasons. The first concerns loss of intensity from absorption in the films.

Table 6.1

R	0·70	0·75	0·80	0·85	0·90	0·95
F	8·3	10·3	13·3	18·3	28·3	58

Writing $T = 1 - R - A$, where A is the absorption, Equation 6.7 becomes

$$I_{max} = I_0 \left(1 - \frac{A}{1 - R}\right)^2$$

The peak intensity remains high only if A is small compared to $1 - R$. This constitutes a restriction on R if metallic reflecting films are used, but in the infra-red and visible regions high reflectivity with negligible absorption can be obtained from dielectric coatings. Two dielectric materials are required, one of high refractive index and one of low index. They are deposited on the interferometer plates in alternate layers, each of optical thickness $\frac{1}{4}\lambda_0$, beginning and ending with the high index material. In these conditions the reflected rays are all in phase for the wavelength λ_0, and the reflectivity is a maximum. As the number of layers is increased, the reflectivity increases for λ_0 but falls off more rapidly for wavelengths on either side. Zinc sulphide ($\mu \approx 2\cdot3$) and cryolite ($\mu \approx 1.3$) are the materials used in the visible region. With 5 or 7 layers reflectivities as high as 98% can be obtained without significant absorption, a great improvement on the best silver films formerly used [4]. In the ultra-violet region materials of high refractive index tend to absorb, but dielectric coatings with reflectivities of ~90% have been successfully used down to about 2500 Å. At shorter wavelengths aluminium films have to be used; reflectivities are in the range 80–85% and the absorption is relatively high. The Fabry–Perot is not a practical proposition for normal spectroscopic purposes much below 2000 Å.

Even when reflecting coatings present no problem, the effective finesse is limited by the imperfections of the interferometer plates. The effect of any departure from complete flatness is that all the rays emerging from the interferometer at a given angle no longer have exactly the same phase difference. A bump of $\lambda/50$ changes the local path difference by $\lambda/25$ and hence changes the phase by $1/25$ of an order. If the fringe maximum can wander about through $1/25$ of an order, the finesse cannot be greater than 25. Although interferometer plates have been worked to $\lambda/100$, or even better over a small area, $\lambda/50$ is a more usual figure. Little is gained by using a reflection finesse appreciably greater than the plate finesse, and a detailed treatment shows that the two should be approximately matched [4]. The effective finesse is then 0·6 of the finesse of either, usually between 25 and 30 in the visible and infra-red.

6.4   Resolving power

To obtain an expression for the resolving power of the Fabry-Perot, we neglect plate imperfections for the moment and return to Equation 6.12 for the half-value width:

$$\delta\sigma = \frac{1-R}{\pi\sqrt{R}}\,\Delta\sigma = \frac{1}{F}\,\Delta\sigma$$

At first sight it would appear that the superposed profiles of two equally intense lines separated by $\delta\sigma$ would show no dip at all at the midpoint where each has intensity $\frac{1}{2}I_{max}$. As shown in Fig. 6.7, however, each of the line peaks is in fact raised by the wing of the other to a value I_p where $I_p = I_{max} + I_w$. I_w is the intensity at a wavenumber interval $\delta\sigma$ from the maximum and can be found from the Airy distribution (Equation 6.10b):

$$I_w = \frac{I_{max}}{1 + f\sin^2 \pi(\delta\sigma/\Delta\sigma)}$$

The sine can again be replaced by the angle, and by analogy with Equation 6.11 we have

$$\sin^2\left(\pi\,\frac{\delta\sigma}{\Delta\sigma}\right) \approx \pi^2\,\frac{(\delta\sigma)^2}{(\Delta\sigma)^2} = \frac{4}{f}$$

$$\therefore \quad I_w = \frac{I_{max}}{1+4} \quad \text{and} \quad I_p = \frac{6}{5}I_{max}$$

The intensity in the dip is therefore 5/6, or 83%, of the peak intensity, which is quite close enough to the value of 81% required by the strict application of Rayleigh's criterion. Consequently the expression

$$\delta\sigma = \Delta\sigma/F \qquad (6.14)$$

holds for the limit of resolution as well as the half-value width. It is independent of the wavenumber and depends only on the reflectivity and the spacing. For $R = 90\%$ (corresponding to

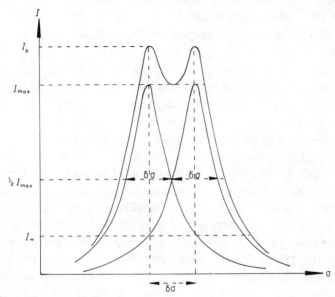

Fig. 6.7. Rayleigh's criterion applied to Fabry–Perot fringes. The peak intensity I_p is the sum of the maximum from one line, I_{max}, and the intensity I_w in the wing of the other line at $\delta\sigma$ from its centre.

$F = 28$) and $t = 1$ cm, the resolution limit is $0\cdot018$ cm^{-1}, which is the order of magnitude of a small hyperfine structure or isotope splitting.

The resolving power is easily found from Equation 6.14:

$$\mathscr{R} = \frac{\sigma}{\delta\sigma} = \frac{\sigma}{\Delta\sigma}F = nF \qquad (6.15)$$

using $n = 2t\sigma = \sigma/\Delta\sigma$ from Equations 6.3 and 6.4. Comparing this with the expression for the resolving power of a grating in the form nN (Equation 5.8), one can interpret the finesse as the effective number of interfering beams. This is consistent with the concept of resolution limit as the reciprocal of the extreme path difference imposed by the instrument, because for F beams the latter becomes $F \times 2t$, or $F/\Delta\sigma$, in agreement with Equation 6.14. The quantity $2tF$ compares with $2W$ for the grating. Obviously tF will exceed the width of even the largest gratings when t is greater than a centimetre or so. Looked at from the point of view of resolving power, n can be made as large as wanted simply by increasing the plate separation. For $t = 10$ cm and $\lambda = 5000$ Å, $n = 4 \times 10^5$, and a finesse of 25 is sufficient to give a resolving power of 10^7, an order of magnitude higher than can be got with a grating.

In practice a resolving power of 10^6 is as high as is useful for most spectroscopic purposes because of the intrinsic width of the spectral lines. Higher resolution is occasionally required to study the line profiles of very narrow lines and for use with lasers, when the line width can be many orders of magnitude smaller than in an ordinary light source (Section 3.9). The expressions (6.14) and (6.15) refer of course to the *theoretical* resolution. The actual resolution depends on the flatness of the plates, but it has already been seen that the effect of plate defects is to put an upper limit of about 30 on F. A further limitation is that with large spacings the resolving power drops off at the edges of the ring pattern because the lateral displacement of successive reflections eventually pushes them right off the plates. The effective number of interfering beams is thus reduced for large θ, an effect known as 'walk-off'.

6.5   Methods of using the Fabry–Perot interferometer

Becasue of its very small free spectral range the interferometer is normally used with an auxiliary spectrograph to provide cross dispersion. This must be stigmatic but not necessarily of large dispersion, and a prism spectrograph is usually suitable. Various arrangements are possible, depending on the particular problem investigated and on whether the interferometer is used photographically or photo-electrically.

For photographic use the interferometer can be put either outside or inside the spectrograph. If outside, the fringes are focused on the spectrograph slit by a high quality lens. If inside, the interferometer is usually put in the parallel beam between the prism and the collimator lens. A discussion of the relative advantages and disadvantages of these positions from the point of view of scattered light, secondary images, etc, is given in Tolansky's book [1]. The actual pattern produced on the photographic plate is a rectangular section, corresponding to the slit image, through the centre of the ring system. The slit may be up

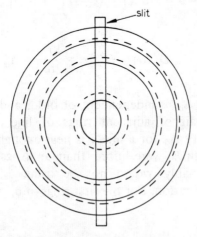

Fig. 6.8 Fabry–Perot ring pattern crossed with spectrograph slit. The full circles are the maxima for wavenumber σ_α and the dashed circles are maxima for σ_β.

to 1 mm wide unless the spectrum has very close lines. The fringe system then consists of short arcs 1 mm long, tending to horizontal lines as one gets farther from the centre of the ring system. (Fig. 6.8). The fringe pattern can be moved up and down the slit by tilting the interferometer.

6.6 Reduction of measurements

To determine wavelength differences from measured fringe positions, two procedures are possible. If the centre of the fringe system is focused on the centre of the slit, complete ring diameters can be measured for each component. The distance between successive rings of one component gives a scaling factor which allows the wavenumber difference between two components to be found from their ring differences. Since θ is always fairly small, Equation 6.2 can be written as $n = 2t\sigma(1 - \theta^2/2)$. Let α be the angular diameter of a ring for wavenumber σ_α, so that $\alpha = 2\theta = D_\alpha/f$ where f is the focal length of the fringe-forming lens. Then the first and second rings of σ_α are given by

$$\left. \begin{array}{l} n_1 = 2t\sigma_\alpha(1 - \alpha_1^2/8) \\[2em] n_2 = n_1 - 1 = 2t\sigma_\alpha(1 - \alpha_2^2/8) \end{array} \right\} \begin{array}{l} \text{Subtracting,} \\[2em] 1 = 2t\sigma_\alpha \dfrac{\alpha_2^2 - \alpha_1^2}{8} \end{array}$$

$$\therefore \quad D_2^2(\alpha) - D_1^2(\alpha) = \frac{8f^2}{2t\sigma_\alpha} = \Delta_\alpha \text{ say.}$$

The scale factor Δ_α is independent of order and is therefore the same for any pair of adjacent rings of line α. Moreover the analogous quantity Δ_β for a line β is near enough the same as Δ_α because if the lines were more than an Ångström apart (say $1:10^4$) one probably would not be using a Fabry–Perot to measure their separation. For the first rings of σ_α and σ_β:

$$\left. \begin{array}{l} n_1 = 2t\sigma_\alpha \left(1 - \dfrac{\alpha_1^2}{8}\right) \\[2em] n_1 = 2t\sigma_\beta \left(1 - \dfrac{\beta_1^2}{8}\right) \end{array} \right\} \begin{array}{l} \text{Subtracting,} \\[1em] 2t(\sigma_\alpha - \sigma_\beta) = 2t\sigma_\alpha \dfrac{\alpha_1^2}{8} - 2t\sigma_\beta \dfrac{\beta_1^2}{8} \\[2em] \simeq 2t\sigma_\alpha \dfrac{\alpha_1^2 - \beta_1^2}{8} \end{array}$$

$$\therefore \quad D_1^2(\alpha) - D_1^2(\beta) = \frac{8f^2}{2t\sigma_\alpha} 2t(\sigma_\alpha - \sigma_\beta) = \delta_{\alpha\beta} \text{ say}$$

where we have again assumed σ_α and σ_β to be very close. The quantity $\delta_{\alpha\beta}$ is also independent of order and can be measured in every order. The wavenumber difference is found from the mean values of $\delta_{\alpha\beta}$ and Δ_α by

$$\sigma_\alpha - \sigma_\beta = \frac{1}{2t} \frac{\delta_{\alpha\beta}}{\Delta_\alpha}$$

t has to be measured to only about 1% in view of the fairly large percentage errors in $\delta_{\alpha\beta}$ and Δ_{α}, which are differences of squares.

The alternative is to tilt the interferometer and measure the off-centre fringes, the centre of the ring pattern being off the slit. The definition of the off-centre fringes is sometimes better and there are more of them on the length of the slit. The reduction of the measurements in this case is not very different in principle and can be found in [1].

Up to now the Fabry–Perot interferometer has been treated as a high resolution instrument for measuring very close splittings of spectral lines such as occur in hyperfine structure and isotope shifts. The interferometer has in fact two other very important applications: the determination of absolute length and the establishment of extremely accurate secondary standards of wavelength. It was first used to determine the absolute wavelength of the red line of cadmium in terms of the International Metre in Paris in 1905 by Fabry, Perot and Benoit. Since the present international standard of length is the wavelength of a krypton line, the comparison must now be considered the other way about, the function of the Fabry–Perot being to determine standard metres in terms of the krypton wavelength. The second function of the interferometer is to measure a fairly large number of other spectral lines relative to the krypton standard. These should be well spread over the spectrum and sufficiently numerous for accurate interpolation between them. A description of the methods of measuring absolute lengths and secondary standards is outside the scope of this book. Accounts may be found in, for example, [5] and [6].

6.7 Photo-electric use

The Fabry–Perot interferometer is now frequently used with photo-electric detection as a spectrometer. A circular aperture isolates the centre of the ring system, and the spectrum is scanned by changing the effective spacing of the interferometer – in other words, θ is fixed ($\cos \theta = 1$) and *I* is recorded as a function of *t*, in contrast to the method of photographic detection when *t* is fixed and *I* is recorded as a function of θ. In fact, it is so difficult to change the physical spacing with sufficient accuracy while keeping the plates parallel to $\lambda/50$ or so that the scanning is usually done by enclosing the interferometer in an airtight box and varying the pressure. Going back to Equation 6.2, at the centre of the pattern $n\lambda = 2\mu t$; with a 1 cm spacer a scan of one complete order in the visible region requires a change of refractive index of $\Delta\mu = \lambda/2t \sim 2 \times 10^5$, corresponding to a change in pressure of the order of 0·1 atmospheres. If the pressure can be changed linearly with time, the output from the photomultiplier behind the circular aperture can also be recorded as a function of time; otherwise the output must be recorded directly as a function of pressure.

The general advantages of photo-electric over photographic detection for intensity measurements were discussed in Section 3.12. In addition, there are two important advantages associated with using the centre of the ring system. First, the dispersion is constant. It was shown in Section 6.2 that the angular dispersion of the Fabry–Perot is anything but constant – indeed it goes infinite at the centre – but if the spectrum is recorded as a function of pressure rather than angle the angular dispersion is irrelevant and the quantity of interest is $d\lambda/dP$. Writing $\mu = 1 + aP$, where P is the pressure and a is constant over the range used, we have

$$\frac{d\lambda}{dP} = \frac{d\lambda}{d\mu} \cdot \frac{d\mu}{dP} = \frac{2t}{n} a,$$

which is effectively constant for a scan over a few orders. Secondly, there is no walk-off. The light entering the central aperture is incident almost normally on the Fabry–Perot, and there is virtually no lateral displacement between the multiple reflections.

The use of a circular aperture does, however, introduce an important new consideration. If resolving power is not to be lost, the aperture must isolate only a small fraction of an order, comparable with $\delta\sigma/\Delta\sigma$, or $1/F$. On the other hand, if the light source is weak it is desirable to have as large an aperture as possible, leading to the usual dichotomy between resolution and intensity. Further discussion of this question will be postponed to Section 6.15, but it may be remarked here that because of the axial symmetry the limiting aperture for a Fabry–Perot can be much larger than the limiting slit area of a grating instrument.

6.8 Double etalon

It has been seen that both the resolution limit $\delta\sigma$ and the free spectral range $\Delta\sigma$ are proportional to $1/2t$, so that at high resolution the range between orders may well be considerably less than a wavenumber. It is often difficult to isolate this waveband with the cross-dispersion instrument without using a long focal length, which decreases the intensity, or a very narrow slit, which makes the fringes difficult to measure. An alternative way of tackling the problem is to use a second Fabry–Perot, having say 1/5 the spacing, in series with the first. Because the angular dispersion is independent of t, the ring patterns of the two etalons coincide except that the longer etalon has 5 rings for every one of the shorter. The latter acts as a filter, transmitting only every fifth order of the longer etalon.

The fine adjustment necessary to obtain the required ratio of the two spacings is achieved by adjusting the pressure in one of the etalons. The ratio need not in fact be an exact integer because if it is out by an integral number of half wavelengths the maxima still occur at the same angle, but only for a rather narrow band of wavelengths. The advantage of using the second etalon in this filtering capacity is that the intensity in the transmitted peaks is scarcely reduced: it was shown earlier that in the ideal case of zero absorption the peak intensity is equal to the incident intensity. On the other hand, the intensity away from the maxima of the filter etalon does not drop to zero, so the 'suppressed orders' are not entirely suppressed and may have to be allowed for in interpreting the spectrum [1, 4].

THE MICHELSON INTERFEROMETER

6.9 Basic ideas

The instrument used in Fourier transform spectroscopy is a Michelson interferometer in which one mirror can be moved in the direction of the incident light. The fixed-mirror Michelson is not an instrument of any importance in spectroscopy, although both it and its variants are widely used in other contexts. However, to understand what happens when one moves one of the mirrors one must start with the interference pattern obtained from fixed mirrors.

Fig. 6.9 shows a conventional Michelson interferometer set up to give fringes of equal inclination localized at infinity. Light from an extended source S is divided at the semi-silvered surface of the beam-splitter B into two perpendicular beams. These are reflected back normally from the mirrors M_1 and M_2 and recombined at B, and the interference pattern is imaged in the focal plane of the lens L. A compensating plate C, identical with B but for the

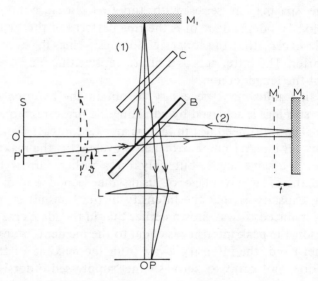

Fig. 6.9 Path of light ray through Michelson interferometer. See text for explanation. Refraction within plates B and C is identical for the two beams and is ignored.

silvering, is introduced into the first beam to equate the extra passages through the beam splitter of the second beam. As in the Fabry–Perot arrangement of Fig. 6.3, a lens L' is usually inserted between S and B so as to form an image of S in the focal plane of L. Again here, L' plays no part in the formation of the interference pattern, but it increases the intensity of illumination of the fringes.

Fig. 6.9 shows the path of a pencil of rays inclined at an angle θ to the axis. If the image M_1' of M_1 in B is exactly parallel to M_2, but displaced from it a distance t, the corresponding rays (1) and

(2) may be represented as in Fig. 6.10. The path difference between them is identical with that between the successive reflections in Fig. 6.1 - i.e., $\Delta = 2\,t\cos\theta$ - but in this case the rays have equal amplitude since they have undergone the same number of transmissions and reflections.

All rays reaching L at the same angle θ are superposed at P in the focal plane of L, and interference maxima occur for

$$\Delta = n\lambda = 2t\cos\theta \qquad \text{or} \qquad n = 2t\sigma\cos\theta \qquad (6.16)$$

The extended source gives a range of angles θ, and there will be a bright point along the line OP whenever θ satisfies Equation 6.16. Because of the axial symmetry the complete interference pattern

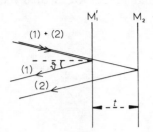

Fig. 6.10 Effective paths for Michelson adjusted for fringes of equal inclination. Rays (1) and (2) correspond to those in Fig. 6.9.

produced by monochromatic light is a set of concentric rings centred on O. The difference between this ring pattern and that of the Fabry-Perot is in the intensity distribution over the fringes, for we are here concerned with two interfering beams instead of, effectively, F beams.

6.10 Intensity distribution in the fringes

The complex amplitude obtained from superposing two beams of equal amplitude A_0 and phase difference δ is

$$A = A_0(1 + e^{i\delta})$$

where in this case

$$\delta = 2\pi\Delta/\lambda = 2\pi\sigma\Delta \qquad (6.17)$$

Multiplying by the complex conjugate to get the intensity gives

$$I = AA^\star = A_0^2(2 + 2\cos\delta) = 2I_0'(1 + \cos\delta)$$

The factor 2 here is quite meaningless and appears only because we have superposed two beams each of arbitrary intensity I_0'. Even if there are no absorption losses in the interferometer, half the light energy incident on B is ultimately reflected back along the path BS. The intensity distribution is in fact usually given in terms of the average of I over δ, I_0, as

$$I = I_0(1 + \cos \delta) = I_0(1 + \cos 2\pi\sigma\Delta) \qquad (6.18a)$$

which may also be written as

$$I = 2I_0 \cos^2 \delta/2 = 2I_0 \cos^2 \pi\sigma\Delta \qquad (6.18b)$$

The condition for a maximum is obviously $\Delta = n\lambda$, or $\sigma = n/\Delta$, and the minima occur at values of Δ or σ half-way between the maxima.

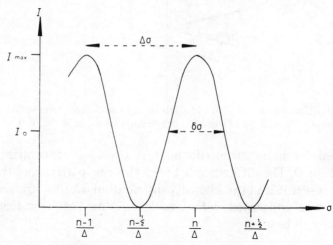

Fig. 6.11 Intensity distribution as a function of wave number for Michelson interferometer. This is the two-beam interference distribution $I = I_{max} \cos^2 \pi\sigma\Delta$.

The $\cos^2$ fringes are shown in Fig. 6.11, where I is plotted as a function of σ for fixed Δ; an exactly similar plot is obtained if σ is fixed and I is regarded as a function of Δ (Fig. 6.12). Since $I = I_0 = \frac{1}{2}I_{max}$ when $\cos 2\pi\sigma\Delta = 0$, or $\sigma\Delta = n \pm \frac{1}{4}$, the half-value width of these fringes is half an order

$$\text{i.e.,} \qquad \delta\sigma = \frac{1}{2} \Delta\sigma \qquad (6.19)$$

By comparison with Equation 6.14 the Michelson can be said to have a finesse of 2, as is to be expected from a two-beam instrument. Since the finesse is an order of magnitude smaller than that of the Fabry–Perot, so, too, is the resolving power for a given plate separation t. The Michelson is not, therefore to be regarded as an extremely high resolution instrument in the sense of the Fabry–Perot. Its importance comes from its peculiar advantages as a scanning instrument [4, 8, 9].

6.11   The Michelson interferometer as a scanning spectrometer

For photo-electric scanning the Michelson interferometer, like the Fabry–Perot, is used with a circular aperture in the focal plane of the lens L to isolate the centre of the ring system ($\cos \theta \approx 1$). A photo-electric detector behind this aperture records I as a function of path difference Δ, where $\Delta = 2t$. The Michelson is scanned by moving one of the mirrors in a direction normal to its surface so as to change t in Figs. 6.9 and 6.10.

Suppose first that the light is monochromatic, of wavenumber σ_1. The trace of I against Δ, which is known as an interferogram, is in this case a $\cos^2$ curve with a maximum whenever $\Delta = 1/\sigma_1$, $2/\sigma_1 \cdots$, as in Fig. 6.12a. If the source also radiates light of a slightly different wavenumber σ_2, the resulting interferogram has a second $\cos^2$ curve superposed on the first, with maxima at $1/\sigma_2$, $2/\sigma_2 \cdots$. As Δ is increased, these two sets of maxima get more and more displaced from one another and the contrast of the fringes decreases, as shown in Fig. 6.12b. When a maximum of σ_1 coincides with a minimum of σ_2, which occurs when

$$\left.\begin{array}{l} \sigma_1 \Delta = n \\ \\ \sigma_2 \Delta = n - \tfrac{1}{2} \end{array}\right\} \quad \text{i.e.} \quad (\sigma_1 - \sigma_2)\Delta = \tfrac{1}{2}$$

the fringes have minimum contrast and disappear entirely if the two components have the same intensity. As Δ increases further, the maxima gradually get in phase again, and the fringes re-appear.

If there is a continuous band of radiation in the range $\sigma_1 - \sigma_2$, the fringes tend to disappear a little sooner, when half the band cancels out the other half, or, in other words, the band occupies about an order

$$\text{i.e.,} \qquad \sigma_1 - \sigma_2 \sim 1/\Delta \qquad\qquad (6.20)$$

In this case the fringes do not reappear at larger Δ. Consequently, the interferogram for a group of spectral lines having wavenumbers in the range $\sigma_1 - \sigma_2$ has the general form of Fig. 6.13.

Fig. 6.13 does not look in the least like a spectrum, but it contains all the information necessary to reconstruct the spectral

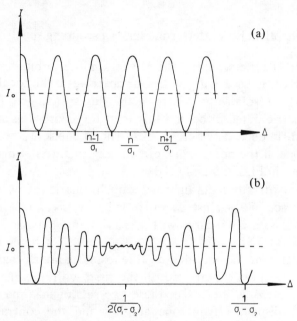

Fig. 6.12 Interferogram from source radiating (a) monochromatic light and (b) light at two sharply defined wavelengths.

distribution of the incident light. This is obvious for a very simple spectrum: given the interferogram of Fig. 6.12a, one could easily infer that the light is monochromatic and of wavenumber σ_1, and from Fig. 6.12b one could deduce from the variation of fringe contrast with Δ the existence of the two components σ_1 and σ_2. Michelson used this variation of contrast, or 'visibility', of the fringes to determine hyperfine structures in the 1890s (see, e.g., [5], [8]) but his methods are applicable only to fairly simple cases. In general, to obtain the spectral distribution from an interferogram such as Fig. 6.13 it is necessary to use the Fourier transform method. One might ask first why the output from the basically similar Fabry–Perot scanning interferometer can be interpreted without Fourier transformations. The reason is that the Fabry–Perot is always used with a very narrow range of wavelengths, isolated by a filter or a monochromator, and is scanned over a few orders only. Provided the wavenumber band is smaller than the free spectral range or distance between orders – that is, $\sigma_1 - \sigma_2 \leqslant 1/\Delta$, cf Equation 6.20 – the intensity at any point on a trace such as that of Fig. 6.5 corresponds to one and only one wavenumber component. Of course, at spacings large

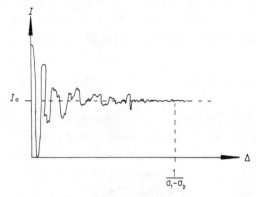

Fig. 6.13 Interferogram from source radiating a finite band of wavelengths. For a band extending from σ_1 to σ_2 the fringes disappear when $\Delta \sim 1(\sigma_1 - \sigma_2)$.

enough that the free spectral range is reduced almost to the intrinsic line width the Fabry–Perot fringes also lose contrast and tend to disappear. Except for a laser source, this limits the useful spacing to about 10 cm and the resolving power to a few million as previously remarked.

6.12 The Fourier transform method

Suppose first that the source radiates a discrete number of sharp lines. The wavenumber of any one of these can be written as $m\sigma_0$, a multiple of an elementary unit of a wavenumber σ_0, and the corresponding intensity as I_m. Its contribution to the total signal at a given value of Δ is found from Equation 6.18a to be

$I_m (1 + \cos 2\pi m\sigma_0 \Delta)$. The interferogram obtained from the source is therefore

$$I(\Delta) = \sum_m I_m (1 + \cos 2\pi m\sigma_0 \Delta) \rightarrow \sum_m I_m \cos 2\pi m\sigma_0 \Delta \quad (6.21)$$

where the constant part, ΣI_m, has been disregarded because it corresponds to the mean level I_0 of Fig. 6.13 and can simply be subtracted off. To obtain the intensity distribution in the source from the experimental interferogram, we have to extract from $I(\Delta)$ the coefficients I_m. This is done by elementary Fourier analysis. A given coefficient I_p is found by multiplying both sides of Equation 6.21 by $\cos 2\pi p\sigma_0 \Delta$ and integrating with respect to Δ over a cycle. All terms in the sum on the right hand side vanish except the pth, which integrates to $I_p/2\sigma_0$, giving

$$I_p = 2\sigma_0 \int_0^{1/\sigma_0} I(\Delta) \cos 2\pi p\sigma_0 \Delta \, d\Delta \quad (6.22)$$

To apply this idea to a continuous spectral distribution $B(\sigma)$ instead of a lot of discrete components, we have to use Fourier integrals instead of Fourier series, but the basic procedure is the same. The interferogram from a source in which $B(\sigma)$ is the relative intensity per unit wavenumber is

$$I(\Delta) = \int_0^\infty B(\sigma) \cos 2\pi\sigma\Delta \, d\sigma \quad (6.23)$$

where the constant part has again been disregarded. The spectral distribution is extracted from $I(\Delta)$ by means of the Fourier transform analogous to Equation 6.22:

$$B(\sigma) \propto \int_0^\infty I(\Delta) \cos 2\pi\sigma\Delta \, d\Delta \quad (6.24)$$

Set out step by step, $B(\sigma)$ is found as follows: $I(\Delta)$ is measured at a number of different values of Δ (sampling points), and $\cos 2\pi\sigma\Delta$ is evaluated at each point for a fixed value of σ. The product $I(\Delta)$ $\cos 2\pi\sigma\Delta$ is then plotted as a function of Δ, and the area under the curve is measured. This gives the relative intensity of the particular component σ. The whole process is repeated for another value of σ, and so on until the spectral distribution is gradually built up. In practice all this tedious calculation is done by a computer. It should be quite clear why Fourier transform spectroscopy was not a practical proposition until computers became readily available.

It is possible to work with frequency rather than wavenumber by treating the interferogram as a function of the delay time τ between the two beams, given by $\tau = \Delta/c$. Then the interferogram is

$$I(\tau) = \int_0^\infty B(\nu) \cos 2\pi\nu\tau \, d\nu$$

and the spectral distribution $B(\nu)$ is obtained from the Fourier transform

$$B(\nu) \propto \int_0^\infty I(\tau) \cos 2\pi\nu\tau \, d\tau$$

Since $\nu = c\sigma$, it is easily seen that the ν, τ pair is equivalent to the σ, Δ pair.

Using yet another representation, one can show that the Michelson interferometer effectively reduces the sinusoidal fluctuations of the original light waves, which are of course far too rapid to be followed by any detector, to a manageable recording speed: if the path difference is changed at constant speed v, successive maxima of wavelength λ occur at time intervals λ/v. The reduced frequency f registered by the detector is therefore $f = v\sigma$. The spectral distribution as a function of f can be obtained from the interferogram as a function of time, $I(T)$, by the Fourier transform

$$B(f) \propto \int_0^\infty I(T) \cos 2\pi f T \, dT$$

6.13 Resolving power and intensity distribution

The resolving power of the Michelson interferometer can be found directly from the maximum path difference imposed by the instrument. For a mirror displacement $t = X/2$, the path difference Δ between the two beams is X, and from Equation 4.4

$$\mathcal{R} = \Delta/\lambda = X/\lambda = \sigma X \qquad (6.25)$$

Since $\mathcal{R} = \sigma/\delta\sigma$, the resolution limit is

$$\delta\sigma = 1/X \qquad (6.26)$$

which could, of course, have been obtained directly from Equation 4.5.

It is interesting to look at the resolving power from the point of view of the Fourier transform process. At the beginning of the last section we considered the interferogram from a number of discrete lines whose wavenumbers could be expressed as multiplies of a unit σ_0. σ_0 must set the limit of precision with which the wavenumber of any line can be specified, and two lines cannot be resolved unless they are separated by at least σ_0. But to obtain the spectral distribution from the interferogram by Equation 6.22 it is necessary to integrate over a complete cycle, $\Delta = 0$ to $\Delta = 1/\sigma_0$. A resolution limit of σ_0 therefore requires a scan over at least $1/\sigma_0$, in agreement with Equation 6.26. An analogous argument applies to the continuous distribution: if the transform (Equation 6.24) is to reproduce $B(\sigma)$ absolutely accurately, the integration must be carried out to $\Delta = \infty$. Cutting it off at $\Delta = X$ has the effect of smearing out each spectral element by an amount $\delta\sigma = 1/X$.

For many purposes (line profile work, for instance) it is not enough to know just the resolution limit of a spectrometer; one requires the complete instrument function – that is, the intensity distribution in the output when the input is an ideal sharp isolated spectral line. For the spectrometers previously discussed the output from the detector gave directly the instrument function: the single slit diffraction pattern (Equations 4.2, 5.5) for the prism and grating and the Airy distribution (Equation 6.10) for the Fabry–Perot. For the Michelson spectrometer the instrument function is not the interferogram itself but the result of the Fourier transform process. This can be found quite easily as follows.

The interferogram from a monochromatic line of wavenumber σ_1 is the cosine distribution of Fig. 6.12a and Equation 6.18

$$I(\Delta) \propto \cos 2\pi\sigma_1\Delta$$

The intensity distribution $B'(\sigma)$ obtained from this interferogram by the Fourier transform, Equation 6.24, for a maximum path difference of X is

$$B'(\sigma) \propto \int_0^X I(\Delta) \cos 2\pi\sigma\Delta \; d\Delta \propto \int_0^X \cos 2\pi\sigma_1\Delta \cos 2\pi\sigma\Delta \; d\Delta$$

$$= \frac{1}{2} \int_0^X \{\cos 2\pi(\sigma + \sigma_1)\Delta + \cos 2\pi(\sigma - \sigma_1)\Delta\} \; d\Delta$$

$$= \frac{1}{2} X \left\{ \frac{\sin 2\pi(\sigma + \sigma_1)X}{2\pi(\sigma + \sigma_1)X} + \frac{\sin 2\pi(\sigma - \sigma_1)X}{2\pi(\sigma - \sigma_1)X} \right\} \qquad (6.27)$$

The first term is negligible compared to the second if $\sigma_1 X \gg 1$ (that is, $X \gg \lambda$) which is always the case. The instrument function is therefore

$$B'(\sigma) \propto \frac{\sin 2\pi(\sigma - \sigma_1)X}{2\pi(\sigma - \sigma_1)X} \qquad (6.28)$$

This has its central maximum at $\sigma = \sigma_1$ and the first zero on either side at $\sigma - \sigma_1 = \pm 1/2X$, as shown in Fig. 6.14a.

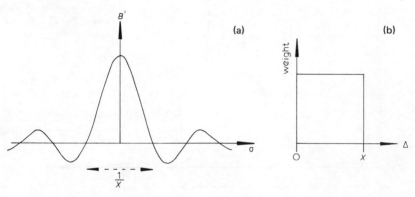

Fig. 6.14 Fourier transform instrument function. (a) shows the intensity distribution B' for monochromatic light obtained with the rectangular cut-off (b).

If a second line centred on the first zero of the first line is assumed to be just resolved, the limit of resolution is $\delta\sigma = 1/2X$, corresponding to a resolving power of $2X/\lambda$. This is twice what we were expecting. It is not, however, a very meaningful value because of the awkward line shape. The secondary maxima have about 15% of the intensity of the central maximum and may well get mixed up with weak neighbouring lines. The intensity appears to go negative only because the constant part has been left out, but the dips are nevertheless inconvenient.

The line shape can be improved at the expense of the theoretical resolving power by a process known as apodization – literally, removing the feet. Equation 6.28 has the form $\sin 2\alpha/2\alpha$ of the *amplitude* of a single slit diffraction pattern as a result of the rectangular weighting function, Fig. 6.14b, that represents the abrupt cut-off at $\Delta = X$. It can be converted to the familiar single slit *intensity* distribution $\sin^2\alpha/\alpha^2$ by using the triangular weighting function $1 - \Delta/X$ of Fig. 6.15b. The instrument function is then (Fig. 6.15a)

$$B'(\sigma) \propto \frac{\sin^2 \pi(\sigma - \sigma_1)X}{(\pi(\sigma - \sigma_1)X)^2} \qquad (6.29)$$

The first zero is at $\sigma - \sigma_1 = \pm 1/X$, and the resolving limit is $\delta\sigma = 1/X$. For most purposes the loss of the notional factor two in resolving power is more than offset by the simpler line shape and

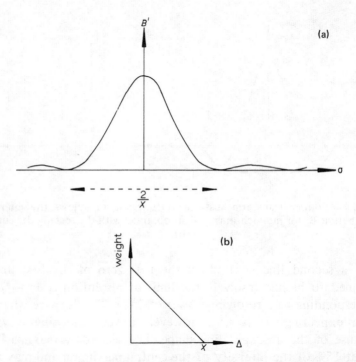

Fig. 6.15 Fourier transform instrument function for triangular weighting function. (a) shows the intensity distribution obtained with the triangular cut-off (b).

reduction of the secondary maxima. The weighting is done mathematically when performing the Fourier transform, not physically by tapering off the output from the detector. One of the advantages of Fourier transform spectroscopy is that different weighting functions can easily be introduced to suit different problems.

6.14   Sampling intervals

In practice $I(\Delta)$ is recorded and evaluated at discrete values of Δ rather than as a continuous function, and we need to know how small these sampling intervals should be. If the resolution limit is $\delta\sigma$ and the spectrum extends from σ_A to σ_B, the maximum number of meaningful intervals into which the spectrum can be divided is $N = (\sigma_A - \sigma_B)/\delta\sigma$. Since the interferogram contains the same information as the spectrum, this suggests that N sampling points should be sufficient. The sampling interval x is then

$$x = \frac{X}{N} = \frac{1}{N\delta\sigma} = \frac{1}{\sigma_A - \sigma_B}$$

The effect of calculating the spectrum from the sum

$$B''(\sigma) = \sum_{p=0}^{N} I(px) \cos 2\pi\sigma px$$

rather than from the integral with the same upper limit of Δ

$$B'(\sigma) = \int_{0}^{X} I(\Delta) \cos 2\pi\sigma\Delta \; d\Delta$$

is that the spectrum is repeated at spectral intervals $1/x$, just as the function $F(z)$ represented by the Fourier series

$$F(z) = \sum_{p} a_p \cos 2\pi pz/L$$

is periodic in z/L and is repeated at intervals L. For example, a single line σ_1 emerges from the transform as the distribution of Fig. 6.14a (or Fig. 6.15a) centred on wavenumbers σ_1, $\sigma_1 + 1/x$, $\sigma_1 + 2/x$, etc. If there is to be no overlapping of these different 'orders', the spectral range $\sigma_A - \sigma_B$ must not exceed $1/x$. Conversely, for a given spectral range, $1/(\sigma_A - \sigma_B)$ is an upper limit to the size of the sampling steps.

The free spectral range $1/x$ of this instrument may be compared with the free spectral ranges of the Fabry–Perot and the grating, which are determined by the path difference between adjacent interfering beams,

$$\Delta\sigma = \frac{1}{2t} = \frac{1}{n\lambda} \qquad \text{(Equation 6.4)}$$

and

$$\Delta\sigma = \frac{1}{d(\sin i + \sin \theta)} = \frac{1}{n\lambda} \qquad \text{(Equation 5.13) respectively.}$$

Moreover, just as the resolving limit of the grating is determined by the *total* path difference across it (Equations 4.5 and 5.10) and hence by the number of interfering beams N according to $\delta\sigma = 1/Nn\lambda$, so the resolving limit of the Fourier transform spectrometer is determined by the number of steps according to $\delta\sigma = 1/X = 1/Nx$. Evidently, sampling the interferogram at N points is equivalent to sampling the wavefront at N points with a grating. By contrast, a prism samples the wavefront continuously and does not give rise to a periodically repeating spectrum.

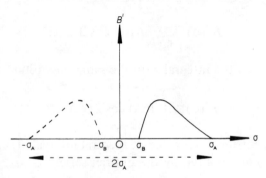

Fig. 6.16 Fourier transform spectrum showing 'negative' wavenumbers. The complete Fourier transform B' includes the negative wavenumber band shown dashed.

There is actually one point of difference in this analogy. Because the interferogram is symmetric about $\Delta = 0$, its Fourier transform is symmetric about $\sigma = 0$, so that the transform process yields not only the true spectrum $B'(\sigma)$ at positive σ but also a mirror image of it, $B'(-\sigma)$ at negative σ, as shown in Fig. 6.16. This mirror image did in fact appear in Equation 6.27, where the first term, which was disregarded in arriving at the instrument function (Equation 6.28), is an identical instrument function centred on $\sigma = -\sigma_1$. To avoid overlapping, the sampling steps must be small enough that the free spectral range $1/x$ covers both the actual band width $\sigma_A - \sigma_B$ and its mirror image. This would suggest (see Fig. 6.16) $1/x = 2\sigma_A$, where σ_A is the largest

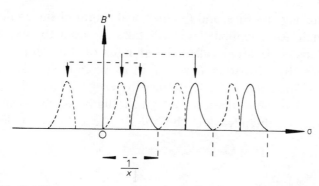

Fig. 6.17 Significance of free spectral range in Fourier transform spectroscopy. B'' is the Fourier transform spectrum obtained from finite sampling steps x. The arrows show the minor image pairs. Each spectrum with its mirror image is repeated at intervals $1/x$.

wavenumber present. Actually, by choosing x appropriately it is possible to fit the mirror spectra into the gaps between successive repetitions of the true spectra, as indicated in Fig. 6.17, and thereby keep the value of $1/x$ down to about twice the band-width –

$$\text{i.e.,} \qquad x \leqslant \frac{1}{2(\sigma_A - \sigma_B)} \qquad (6.30)$$

The number of sampling points N' is then given by

$$N' \geqslant 2N = \frac{2(\sigma_A - \sigma_B)}{\delta\sigma} \qquad (6.31)$$

For a given resolution, the sampling requirements obviously get much more severe as one goes to shorter wavelengths unless the band-width is very narrow.

6.15 Light-gathering power (etendue or through-put)

It was shown in Section 4.3 that with optimum illumination from a given source the light flux reaching the detector of a spectrometer is proportional to the area of the slit and to the solid angle accepted by the instrument – that is, to the solid angle subtended by the collimator or other limiting aperture at the slit. The light-gathering power of a spectrometer can therefore be defined by $L = S\,A/f^2$, where S is the area of the slit, A is the area of the limiting aperture, and f is the focal length of the collimating lens, which for simplicity we will take as equal to that of the camera lens. L is often called the etendue or through-put of the instrument. By re-writing L as $(S/f^2)A$, we can express it as the product of the solid angle Ω defined by $\Omega = S/f^2$ and the area A. Ω is known as the admission angle; it should be emphasized that it is the angle subtended *by* the slit *at* the collimating lens, as

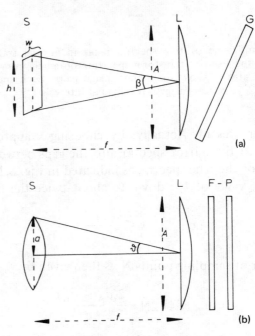

Fig. 6.18 Admission angles for (a) diffraction grating and (b) interferometer. In (a) the admission angle is given by $\Omega = wh/f^2 = w/f\,\beta$ and in (b), where S is a circular aperture, it is given by $\Omega = \pi a^2/f^2 = \pi\theta^2$.

opposed to the angular aperture, which is the angle subtended *by* the lens or dispersing element *at* the slit. The area A is usually fixed by the size of grating or interferometer plate available and does not vary greatly between one type of spectrometer and another, although it is usually rather smaller for a prism or a Fabry–Perot than for a grating or a Michelson. L therefore depends mainly on Ω, and it turns out that Ω is inversely proportional to the resolving power $\mathcal{R}$, according to the rule $\mathcal{R}\Omega$ = constant. The constant is, however, different for prisms and gratings on the one hand and interferometers of axial symmetry on the other, as we shall now show (see also [4,8]).

In Section 5.9 it was found that the resolving power of a grating used with a slit of width w greater than the optimum width is approximately f/w. The admission angle is given by $\Omega = wh/f^2$, where h is the slit height (Fig. 6.18).

$$\therefore \qquad \mathscr{R}\Omega = h/f = \beta \text{ say} \qquad\qquad (6.32)$$

The allowable admission angles for the Michelson and Fabry–Perot spectrometers can be considered together. It was mentioned in section 6.7 that the circular aperture isolating the centre of the Fabry–Perot ring system should be chosen to admit a wavenumber spread comparable with the resolution limit $\delta\sigma$. By symmetry, the aperture can equally well be put at S to limit the area of the source, assuming the optical arrangement to be that of Fig. 6.3. An aperture of radius a limits the angle of incidence on the interferometer to θ where $\theta = a/f$ (Fig. 6.18). The corresponding admission angle is $\Omega = \pi a^2/f^2 = \pi\theta^2$. If the interferometer plate separation is t, the path difference Δ changes from $2t$ to $2t\cos\theta$ in going from normal incidence to incidence at angle θ.

$$\therefore \qquad \delta\Delta = 2t(1 - \cos\theta) \approx t\theta^2$$

The corresponding wavenumber spread is obtained by differentiating the equation

$$\Delta = n\lambda = n/\sigma$$

$$\text{i.e.,} \qquad \frac{\delta\Delta}{\Delta} = -\frac{\delta\sigma}{\sigma}$$

The left-hand side of this equation is given by

$$\frac{\delta\Delta}{\Delta} = \frac{t\theta^2}{2t} = \frac{\theta^2}{2}$$

and if $\delta\sigma$ is to be equal to the resolution limit the right hand side is $1/\mathscr{R}$. The limiting value of θ is therefore set by $\theta^2 = 2/\mathscr{R}$ and we have finally

$$\mathscr{R}\Omega = 2\pi \qquad\qquad (6.33)$$

This holds for both the Michelson and the Fabry–Perot spectrometers.

Comparison of Equations 6.32 and 6.33 shows that for a given resolution Ω can be a factor $2\pi/\beta$ larger for the interferometers than for the grating. The useful slit height is limited by practical considerations to about 2 to 5 cm, even if curved slits are used to compensate for the spectral line curvature, because of the practical difficulties of making, positioning and aligning long slits a few microns wide, not to mention producing a long enough image of the source to cover them. For a high resolution instrument $\beta \sim 1/50$ would be an upper limit. The light-gathering power of an interferometer is therefore some two orders of magnitude greater than that of a grating of the same effective area.

6.16   Signal/noise ratio and multiplex advantage

In addition to its high light-gathering power there is another feature that distinguishes the Fourier transform spectrometer from the grating spectrometer, and in this case also from the Fabry–Perot spectrometer. In each of the last two instruments each spectral element is recorded in turn over a suitable integration time t. The total time taken to scan a spectrum of N elements is Nt. In Fourier transform spectroscopy, on the other hand, all N elements contribute to the signal from the detector at each setting, and the integration time needed for the same signal is only t/N. The entire scan would take only time t if the number of settings were the same. Actually 2N settings are required (Equation 6.31) so the total scan takes time $2t$, giving a time advantage of $N/2$.

The principal importance of this time advantage is the improvement in signal/noise ratio with which it is associated in the infra-red region. In this region the noise is primarily detector noise (Section 3.14) arising from fluctuations of temperature or current in the detector. The noise equivalent power is inversely proportional to the square root of the integration time but is independent of the signal. Now each spectral element is recorded over the entire scanning time by the Fourier transform spectrometer but only over $1/N$ of the total time by the grating and Fabry–Perot spectrometers. Consequently the noise equivalent power is a factor $1/N^{1/2}$ lower in the Fourier transform method and the signal/noise ratio is improved by $N^{1/2}$. This improvement is known as the multiplex advantage (or sometimes the Fellgett advantage). It does not hold in the visible and ultra-violet regions where the principal source of noise is photon noise due to fluctuations in the size of the signal. The N-times larger signal obtained from the simultaneous recording of all N elements leads to an increase in noise of a factor $N^{1/2}$,, exactly cancelling the advantage obtained from increased integrating time. Indeed, there may actually be a multiplex disadvantage in this case because of the contribution to signal noise from spectral elements outside the band-width under study.

It should be remembered that the multiplex advantage is also exploited by any instrument using a photographic plate as a detector. A 10-inch plate with a resolution limit of 10 μm can record 25 000 pieces of information simultaneously. However, the long wavelength limit for photographic emulsions is only a little over 1 μm, and as this is close to the cross-over point from photon to detector noise the Fourier transform spectrometer can be regarded as taking over the multiplex advantage from a conventional spectrograph in this region.

6.17 Comparison of grating, Fabry–Perot and Fourier transform spectroscopy

It seems useful at this stage to summarize the principal advantages and disadvantages of these three types of spectroscopic instrument. By restricting the comparison to fairly high resolution it is reasonable to exclude prisms. The particular advantages of prism spectrographs and monochromators at medium to low resolution were mentioned in Section 4.10.

The discussion is somewhat simplified by taking separately the three regions (a) the infra-red at wavelengths above a few micrometres, (b) the region from a few micrometres to about 2000 Å and (c) the far ultra-violet below 2000 Å. Fourier transform spectroscopy can then be eliminated from the contest in regions (b) and (c) because the multiplex advantage has vanished, the high light grasp is shared by the Fabry–Perot, and the number of sampling points becomes prohibitively high – to go from 4000 Å to 5000 Å even at the very modest resolution limit of 1 cm^{-1} ($\mathscr{R} \sim 2 \times 10^4$) requires 10 000 sampling points. The Fabry–Perot is not a practical proposition in region (c) because of the requirements on plate flatness and reflectivity. It is also very difficult to make flat enough plates from the crystals that transmit in region (a) above the quartz cut-off. Beam-splitters for Michelson interferometers in the infra-red do not present such a problem because the criterion for flatness is set by the finesse required, which is about 25 for the Fabry-Perot and only 2 for the Michelson. This degree of latitude makes the Michelson usable all through the infra-red, with polythene or other suitable thin films as beam-splitters beyond the cut-off wavelength of crystals and with mirrors replacing the lenses L and L'. As a matter of fact, Fabry-Perot spectrometers with metallic mesh 'plates' have been used in the submillimetre range [7], but it will simplify the discussion to compare the grating with the Fabry-Perot only in region (b) and with the Fourier transform spectrometer only in region (a), leaving it in undisputed possession of region (c).

The main advantages of the Fabry-Perot over the grating are its higher potential resolving power, the ease with which this may be varied, and its higher light-gathering power; the principal disadvantages are the necessity for cross-dispersion, the more complex instrument function, and (in photographic use) the non-linear dispersion. The Fabry-Perot is not convenient for scanning a wide spectral range, and its free spectral range is fixed by its resolving power. The residual background intensity I_{min} between orders (Figs. 6.5 and 6.6) is a serious difficulty in absorption spectroscopy unless the background source is itself an emission line of limited width. At wavelengths up to just over 1 μm both instruments may be used with either photo-electric or photographic detection. The higher etendue of the Fabry-Perot is a clear advantage in the first case, but at first sight this advantage is lost in photographic use because the greater light flux is simply spread over a larger area of plate. However, it can be shown [8] that if the resolution limit is properly matched to the grain size of the emulsion the ring pattern can be shrunk (by using a lens of shorter focal length) without loss of resolution in proportion as the resolving power is increased. This is possible because high $\mathscr{R}$ means small angular radius for a given ring and hence large angular dispersion. The net gain in illumination is of order 10 to 100 when the resolution is fairly high.

In region (a), the infra-red, we come to the comparison between the grating and Fourier transform spectrometers, which has engendered considerable discussion, sometimes somewhat controversial (see, for example, [3], [7], [8]). The two principal advantages of Fourier transform spectrosocpy are those discussed in the last two sections, the superior light-gathering power and the multiplex advantage. For a given resolving power the light grasp of the Michelson was shown to be about 100 times that of the grating, and this advantage can be used either to reduce the scanning time by a factor of 100 or to prove the signal/noise ratio by a factor 10 (assuming detector noise to be dominant). Alternatively, if the Michelson aperture were reduced by a factor 100 to give the same signal/noise ratio, the resolving power could be correspondingly increased. The multiplex advantage produces a further large gain in signal/noise ratio, scanning time or resolution. For example, suppose we wish to scan over a spectral range of 50 cm^{-1} centred on 100 cm^{-1} (i.e., on 100 μm). With a resolving power of 100, the resolution limit is 1 cm^{-1} and the number of spectral elements N is 50. In principle the multiplex advantage alone would enable the Michelson to do in a couple of minutes what the grating could do in an hour. Taking both advantages together, the available factor of about 5000 could be split so as to scan the band in say 20 minutes with a resolving power about 10 times that of the grating and with a signal/noise ratio also about 10 times better. Generally speaking, the advantages of the Michelson increase towards longer wavelengths. It is probably superior to the grating for most purposes above 100 μm, but at very long wavelengths its light grasp may be unnecessarily large in that the sources and detectors may be too small to take full advantage of it. Below 10 μm, on the other hand, the number of sampling points gets inconveniently large if the resolving power is at all high, and the light grasp is less important, thus favouring the grating. Between these limits, and indeed outside them in particular cases, other factors may sway the balance: the Michelson interferometer is a relatively compact instrument and its instrument function is accurately known and controllable; on the other hand the grating can produce 'instant spectra' and does not require a computer. There are also certain jobs that one or other instrument simply cannot do; for instance, the grating is impossibly slow for many astrophysical applications, and the Fourier transform spectrometer cannot be used as a monochromator.

*SOME OTHER INTERFEROMETERS USED IN
SPECTROSCOPY*

6.18   The wedge etalon

The Fabry-Perot used with a point source produces a uniforml
illuminated field of view if the plates are exactly parallel, sinc
there is no variation of θ over the field. If one plate is tilte
slightly, t becomes the variable, and the result is a set of straigh
fringes parallel to the edge of the wedge formed between the tw

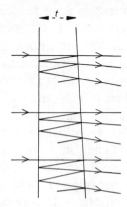

Fig. 6.19 Path of rays through wedge etalon. The light is incident almo
normally, and maxima occur where t satisfies $2t = n\lambda$.

plates (Fizeau fringes), which have a high finesse because of th
multiple reflections (see Fig. 6.19). The principal advantages c
this interferometer are its linear dispersion and the less stric
requirements on the flatness of the plates, for a local bump merel
produces a local kink in the fringe instead of contributing t
smearing out the whole pattern. On the other hand, because of th
departure from cylindrical symmetry much of the light-gatherin
power is lost.

6.19   The spherical Fabry-Perot

This device consists of two identical spherical mirrors separated b
their radius of curvature r [4], [8]. As shown in Fig. 6.20, a ra
ABCDAB having made five crossings emerges not only parallel t
but also co-incident with the original ray AB, with a pat

difference $\Delta = 4r$ if the aperture d/r is sufficiently small for spherical aberration to be neglected. With Δ independent of θ, the angular restriction on the ordinary Fabry–Perot no longer applies. The restriction on d/r which replaces it makes for a smaller light-gathering power than that of the plane Fabry–Perot at plate separations of a few cm; but since d can be increased proportionately to r, and hence to the resolving power, the device has the unusual property of etendue proportional to $\mathcal{R}$ instead of to $1/\mathcal{R}$. At resolving powers above about 10^7 ($t > 10$ cm in the plane Fabry–Perot, corresponding to $r > 5$ cm) there is an appreciable

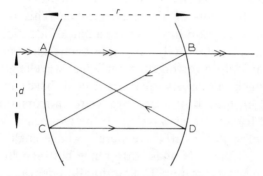

Fig. 6.20 Path of rays through spherical Fabry–Perot etalon. The reflected part of AB follows the path BCDAB and emerges co-incident with the directly transmitted part.

light gain in using the spherical rather than the plane Fabry–Perot. Such high resolving power is useless with ordinary spectral lines because the Doppler width merely smears the line over a whole order, but it may be useful for atomic beam sources. The most obvious application is to laser sources, where the band width $\delta\sigma$ may correspond to path differences $\Delta = 1/\delta\sigma$ of several metres.

6.20   SISAM

This device gets its weird name from the initial letters of its description in French, which, apart from the word order, is the same in English: spectroscopy interferometric with selection by amplitude of modulation. Lest this sound too formidable, it should be pointed out that the Fourier transform spectroscopy we have been describing in this chapter could be designated SISFM –

simply substituting frequency for amplitude – if that were pronounceable. From Equation 6.18, the intensity of the signal from a Michelson interferometer due to a single component σ_i is

$$I = I_i(1 + \cos 2\pi\sigma_i\Delta),$$

which may be written

$$I = I_i(1 + \cos 2\pi\sigma_i Vt)$$

where $V/2$ is the velocity of the moving mirror. Thus the signal is modulated at a frequency $\sigma_i V$ which depends on σ_i, and with unit depth independent of σ_i. In SISAM the depth of modulation is also made a function of the wavenumber. This is achieved by replacing the Michelson mirrors with a pair of identical diffraction gratings, which can be rotated together about axes parallel to their rulings. At any given setting, the particular wavelength σ_i which is diffracted back along its own path will have unit depth of modulation, just as if the gratings were mirrors set accurately parallel. But for any other wavelength the two beams will return at different angles, just as if there were a wedge angle between the two mirrors, so that the phase difference between the two beams varies over the cross section. The modulation depth is

$$\frac{\sin 2\pi(\sigma - \sigma_i)D}{2\pi(\sigma - \sigma_i)D}$$

[4], [8], where D is the maximum path difference across the beam. Only a narrow frequency band around σ_i is modulated when the path difference Δ is changed. As the gratings are rotated to different angles, a change of Δ results in a modulated signal whenever the angle is right for one of the frequencies in the source. Since it would obviously be difficult to change Δ by displacing one of the gratings in this device, the compensating plate C of Fig. 6.9 is rotated to achieve the same effect. The resolving power of the instrument depends on the band-width of the modulation function

$$\frac{\sin 2\pi(\sigma - \sigma_i)D}{2\pi(\sigma - \sigma_i)D}$$

and hence on D, the path difference across either beam, so that it is essentially the same as for a grating spectrometer using the same grating. To use two gratings and record modulation depth as a function of rotation angle, instead of simply recording intensity as a function of angle with a single grating in a conventional spectrometer, might seem a pointless exercise. The advantage is that the admission angle is characteristic of the axially symmetric rather than the slit type of instrument, so that there is a gain of some two orders of magnitude in light-gathering power.

6.21   The Mach–Zehnder interferometer

The Mach–Zehnder interferometer is a refractometer rather than a spectrometer. Its connection with spectrosocpy is that the variation of refractive index of a gas or a vapour near an absorption line – that is, the anomalous dispersion – is determined by the

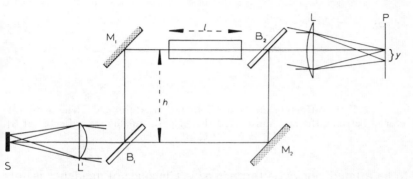

Fig. 6.21 Mach–Zehnder interferometer. See text for explanation.

transition probability of the absorption line. This relation is exploited in the so-called hook technique for the experimental determination of transition probabilities, as described in Chapter 10. The Mach–Zehnder interferometer is the instrument used in this method.

The interferometer is a modification of the Jamin refractometer, which is itself a variant of the Michelson. As shown in Fig. 6.21, the beam-splitter B_1 divides the incident light into two beams which, instead of retracing their paths after reflection, travel parallel to one another and are recombined at a second beam-splitter B_2. In order that $M_1 B_2$ be parallel to $B_1 M_2$, M_1 must be parallel to B_1, as shown in Fig. 6.22. The path difference between the two beams is the same as that between successive reflections in Fig. 6.1 – that is, $2t \cos \theta$, as given in Equation 6.1, where t is the perpendicular distance between the plates (Fig. 6.22). If B_2 and M_2 are also parallel to B_1, the second pair of reflections introduces an equal path increment into the lower beam, and the net path difference after recombination, Δ, is zero for any angle of incidence. Now suppose each of the plates B_1 and M_1 to be rotated anti-clockwise about an axis perpendicular to the page through a small angle $\alpha/2$ and each of the plates B_2 and M_2

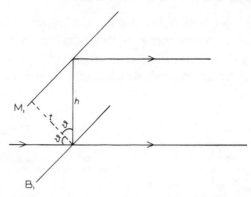

Fig. 6.22 Path difference in Mach–Zehnder interferometer. The path difference between the ray transmitted through B_1 and the ray reflected at B_1 and M_1 is $\Delta = 2t \cos \theta = 2h \cos^2 \theta$.

to be rotated clockwise through $\alpha/2$. The angle of incidence is now $\theta - \alpha/2$ for the first pair and $\theta + \alpha/2$ for the second. The net path difference becomes:

$$\Delta = 2t \cos (\theta - \alpha/2) - 2t \cos (\theta + \alpha/2)$$

$$= 2t\{\cos \theta \cos \alpha/2 + \sin \theta \sin \alpha/2 - \cos \theta \cos \alpha/2 + \sin \theta \sin \alpha/2\}$$

$$\simeq 2t\alpha \sin \theta$$

for small α. The path difference now depends, for a given α, on θ, the angle of incidence in the plane of the diagram. The two beams interfere constructively for angles θ such that

$$\Delta = 2t\alpha \sin \theta = n\lambda \qquad (6.34)$$

Since they are both travelling in the same direction after recombination, the interference fringes are localized at infinity, or in the focal plane P of the lens L. Rays of different θ are focused at different points y, as shown in Fig. 6.21. In contrast to the Fabry–Perot and the Michelson, this system is not axially symmetric because Δ depends on the wedge angle α; the interference fringes are straight lines parallel to the edge of the wedge (i.e., perpendicular to the page). Since the Mach–Zehnder is a two-beam interferometer, the fringes have the same $\cos^2$ intensity distribution as those of the Michelson.

The angular spacing of the fringes is found by differentiating Equation 6.34 and putting $\Delta n = 1$ to give

$$2t\alpha \cos \theta \ \Delta\theta = \lambda\Delta n = \lambda$$

The linear spacing with a lens of focal length f is

$$\Delta y = f\Delta\theta = f\lambda/2t\alpha \cos \theta$$

Usually $\theta = 45°$, and Δy can be rewritten in terms of the separation h of $B_1 - M_1$ and $B_2 - M_2$ (see Fig. 6.22) as

$$\Delta y = f\lambda/2h\alpha \cos^2 \theta = f\lambda/h\alpha$$

Similarly, the path difference (Equation 6.34) can conveniently be written in terms of y rather than θ as $\Delta = a + by$ where $b \equiv h\alpha/f$.

For use as a refractometer an absorption tube of length l is inserted in one arm as shown in Fig. 6.21. If the tube contains a gas or vapour of refractive index μ the path difference between the beams is

$$\Delta = a + by + (\mu - 1)l \qquad (6.35)$$

and the refractive index can be measured from the fringe shift. If the fringes are focused across the slit of a spectrograph, the y-value of a given fringe – the pth – will vary with wavelength according to $by + (\mu_\lambda - 1)l = p\lambda$ and the fringe can be regarded as tracing the curve of $(\mu_\lambda - 1)$ versus λ.

References

1. Tolansky, S. 'High Resolution Spectroscopy', Methuen, 1947
2. Francon, M. 'Optical Interferometry', Academic Press, 1966
3. Steel, W. H. 'Interferometry', Cambridge University Press, 1967
4. Bousquet, P. (a) 'Spectroscopie Instrumentale', Dunod, 1969; (b) (Trans. Greenwood, P.), 'Spectroscopy and its Instrumentation'; Hilger, 1971

5. Born, M. and Wolf, E. 'Principles of Optics', Pergamon, 1965
6. Ditchburn, R. W. 'Light', Blackie, 1963
7. Martin, D. H., ed. 'Spectroscopic Techniques for far Infra-red, Sub-millimetre and Millimetre Waves', North Holland, 1967
8. Jacquinot, P. New Developments in Interference Spectroscopy, *Rep. Prog. Phys.* **23**, 267, 1960
9. Gebbie, H. A. and Twiss, R. Q. Two Beam Interferometric Spectroscopy, *Rep. Prog. Phys.* **29**, 729, 1966

CHAPTER SEVEN

Microwave and radiofrequency spectroscopy

7.1 Introduction

Microwave and radiofrequency spectroscopy measure relatively small energy changes, between about 10 cm^{-1} and a few thousandths cm^{-1} (300 GHz to 30 MHz). At the short wavelength end of this range (1 mm $< \lambda <$ 10 cm, say) come pure rotational transitions of molecules, Zeeman and Stark splittings and the hyperfine structure of heavy atoms. Towards longer wavelengths (lower frequencies) one can measure the hyperfine structure of light atoms, the fine structure of hydrogen-like ions, and Zeeman splitting from the nuclear magnetic moment.

All these effects can be observed as small perturbations (shifts or splittings) of transitions in the optical region – by resolving, for example, the rotational structure of molecular bands or the hyperfine structure of atomic lines in the visible – but they then appear as very small differences in large quantities. A typical hyperfine splitting in a light atom of a few hundredths cm^{-1} represents a difference of one part in 5×10^5 in the visible transitions. This is very close to the limit of accuracy of the measurements, which is usually set by the Doppler width at about $1 : 10^6$, or about $1 : 10^7$ in heavy atoms. However, the Doppler width is a constant fraction of the frequency (Section 8.3), so the frequency spread is very much smaller at low frequencies. It follows that the direct determination of small energy differences is many orders of magnitude more accurate than the indirect.

217

Such direct determinations are not always possible, because radiative transitions depend on the existence of a transition moment, electric dipole, magnetic dipole, electric quadrupole or higher order. It was seen in Section 2.8 that, since an atom has no permanent electric dipole moment, electric dipole transitions can occur only between states of opposite parity. Of the effects listed above, only the fine structure of hydrogen falls into this category; all the other types of transition involve different states of the same electron configuration and can be observed directly only if there is a permanent electric or magnetic dipole moment. In general, magnetic interactions are more characteristic of free atoms and electric interactions of molecules. Any atom with a resultant electronic angular momentum ($J \neq 0$) has a magnetic moment, but most stable molecules in their ground states do not, because the electrons are paired off to give zero resultant spin and orbital angular momentum. On the other hand, most heteronuclear molecules have an asymmetric electron distribution and hence an electric dipole moment; this allows investigation of the splitting between rotational levels and between Stark sub-levels in an external electric field. Pure rotational radiative transitions do not occur, however, in homonuclear molecules or spherical top molecules (which have spherical symmetry). The same general considerations apply to hyperfine structure. The interaction between the nuclear magnetic dipole moment and the internal magnetic field from the unpaired electrons is the more important in free atoms, and the interaction between the electric quadrupole moment of the nucleus and the gradient of the internal electric field is the more important in most molecules.

The short wavelength limit of the microwave region is set by the range of klystron oscillator harmonics at a fraction of a millimetre, thus overlapping the infra-red region; the transition to radiofrequency is usually taken to be at about 30 cm, so that in round figures the limits of the microwave region are $10^6 - 10^3$ MHz ($30 - 0.03$ cm^{-1}). The availability of narrow-band tunable sources entirely eliminates the need for dispersers. An absorption spectrum is scanned simply by sweeping the source frequency over the relevant range and recording the signal from a tuned detector as a function of frequency. One problem that arises, however, is that the amount of energy absorbed by a transition between two states decreases as the energy difference between the states decreases, and it may become exceedingly small as radiofrequencies are approached. To overcome this difficulty ways of detecting transitions other than by measuring the energy absorbed have been devised – for example, the change in deflection of an atomic beam or the change in polarization of resonance radiation in the visible region.

The first kind of experiment to be discussed in this chapter is the straightforward microwave absorption spectroscopy of gases. The effects studied are primarily rotational transitions of molecules, in which electric quadrupole effects may appear as a fine structure, and Stark shifts in an external electric field. The classic experiment of Lamb and Retherford on the fine structure of hydrogen also falls into the category of electric dipole transitions in the microwave region.

The other types of experiment are all concerned with magnetic dipole transitions between the sub-levels into which a given energy level is split in a magnetic field, external or internal. The splitting of the sub-levels, ΔE_B, depends on the strength of the field B, and transitions are observed when the frequency of the oscillator passes through the 'resonance' value $h\nu = \Delta E_B$. Experiments of this type are collectively known as magnetic resonance experiments for this reason. In external magnetic fields of a convenient size, say 0.1 Wb m^{-2} (1 kilogauss), the energy splittings from the interaction of the field with the electronic magnetic moment, which is of order μ_B, fall in the microwave region; the resonance frequencies are investigated by electron spin resonance in the case of solids and liquids and by beam deflection experiments in the case of free atoms. The energy splittings from the nuclear magnetic moment, being some three orders of magnitude smaller, fall in the rf region even in high external fields. The technique used to measure them in solids and liquids is known as nuclear magnetic resonance, and in free atoms and molecules they may be investigated in beam deflection experiments. Rather closely linked with the atomic and molecular beam magnetic resonance experiments are the double resonance experiments, described in Section 7.11, which use optical resonance radiation to detect rf resonances. The final section describes a type of experiment known as level-crossing spectroscopy, which does not involve any rf radiation at all, but which is included in the present chapter rf reasons that are more logical than might at first appear.

7.2 Microwave spectroscopy

Microwave spectroscopy is a direct result of the development of radar during the 1939-45 war. The short wavelength limit has been steadily pushed down since then, to reach a few tenths of a millimetre, but the region around 1 cm is still the easiest to operate in. Rotational transitions in molecular spectra fall in this wavelength region because the rotational constants of most molecules are of order 1 cm^{-1}. Much of microwave spectroscopy is concerned with measurements in absorption of rotational spectra. Emission spectra from laboratory sources are usually too weak to be of practical importance. Not all molecules can be investigated by microwave absorption, however: molecules such as the diatomic hydrides have such large rotational constants, owing to their small moments of inertia, that their rotational spectra fall in the infra-red, and molecules without a permanent electric or magnetic moment have no pure rotational spectrum. This last category includes most homonuclear diatomic molecules (except a few like O_2 with a permanent magnetic dipole moment) and symmetric top molecules.

The rotational constants are important because molecular bond lengths and angles may be deduced from them. In addition, microwave absorption spectroscopy gives information on the magnitudes of electric dipole and nuclear quadrupole moments, as will be shown below.

We are concerned here only with the absorption of gases. In solids and liquids the rotational structure is smeared out, and the microwave spectrum, like the optical, consists of bands whose interpretation must be sought in solid- and liquid-state physics.

7.3 Experimental aspects of microwave spectroscopy

Fig. 7.1 shows schematically a typical lay-out for microwave absorption. The source is normally a klystron oscillator of appropriate working frequency, tunable over a frequency range of about 15%. Below 5 mm it is necessary to use harmonic multiplication. The radiation passes through a waveguide of suitable dimensions to a crystal detector. Part of the waveguide is isolated by mica windows and filled with gas at a pressure usually in the range 10^{-1} to 10^{-4} torr to form the absorption cell. A small fraction of the input is tapped off to a wavemeter to measure the frequency. The detector itself is usually a silicon-tungsten crystal; this acts as a rectifier whose output is proportional to the incident power (that is, a square law detector), which is usually of the order of a milliwatt. The simplest way of increasing the signal/noise ratio over that from d.c. detection is the crystal video system shown in Fig. 7.1. The klystron frequency is swept over a small range (say 1% of the central frequency) at an audio frequency. If

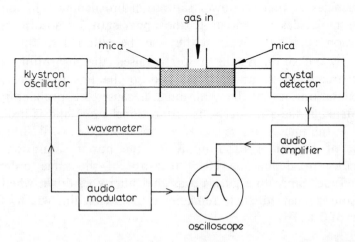

Fig. 7.1   Schematic arrangement for microwave absorption spectroscopy. A section of waveguide isolated by mica windows acts as an absorption cell. The source is swept at audio frequency over a small frequency range.

the audio signal is applied to the X-plates of an oscilloscope and the amplified signal to the Y-plates, any change of signal with frequency will show up on the oscilloscope. This will, in fact reproduce the line shape provided that the line is narrow compared to the frequency sweep and that the amplifier bandwidth is adequate.

This basic technique can be refined or modified in a large number of ways – for example, source modulation and phase sensitive detection, superheterodyne detection. For a full description of the various methods and their relative advantages a textbook such as [1, 2] should be referred to.

The accuracy with which microwave frequencies can be measured is of order $1 : 10^7$, but, as in optical spectroscopy, the precision with which the frequency of a transition can be determined is usually limited by the intrinsic line width. Of the line-broadening mechanisms discussed in Chapter 8, pressure broadening is the most important in the microwave region. At 10^{-2} torr the half-value width of a pressure broadened line is about 0·1 MHz. With a central frequency of order 30 000 MHz (1 cm wavelength), one can expect a resolving power in the $10^5 - 10^6$ range, comparing well with the optical region.

A line-broadening effect important at low, but not at optical, frequencies is that of power saturation broadening. This occurs when molecules are raised to the upper state by absorption of radiation more quickly than they can be returned to the lower state by collisions. As the power is raised, the absorption at the peak frequency saturates first, and so the line broadens. The de-excitation has to be collisional because the probability of spontaneous radiative decay is proportional to ν^3 and is insignificant in the microwave region (Sections 3.4 and 9.3). With power input of the order of a milliwatt, the power saturation and pressure broadening at 10^{-2} torr are of the same order of magnitude. Since one goes up as the other goes down when the pressure is changed, there remains an irreducible line width of the order of 0·1 MHz.

7.4 The principal information obtained from microwave spectroscopy

7.4.1 *Rotational constants*

The rotational energy of a diatomic or linear polyatomic molecule is given (equation 2.34) by

$$F(J) = B_v J(J + 1) - D_v J^2 (J + 1)^2$$

where B is inversely proportional to the moment of inertia, D is a small correction for centrifugal stretching, and both B and D are functions of the vibrational quantum number v as well as the electronic state. The selection rule for electric dipole radiation is $\Delta J = \pm 1$ so that absorption can occur for wave numbers σ given by

$$\sigma = 2B(J + 1) - 4D(J + 1)^3 \ \text{cm}^{-1}$$

where B and D are in cm^{-1} and J refers to the lower state. Since $D \ll B$, the rotational absorption consists of a set of lines of almost equal spacing 2B (Section 2.18). In diatomic molecules B gives directly the moment of inertia I and hence the bond length. In linear polyatomic molecules it is necessary to do experiments with different isotopes of one or more of the atoms to obtain the individual bond lengths unambiguously. The analysis of non-linear polyatomic molecules is naturally a good deal more complicated. With suitable isotopic substitutions, however, bond lengths and angles can usually be extracted to accuracies of about ±0·005 Å and ±30 minutes of arc respectively.

7.4.2 Hyperfine structure and quadrupole moments

Magnetic hyperfine structure is negligible in the ground states of most molecules because of the pairing of the electron spin and orbital angular momenta. On the other hand, the departure from spherical symmetry of the electron charge distribution can produce an electric field gradient at the nuclei. Taking the z-axis in the direction of this gradient, one can write

$$-\frac{\partial F_z}{\partial z} = \frac{\partial^2 V}{\partial z^2} = q$$

When the nuclear spin I is greater than or equal to 1, the nuclear charge distribution also lacks spherical symmetry, being either stretched out or squashed in along one axis (prolate or oblate spheroid) by an amount which is measured by the nuclear electric quadrupole moment eQ. The energy of the molecule then depends on the orientation of Q to the z-axis, as given by Equation 2.27:

$$E_Q = eQq \, f(F, I, J)$$

The number of hyperfine sub-levels is determined by the quantum numbers I, J and F and their separation by the quadrupole coupling constant eQq. q is very difficult to calculate, except in especially favourable cases, but if Q is known from atomic beam experiments the microwave results can be used to evaluate q, and this is a very useful constant for determining bond types and testing wave functions. Even if Q is not known, useful information can be obtained from the relative values of q in different compounds containing the same nucleus.

7.4.3 Stark effect

Most molecules show appreciable Stark splitting in an external electric field. As with the quadratic Stark effect in atomic spectra (Section 2.9), each level of given J splits into $J + 1$ sub-levels, with $+M_J$ and $-M_J$ having the same energy. In linear molecules the splitting is proportional to $\mu^2 F^2$, where μ is the electric dipole moment and F the field strength. It is not linear in μF, as one might at first expect, because of the rotation. In the absence of any interaction the dipole rotates uniformly and has no average component in the field direction. The first effect of the interaction is to produce a net orientation in the field direction proportional to μF, so that the average effective dipole is proportional to $(\mu F)\mu$ and the interaction energy to $(\mu F) \cdot \mu F$. μ can therefore be determined from the Stark splitting. In a high field this may be as large as 20 MHz and can be measured to about $1 : 10^3$, so the method is very accurate if F is accurately known. In practice the field is usually calibrated by measuring the Stark splitting for a molecule of known μ.

7.4.4 Particular cases

The inversion spectrum of ammonia was the first microwave absorption ever studied and should certainly be mentioned here. The ammonia molecule NH_3 is a pyramid with N at the peak of a base of H atoms, and the two configurations with the molecule turned inside out with respect to one another are both states of stable equilibrium. The resonance interaction between these two states leads to a splitting of about 24 000 MHz. The transition between the two states at 8 mm wavelength was not only the first to be measured with microwaves but also the first to be used in a maser.

The Lamb shift in hydrogen-like spectra also falls in the microwave region. In hydrogen and H-like ions levels of the same quantum number j are almost but not quite degenerate. The splitting can just be measured as a very small difference between two optical transitions, with an error comparable to the size of the effect. The $2s_{1/2} - 2p_{1/2}$ splitting, for example, produces extra components in the red H_α line, but their separation of about 0.03 cm^{-1} is very close to the resolution limit in the best experimental conditions. Lamb and Retherford measured the energy difference directly in a type of beam experiment, the transition frequency being found from the sudden change in the number of metastable $2s_{1/2}$ atoms, and achieved an accuracy of about $1 : 10^4$. Lamb shifts have since been measured for a considerable number of levels in hydrogen and H-like ions.

7.5 Magnetic resonance spectroscopy: general considerations

Magnetic resonance spectroscopy is concerned with radiative transitions between magnetic sub-levels. The energy separation of these sub-levels rarely exceeds about a wavenumber, and the relevant spectral region therefore extends from the long wavelength half of the microwave region into the radiofrequency region. For a given field B, the resonance frequency defined by

$$h\nu_0 = \Delta E_B \tag{7.1}$$

can be found by recording the signal as a function of frequency, as in Fig. 7.2a. In practice it is usually more convenient to hold the frequency at some fixed value and record the signal as a function of field, as in Fig. 7.2b.

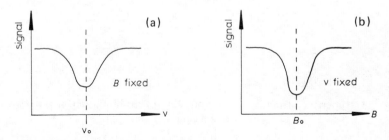

Fig. 7.2 Signal in magnetic resonance spectroscopy as a function of (a) frequency and (b) magnetic field.

Let us start with the energy of a free atom in an external magnetic field. Unless the atom is an a $^1S_0$ state, it has a magnetic moment μ_J associated with the electronic angular momentum $J\hbar$, and the interaction of μ_J with the external field B gives rise to Zeeman splitting of the energy levels (Section 2.9). The interaction energy is given by Equation 2.19, from which we have for the separation of adjacent sub-levels

$$\Delta E_B = g_J \mu_B B \tag{7.2}$$

where μ_B is the Bohr magneton ($9 \cdot 273 \times 10^{-24}$ J per weber m^{-2}) and g_J is the Lande g-factor, of order unity. Taking the largest easily obtainable value of B to be about 2 weber m^{-2} (20 kilogauss), we have

$$\Delta E_B \leqslant 2 \times 10^{-23} \text{ J} \sim 1 \text{ cm}^{-1}$$

Transitions satisfying Equation 7.1 therefore correspond to wavelengths from 1 cm upwards.

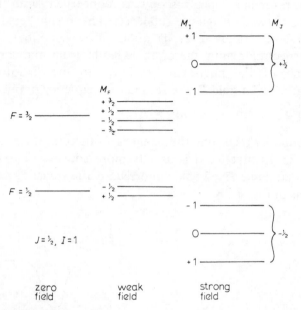

Fig. 7.3 Hyperfine structure of ground state of lithium atom in zero, weak and strong magnetic field.

Next we have to consider the interaction of any nuclear magnetic moment μ_I with the internal field B_J attributable to the electronic spin and orbital motion – that is, hyperfine structure, Section 2.13 – and the effect on this of the external field. μ_I is some three orders of magnitude smaller than μ_J (Equation 2.25), but the observed hyperfine splittings are in the range $0 \cdot 01$ cm^{-1} to 1 cm^{-1}. It follows that the internal field B_J must be of order 100 weber m^{-2}. Fig. 7.3 shows on the left-hand side the two hyperfine levels of an atom such as deuterium or lithium, which have $J = \frac{1}{2}$ in the ground state and nuclear spin $I = 1$. The total angular momentum quantum number F, defined by $F = I + J$ (Section

2.13), can take the two values $3/2$ and $\frac{1}{2}$. In an external magnetic field F remains a good quantum number provided the field is weak enough that the $\mu_J - B$ interaction is small compared with the $\mu_I - B_J$ interaction – that is, $B \leqslant 10^{-2}$ weber m^{-2}. The field then splits each hyperfine level into $2F + 1$ sub-levels, as shown in the centre of Fig. 7.3: Zeeman effect of hyperfine structure. The right-hand side of the figure shows the opposite extreme, $B \gg 1$ weber m^{-2}, where the $\mu_J - B$ interaction is the dominant one. The $\mu_I - B_J$ interaction then splits each Zeeman level into a number of hyperfine levels – i.e., hyperfine structure of Zeeman effect. Since $B_J \sim 100$ weber m^{-2}, it is still generally true that $\mu_I B_J \gg \mu_I B$, and the latter interaction can usually be ignored. However, in the case of most free *molecules*, as well as atoms in $^1S_0$ states, when μ_J and B_J are zero, the interaction of μ_I with the external field is the dominant effect. The transitions between the hyperfine and Zeeman sub-levels of free atoms and molecules can be investigated at various field strengths by atomic and molecular beam experiments and by the double resonance method.

In solids and liquids, as in molecules, the valence electrons are usually paired off, leaving only μ_I to interact with the external field. There are two important types of exception, however: atoms or ions with partly filled inner shells (transition elements and rare earths) and free radicals *do* have unpaired bound electrons. The separation of their magnetic sub-levels is therefore of order $\mu_B B$ as with free atoms. Magnetic resonance spectroscopy with such substances is known as electron spin resonance, or ESR. As the unpaired electrons are responsible for paramagnetism, it is also known as paramagnetic resonance. In all other types of solid and liquid the separation of the magnetic sub-levels is of order $\mu_I B$, and transitions between them are investigated by nuclear magnetic resonance (NMR). In both ESR and NMR the energy levels are influenced by the surrounding atoms as well as by the external magnetic field. Although solid state spectroscopy is beyond the scope of this book, a few remarks on ESR and NMR are included in this chapter because of the close links with atomic and molecular beam magnetic resonance spectroscopy.

As already remarked, measurement of absorbed energy becomes an ever less sensitive method of detecting radiative transitions as the wavelength increases from the microwave towards the radio-frequency region. In thermal equilibrium the relative populations of two states separated in energy by ΔE is given by Boltzmann's formula (Equation 3.6):

$$N_2/N_1 = g_2/g_1 \; e^{-\Delta E/kT}$$

where the g's are the statistical weights of the two states and are of the same order of magnitude. At room temperature $kT \approx 200 \; \mathrm{cm}^{-1}$, so for ΔE less than about a wavenumber (wavelengths from 1 cm upwards) the exponential factor is almost unity, and the populations of the two levels are nearly equal. Consequently stimulated emission is almost as probable as absorption, and the absorption of radiation from upward transitions is largely cancelled by the emission from downward transitions. Spontaneous radiation can be neglected at these frequencies because it falls off with ν^3 (Equation 3.8). Quantitatively, the *net* rate of upward transitions can be written in terms of the Einstein *B*-coefficients defined in Section 3.4 as

$$B_{12}N_1\rho(\nu) - B_{21}N_2\rho(\nu)$$

per sec per unit volume where ρ is the radiation density at the frequency ν defined by $h\nu = \Delta E$. Using Boltzmann's equation and the relation between the *B*-values (Equation 3.8), the net rate of upward transitions is

$$B_{12}N_1\rho(1 - e^{-h\nu/kT})$$

which for $h\nu \ll kT$ becomes

$$B_{12}N_1\rho h\nu/kT$$

Since the energy absorbed in each upward transition is $h\nu$ the rate of absorption of energy is proportional, other things being equal, to ν^2. In ESR and NMR experiments it usually pays to work with the largest practicable magnetic fields so as to make ν as high as possible.

Absorption in magnetic resonance experiments is in any case weaker than in the microwave experiments described in Section 7.2 because it is governed by the magnetic dipole transition probability, which is only about $10^{-4} - 10^{-5}$ times as large as the electric dipole transition probability (Sections 2.8 and 9.3). Transitions involving the nuclear magnetic moment only are down by a further factor $(\mu_I/\mu_J)^2$, or about 10^{-6}. Nevertheless, energy absorption at resonance is detected by both ESR and NMR techniques, where the use of solid or liquid samples provides a high concentration of absorbers. The atomic and molecular beam and double resonance experiments depend on changes of beam deflection and polarization respectively to detect transitions.

7.6 Electron spin, or paramagnetic, resonance

The discussion of the Zeeman effect in Section 2.9 made the point that the interaction between μ_J and the external field was small compared to the magnetic interaction between the spin and orbital parts of the angular momentum Lh and Sh, so that J remained a good quantum number. In the solid state, however, the electric fields in the crystal can interact with L strongly enough to uncouple L and S and split a level of given L into $2L + 1$ sub-levels separated by several thousand cm^{-1}. To a first approximation each of these sub-levels is $(2S + 1)$-fold degenerate because S interacts with the crystalline field only through the residual $L-S$ coupling. At normal temperatures only the lowest orbital sub-level is occupied. Its degeneracy is removed by an external magnetic field, and the separation of the $2S + 1$ sub-levels is determined primarily by the interaction of the spin magnetic moment μ_S with the field: $E_B = - \mu_s . B$. Using equation 2.15,

$$E_B = g_s\mu_B S . B = g_s\mu_B B M_s \qquad (7.3)$$

Fig. 7.4 illustrates the splitting as a function of B for Mn^{++}, which has a $^6S_{5/2}$ ground state. The magnetic dipole selection rule $\Delta M_J = \pm 1$ allows transitions between adjacent sub-levels only, and Equation 7.1 becomes

$$h\nu_0 = g_s\mu_B B \approx 2\mu_B B$$

With $B = 1$ weber m^{-2} the transitions fall at about 1 cm wavelength.

When second-order effects are taken into account, Equation 7.3 no longer describes the energy levels exactly. Although g is indeed very close to the free electron value g_s in the case illustrated above, where $L = 0$, and in many molecular radicals, the spin-orbit coupling cannot be entirely neglected in most paramagnetic crystals. Fig. 7.5 shows the energy levels of the lowest orbital state

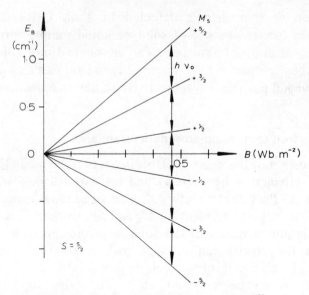

Fig. 7.4   Electron spin resonance: energy levels of Mn^{++} as a function of
field. At a given frequency ν_0, transitions between adjacent levels all occur at
the same field.

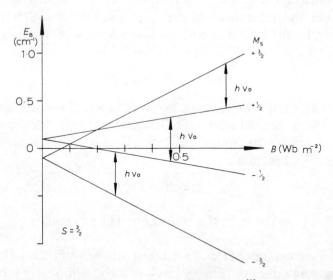

Fig. 7.5 Electron spin resonance: energy levels of Cr^{+++} as a function of field.
The levels $M_s = \pm\frac{1}{2}$ are slightly separated from $M_s = \pm 3/2$ by the crystalline
field. The four M_s levels are therefore not equally separated in the magnetic
field, and transitions between adjacent levels at a given frequency ν_0 occur at
three different values of B.

of Cr^{+++}, which has a $^4F_{3/2}$ ground state, as a function of B, for a chrome alum crystal. The spin degeneracy is partly removed by the crystal field, and the magnetic sub-levels are no longer equally spaced. Resonances for a fixed frequency ν_0 are observed at three different field strengths, as shown. In general, the number and positions of the resonances depend not only on the type of ion but also on the type of crystal – and indeed on the orientation of the magnetic field relative to the crystal axis. Further consideration shows that the *widths* of the resonances also depend on the interactions of the paramagnetic ions with each other and with other atoms. ESR spectroscopy therefore gives much useful information on solid state physics. For a fuller account, including the application of ESR to free radicals, which has not been touched on here, a reference such as [1] or [2] should be consulted.

If the paramagnetic ion also has a nuclear spin, each of the M_S levels is further split by hyperfine structure (Fig. 7.3). If the hfs can be resolved, the nuclear spin and hyperfine interaction constant can be determined from the number and separation of the components. The hfs patterns and the information deduced from them are essentially the same for paramagnetic ions as for free atoms and will be discussed in Section 7.9.

7.7 Nuclear magnetic resonance

When a magnetic field is applied to a diamagnetic substance the dominant interaction is that of the nuclear magnetic moment μ_I with the external field. Compared with paramagnetic resonance, the nuclear magnetic resonance frequencies are down by a factor of order μ_I/μ_B, $\sim 10^{-3}$, and the transition probabilities by the square of this, or about 10^{-6}, in addition to the ν^2 factor for energy absorption. The basic principles are, however, similar. A given level is split by the external magnetic field into $2I + 1$ sub-levels whose energy is given by $E_B = -\mu_I.B$. Using Equation 2.25,

$$E_B = - g_I \mu_N B M_I \tag{7.4}$$

g_I, the nuclear g-factor is of order unity, and μ_N, the nuclear magneton, is $1/1836\ \mu_B$ (see Section 2.13). Fig. 7.6 illustrates the splitting for $I = 3/2$. The selection rule for magnetic dipole radiation is $\Delta M_I = \pm 1$; M_I changes because the radiation interacts directly with the nuclear moment. The frequency for transitions at field B is therefore $h\nu_0 = g_I \mu_N B$. For $B = 1$ weber m^{-2} (10 kilogauss) the resonant frequency ν_0 is a few MHz.

The most obvious application of NMR is to the determination of nuclear g-factors and hence, if I is known, nuclear magnetic moments. As in ESR the exact position and the width of a resonance are in general influenced by the presence of other particles. In particular, the diamagnetic screening of the external field by the paired electrons, which is of order $1 : 10^5$ or 10^6, depends on the particular chemical compound and can be used in

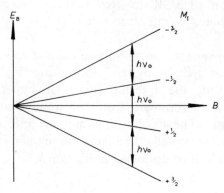

Fig. 7.6 Nuclear magnetic resonance: energy levels for $I = 3/2$ as a function of field. The levels are equally spaced, and for a given frequency ν_0 all transitions $\Delta M_I = \pm 1$ occur at the same field.

studying the chemical bonding. The width of the resonance is determined by the interactions of the nuclear spins with other nuclear spins and with the lattice.

To the order of accuracy of diamagnetic shifts and line widths – say $1 : 10^5$ – it is evidently possible to use a nucleus of known g_I to measure the strength of an unknown magnetic field, and this does indeed constitute an important application of NMR. If a sample of hydrogenous material (water or paraffin wax) is inserted, the field strength can be found from the resonant frequency ν_0 and the known magnetic moment of the proton by the relation $B = 2 \cdot 3487 \times 10^{-8} \ \nu_0$ weber m^{-2}, provided the field is greater than about 10^{-2} weber m^{-2}, or 100 gauss. For fields lower than this the resonant frequency and the signal strength are very low, and the field is better measured by ESR. In this case the hydrazyl free radical is used to give a sharp resonance, and B is found from $B = 3 \cdot 566 \times 10^{-11} \ \nu_0$ weber m^{-2}.

Electric quadrupole interactions may also be important for nuclei having $I \geqslant 1$, for the nuclear quadrupole moment can interact with the inhomogeneous electric field of the crystal when there is no external field. Transitions between the resulting hyperfine levels fall in the rf region. As with microwave experiments, it is the product eQq that is obtained from the splitting, and this is generally used to evaluate the crystal field gradient q for nuclei of known Q.

7.8   Atomic beam magnetic resonance

Atomic beam magnetic resonance experiments measure directly the separations between the hyperfine levels of free atoms that can be obtained from optical spectroscopy only as very small differences in large quantities. By contrast to ESR experiments they give information on interactions within atoms rather than on interactions of atoms with their neighbours. Since the density of atoms in a vapour or beam is many orders of magnitude lower than the density in a solid or liquid, it is impracticable to detect transitions by absorption of energy. Atomic beam experiments exploit the fact that an atom with a magnetic moment passing through an inhomogeneous magnetic field is deflected in the direction of the field gradient by an amount that depends on the orientation of its magnetic moment. Transitions affecting the latter therefore vary the deflection.

The basic experimental arrangement is shown in Fig. 7.7. The force in the z-direction acting on an atom with magnetic moment μ_J in an inhomogeneous field B is given by

$$F_z = \mu_z \frac{\partial B}{\partial z}$$

where μ_z is the component of μ_J in the field direction. In Fig. 7.7 the field in the three regions A, B and C is in the same direction. A and B are of equal length and have equal and opposite field gradients, while C is a region of uniform field. Any atom leaving the source at a slight angle to the axis which is deflected up by the A-field will be deflected down an equal amount by the B-field,

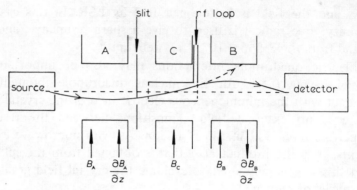

Fig. 7.7 Schematic arrangement for atomic beam magnetic resonance. The diagram is not to scale; distances perpendicular to the beam axis are greatly exaggerated.

finishing up on the axis again. But if μ_z is changed while the atom is in the C-field region, the B-field will no longer bring it back to the detector (dotted line path), and so there will be a change in signal. Since the field *direction* (as opposed to its gradient) is the

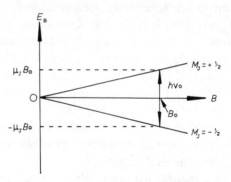

Fig. 7.8 Resonance condition for an atom having $J = \frac{1}{2}$.

same in all three regions, the space quantization, which determines μ_z, does not change unless some other field is applied. This other field takes the form of an rf field in the C-region. If we take the simplest possible case of $J = \frac{1}{2}$, μ_z has the two possible values $\pm\mu_J$ as shown in Fig. 7.8. At the resonance frequency given by

$$h\nu_0 = 2\mu_J B_0 = 2(\tfrac{1}{2}g_J\mu_B)B_0 \, g_J\mu_B B_0$$

an atom in either state may undergo a magnetic dipole transition to the other state and be lost from the beam when it gets into the

B-field. One can see that, in contrast to the techniques measuring energy absorption, transitions either way will contribute to the signal: all that matters is that the magnetic moment should be reversed so as to 'flop-out' the atom. The experiment is in fact frequently run the other way round, with the A- and B-field gradients in the same direction, so that transitions are detected by an increase in signal – the 'flop-in' method.

The source of atoms is usually an oven of some sort, producing atoms in the ground or a very low-lying excited state (< 1 eV), but

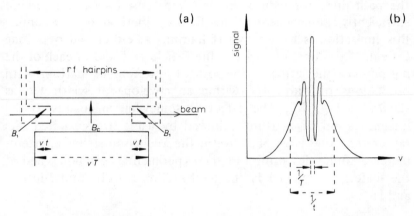

Fig. 7.9 (a) 'Ramsey hairpin' arrangement of rf field and (b) its effect on the signal. (a) is a view from the top in the plane of the beam. The dashed parts of each hairpin are above this plane. B_c is horizontal and B_1 is vertical (perpendicular to the plane of the page). In (b) the signal is an interference pattern (width of central maximum $1/T$) superposed on a Lorentzian of width $1/t$.

in some cases more highly excited states have been produced by discharge techniques. The lifetime of an excited state must be of the order of a millisecond for the atom to get through the apparatus before it decays, which effectively limits the method to metastable states. The detector takes different forms: for a non-condensable gas it may be simply a pressure measuring device; atoms such as the alkalis may be ionized at the surface of a hot filament and detected from the ionization current; other species may be ionized by electron bombardment and passed through a small mass spectrometer to sort out the beam from the background, which, of course, is also ionized. Finally, radioactive atoms are detected simply by condensing them on a bit of foil and then measuring the radioactivity.

The rf loop in the C-field is arranged to produce a magnetic component B_1 perpendicular to B_C, and it is B_1 that actually induces the transitions. The fundamental width of the resonance is determined by the apparent duration of the radiation pulse – that is, by the time T that the atom spends in the rf field – according to the relation $\delta\nu \approx 1/T$. $\delta\nu$ is just the band-width, or frequency spread, associated with a pulse lasting time T. For a velocity of about 10^5 cm sec^{-1} and a 10 cm path, $\delta\nu \approx 10$ kHz. Increasing the path does not reduce this much because B_C cannot be kept sufficiently homogeneous. The Ramsey method of overcoming this limitation is to use two rf hairpins at either end of a long C-field. If the time of flight in the C-field is T and in each of the hairpins t, the effect is a sort of interference pattern with oscillations of width $1/T$ within an envelope of width $1/t$, as shown in Fig. 7.9. The effect is somewhat analogous to the 'magnification of aperture' achieved by using two separated slits rather than one large objective in the Michelson stellar interferometer. A fuller description of the experimental techniques and of the material contained in the next section is to be found in [4] and [5].

7.9 Types of transition in atomic beam magnetic resonance

It is evident from Fig. 7.3 and the associated discussion that the energy levels, and hence the possible transitions between them, depend on the values of J and I and on the strength of the external field B in the C-region relative to the internal field B_J. B_J must be of order 100 weber m^{-2} to account for the observed hfs in zero field. The general form of the energy level diagram depends on whether B is weak enough that

$$\mu_B B \ll \mu_I B_J \tag{7.5}$$

or strong enough that this inequality is reversed. We shall illustrate this here for an atom having $J = \frac{1}{2}$, $I = 3/2$. This is a more general case than it might seem, because most atomic beam work has in fact been done on atoms with a $^2S_{1/2}$ or $2P_{1/2}$ ground state (for example, hydrogen, the alkalis, and the aluminium and copper groups), and quite a number of these have stable isotopes of spin $3/2$ (B^{11}, Na^{23}, K^{39}, Cu^{63}, Ga^{69}, Rb^{87}, Au^{197}).

Fig. 7.10 shows for this case the transition from the weak field to the strong field grouping as B is increased. In zero field the splitting of the two hyperfine levels $F = 1$ and $F = 2$ is given by Equation 2.26 as $\Delta E_{IJ} = 2A$, where A, the hfs constant, is proportional to B_J. In a weak field the hyperfine levels retain their

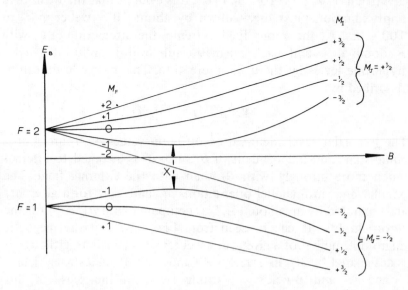

Fig. 7.10 Hyperfine energy levels as a function of field for $J = \frac{1}{2}$, $I = 3/2$. X marks the field – independent transition (see text).

identity (F remains a good quantum number) and are split into $2F + 1$ sub-levels defined by the magnetic quantum number M_F. Each sub-level is shifted by an amount

$$E_B = \mu_F \cdot B = g_F \mu_B F \cdot B = g_F \mu_B M_F B \qquad (7.6)$$

where μ_F is the magnetic moment associated with the total angular momentum $F\hbar$, and g_F, the hfs g-factor, is a function of I, J and F analogous to the Landé g_J-factor of Equations 2.17 and 2.18. To a sufficient approximation

$$g_F = g_J \frac{F^{\star 2} + J^{\star 2} - I^{\star 2}}{2F^{\star 2}}$$

which for $J = \frac{1}{2}$ simplifies to

$$g_F = + \frac{g_J}{2I + 1}, \qquad - \frac{g_J}{2I + 1} \tag{7.7}$$

for $F = I + \frac{1}{2}$ and $F = I - \frac{1}{2}$ respectively. The spacing of both sets of sub-levels is therefore the same – $g_J \mu_B B/4$ for $I = \frac{3}{2}$ – but the order of the M_F values is reversed. If the restriction (Equation 7.5) is written as $\mu_B B \ll (10^{-3} \mu_B) 10^2$, it is evident that the weak field approximation must break down by about 10^{-2} weber m^{-2}, or 100 gauss. At the strong field extreme, the interaction of μ_J with B dominates, and each magnetic sub-level is split into $2I + 1$ hyperfine levels by the $\mu_I B_J$ interaction. The energy levels can be described by

$$E_B = g_J \mu_B B M_J + A M_I M_J \tag{7.8}$$

The hyperfine levels denoted by different values of M_I are almost equidistant and independent of B, since μ_I is assumed to interact much more strongly with B_J than with the external field. The parallel and anti-parallel orientations M_I and $-M_I$ for a given M_J are separated by an amount $A\,I$, as compared with $A(I + \frac{1}{2})$ for the zero-field hfs. It can be seen from Fig. 7.10 that the magnetic quantum number of a given sub-level remains constant, although it is designated by M_F in a weak field and $M_J + M_I$ in a strong field.

The A- and B-fields are strong fields in the sense of this discussion, so that transitions, to be detectable, must occur between two sub-levels that correspond to different M_J on the right-hand side of Fig. 7.10. In weak C-fields the possible transitions obeying the selection rules $\Delta F = \pm 1$, $\Delta M_F = 0, \pm 1$ are as shown in Fig. 7.11. The double lines in this figure are coincident when the spacing in the two sets of sub-levels is exactly equal. The dotted lines correspond to $\Delta M_F = 0$ (σ components) and the full lines to $\Delta M_F = \pm 1$ (π components). σ and π refer to the orientation of the *electric* vector of the rf field relative to the C-field, whereas magnetic dipole transitions are induced by the magnetic vector, which is perpendicular to this. σ transitions therefore require B_1 parallel to B_C, and in the more usual case of B_1 perpendicular to B_C only the π transitions are induced (see Fig. 7.12 and accompanying explanation). The value of I can be determined either from the number of components or from their separation by way of g_F and Equation 7.7. The centre of gravity of the pattern, which coincides with the transition $\Delta F = \pm 1$, $\Delta M_F = 0$ gives the zero-field hfs splitting, and this can also be obtained by measuring the frequency of any of the components as a function of B and extrapolating back to zero field. For these 'weak-field transitions' the C-field may be 10^{-4} weber m^{-2} or less. In strong fields the selection rules are $\Delta M_J = \pm 1$, $\Delta M_I = 0$. If the M_I sub-levels were exactly equally spaced, the resonances would also be equally spaced, giving four lines separated by intervals equal to A, but the small direct interaction between μ_I and the

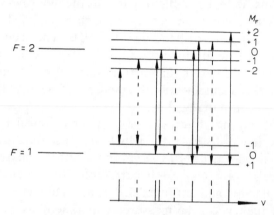

Fig. 7.11 Weak field transitions for $J = \frac{1}{2}, I = 3/2$. Transitions ($\Delta M_F = \pm 1$) are denoted by solid lines and transitions ($\Delta M_F = 0$) by dashed lines. The double lines co-incide when the spacing in the two sets of levels is equal; all components are then equally spaced at intervals given by $h\nu = g_F \mu_B B$.

external field causes small shifts, from which μ_I can be determined. Finally, there are certain transitions, such as the one marked X on Fig. 7.10 ($F = 2, M_F = -1 \leftrightarrow F = 1, M_F = -1$) for which the frequency passes through a minimum as a function of B, so that over a small range it is independent of B. For $J > \frac{1}{2}$, there is usually more than one such transition. These field-independent transitions allow particularly accurate measurements to be made because they are independent of inhomogeneities in B as well as of its absolute magnitude.

In addition to I, μ_I and A, atomic beam experiments can also be used to determine nuclear electric quadrupole moments for nuclei with $I \geqslant 1$, provided that the electronic angular momentum $J > \frac{1}{2}$ (for $J = \frac{1}{2}$ the electric field is spherically symmetric at the nucleus). The quadrupole interaction shows up as a shift in the resonant frequencies and takes the usual form eqQ, where q represents $\partial^2 V/\partial z^2$ as in Section 7.4.2. q can be calculated more reliably for a free atom than for a molecule or solid, so that it is possible to obtain Q itself with fair accuracy.

Atomic beam magnetic resonance has found an important practical application in the atomic clock, now the established frequency standard. The 'clock' is an atomic beam of Cs^{133}, for which $I = 7/2$ and J in the ground state is $\frac{1}{2}$. The resonance frequency for transitions between the two hyperfine states $F = 4$ and $F = 3$ falls in a convenient part of the microwave region ($\lambda \sim 3$ cm). The second, or rather its reciprocal, is defined by taking this frequency to be 9192·631 770 MHz. The accuracy with which this can be determined is about $1 : 10^{12}$, which is several orders of magnitude better than astronomical time standards and exceeds even the accuracy of the standard of length, the wavelength of the krypton orange line.

The hfs of atomic hydrogen also deserves special mention because it is the only case for which very accurate theoretical calculations can be made. In the ground electronic state the two hyperfine levels $F = 1$ and $F = 0$ correspond to the electron and proton spins parallel and antiparallel. The accuracy with which the transition frequency can be measured experimentally ($\sim 1 : 10^7$) is considerably greater than that with which it can be calculated, and the elimination of discrepancies of order $1 : 10^5$ is a good test of the theories of quantum electrodynamics. The transition falls at 1420 MHz in H^1 and is perhaps better known by its wavelength, for the 21 cm line radiated by interstellar hydrogen is of the utmost importance in astrophysics.

7.10 Molecular beam magnetic resonance

In this context, molecular beams are to be understood as comprising molecules and atoms with $J = 0$ in the ground state – that is, atoms in $^1S_0$ states and most molecules. The dominant magnetic interaction is between the nuclear moment μ_I and the external field. There are three main differences between atomic and molecular beam experiments. First, the splitting of the energy levels is as shown in connection with NMR in Fig. 7.6 and as given by Equation 7.4:

$$E_B = - \mu_I . B = - g_I\mu_N BM_I$$

The transition frequencies, given by

$$h\nu = g_I \mu_N B$$

i.e., $\quad \nu \sim 10\,\text{MHz}$ for $B = 1$ weber m^{-2}

fall well in the rf region even at strong fields. Secondly, transitions are induced by the direct interaction of μ_I with the applied rf field B_1, so that the transition probabilities are weaker by a factor $(\mu_N/\mu_B)^2 \sim 10^{-6}$ than in the atomic case and are governed by the selection rule $\Delta M_I = \pm 1$. Thirdly, the beam deflections are only 10^{-3} as large for a given field gradient. For $\partial B/\partial z \sim 10^3$ weber m^{-2} per m (10^5 gauss per cm) over a $\frac{1}{2}$m path, the deflection is only about $\frac{1}{2}$ mm. Consequently, the whole apparatus has to be much longer, 4 metres being a typical length.

The information derived from the resonances is similar to that obtained from NMR: the basic spacing of the sub-levels gives the nuclear g-factor, and deviations from equal spacing give the quadrupole coupling. Molecular rotation and diamagnetic effects must also be taken into account in interpreting the results.

7.11 Applications of tunable lasers to atomic beams

The recent development of tunable dye lasers offers considerable scope for widening the range of atomic beam experiments. In particular, the structure of an excited state can be studied. The conventional magnetic resonance method does not work with an excited state, unless metastable, even if an appreciable population can be attained, because during the radiative lifetime of about 10^{-8} sec the atom travels only a fraction of a millimetre. If, however, the atom has its ground state magnetic moment changed by the excitation – re-emission process, its final deflection will be changed just as in the magnetic resonance technique. The intensity of radiation from a laser can easily be made high enough to pump atoms out of any particular sub-level of the ground state in this way. If, therefore, a tunable dye laser is directed at the beam in the region between the inhomogeneous magnetic fields and is scanned through the wavelength range covered by all the hyperfine components of an optical transition from the ground state, the signal will change whenever the laser frequency coincides with that of one of the components. This is a magnetic deflection but not a magnetic resonance experiment; the various sub-levels of the upper state are excited one by one directly from the sub-levels of the lower state, and there is no rf field present to induce transitions directly between the sub-levels. The change in signal is due to those atoms that return (by spontaneous radiation) to a sub-level different from the one they began in.

A second way of using the dye laser replaces magnetic deflection by collisional deflection. An atom absorbing a photon from a light beam perpendicular to its line of flight acquires transverse momentum $h\nu/c$. When the atom decays to the ground state the photon is emitted in an arbitrary direction, leaving the transverse momentum unchanged on average. The atom therefore suffers a net deflection, and if the atomic beam is sufficiently highly collimated this is detected as a drop in signal. For example, the transverse velocity acquired by a sodium atom is about 3 cm sec^{-1} per collision; with a collimation of order $1 : 10^4$ (20 μm slits, say, separated by 20 cm) it is just possible to detect deflection from a single photon absorption. As in the first variant, the deflection changes only when the laser frequency corresponds to one of the hyperfine components of the optical transition. There is one important difference, however, in that the collision method can be used for atoms with a ground state of $J = 0$ since no magnetic deflection is involved. Indeed, it is actually more efficient in this case because each atom must necessarily return to the state from which it started and so can undergo a large number of collisions during its passage through the laser beam.

7.12 Double resonance experiments

In this technique rf transitions between hyperfine or magnetic sub-levels of an excited state are detected from changes in polarization of a resonance line in the optical region of the spectrum – that is, a line connecting the excited state to the ground state. The principle is best explained by a specific example, for which we take the first experiment of this type ever performed, that by Brossel and Kastler on mercury. The ground state of mercury is $6s^2 \, {}^1S_0$. This state has no hfs and no Zeeman splitting in a magnetic field because $J = 0$. The strong intercombination line at 2537 Å connects the ground state to the excited $6s6p \, {}^3P_1$ level, for which $J = 1$ and $M_J = +1, 0, -1$. In a magnetic field B_0 the 2537 Å line has three components according to the selection rules for electric dipole radiation: one π and two σ components corresponding respectively to $\Delta M_J = 0$ and $\Delta M_J = \pm 1$. The relation between ΔM and the π, σ components in this case (electric dipole transition) is the other way round from that in the atomic beam magnetic resonance (magnetic dipole transition) discussed in the last section. It seems worth attempting to resolve the probable confusion over this point. By the usual convention π

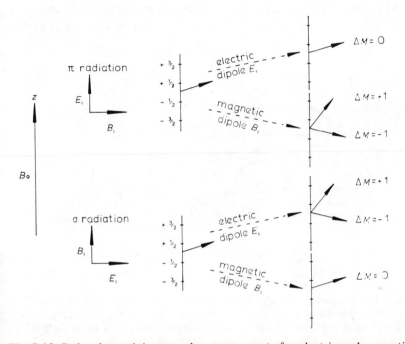

Fig. 7.12 Rules determining π and σ components for electric and magnetic dipole transitions. M can be changed only by the component of field perpendicular to B_0 - that is, by B_1 for π-radiation and E_1 for σ-radiation. The diagram shows schematically the transitions for an atom having J (or F) = 3/2, initially in the state $M = +\frac{1}{2}$.

means that the *electric* vector of the radiation is parallel to B_0 and σ that it is perpendicular to B_0, whatever the type of transition. In Fig. 7.12 the external field B_0 is in the z-direction, and E_1 and B_1 are the electric and magnetic vectors of the radiation emitted or absorbed. The top line, where E_1 is in the z-direction, therefore represents π radiation and the bottom line σ radiation. The corresponding electric dipole and magnetic dipole transitions are shown schematically for the case J (or F) = 3/2, ΔJ (or ΔF) = 0.

The selection rules for ΔM follow from the fact that M can change only through interaction with a vector perpendicular to B_0 – that is, B_1 in the π case and E_1 in the σ case.

In the double resonance experiment mercury vapour in a magnetic field B_0 is illuminated with π radiation from a mercury lamp followed by a polarizer, as shown in Fig. 7.13. This excites only the $M_J = 0$ sub-level of the excited state (Fig. 7.14), and only π radiation can be re-emitted. A detector in the position shown in Fig. 7.13 receives no signal because the direction of propagation

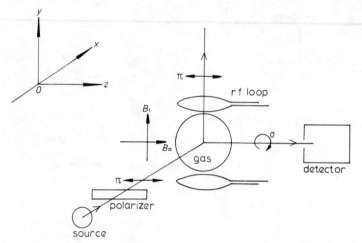

Fig. 7.13 Schematic arrangement for double resonance experiment. Resonance radiation plane – polarized in the z-direction is incident in the x-direction. Re-emission in the z-direction is possible only if the rf field B_1 induces a transition as shown in Fig. 7.14.

of an electromagnetic wave must be perpendicular to the direction of polarization. Now if an rf field of the 'resonance frequency' given by $h\nu = g_J \mu_B B_0$ is applied with B_1 perpendicular to B_0, magnetic dipole transitions to $M_J = \pm 1$ can be induced. The 2537 Å 'resonance line' will then contain σ components (Fig. 7.14) which *can* travel in the direction of the detector. The rf resonance is therefore detected by the optical resonance – hence the name double resonance – as the appearance of a signal. In practice the field B_0 is usually varied for a fixed rf setting, and the resonance is detected by the change in the difference between the π and σ light intensities. Several different arrangements of geometry and polarization are in fact possible.

Although the experiment has been described with reference to the Zeeman sub-levels of the *fine* structure, it can equally well be applied to the Zeeman sub-levels of the *hyperfine* structure shown in Figs. 7.10 and 7.11. Magnetic dipole and electric quadrupole hyperfine interactions have both been measured by this method. It has the advantage over atomic beam magnetic resonance that it can be applied to the structure of excited states so as to determine nuclear spins and quadrupole moments of atoms having no structure in the ground state. However, the intrinsic accuracy is not as great as for measurements made on the ground state, partly because the population of an excited state is relatively small,

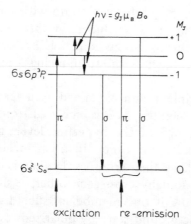

Fig. 7.14 Transitions in double resonance experiment in mercury.

resulting in fewer transitions, and partly because the sharpness of the resonance is limited by the natural lifetime of the excited state. A transition between two levels of lifetime 10^{-8} sec has a natural half-value width of about 30 MHz, or 10^{-3} cm^{-1}. This is still a great improvement on the line widths normally obtainable in optical spectroscopy, where Doppler broadening is perhaps 10 times larger than this. The width of the resonance may in fact be used to measure the lifetime of the excited state provided that suitable corrections can be made for the power broadening produced by high rf field intensity and for a phenomenon known as 'coherence narrowing'. The effect of the latter is to reduce the natural width when the gas pressure is high enough for more than one atom to interact with one photon. A full discussion of this effect, as well as the other aspects and applications of the double resonance method, is to be found in [7].

It is worth remarking that the double resonance and atomic beam magnetic resonance methods have been combined in order to study excited states in atomic beams. A change of M_J while the atom is in the C-field can be achieved by exciting a higher state and inducing an rf transition in this excited state. The atom then decays to a different sub-level of the ground state and can be detected by a change in deflection in the usual way.

7.13   Level crossing experiments

This section has nothing whatever to do with radio-frequency or microwave radiation. It is included in this chapter because it describes another means of investigating hfs energy levels in a magnetic field, although without inducing transitions between them.

Again, the principle is best explained by a specific example. In Fig. 7.10 it can be seen that as B is increased from the weak to the strong field limits, one of the hyperfine levels moves over from one group of levels to the other. In a case where $J > \frac{1}{2}$, two or more sub-levels change their group as the field is increased, so that one or more pairs actually cross each other, as shown in Fig. 7.15 for $J = 1$, $I = \frac{1}{2}$. Now if one member of such a pair is selectively populated from the ground state by polarized resonance radiation in an arbitrary magnetic field, the re-emitted light will have the same polarization. But if B has the critical value B_0 the two sub-levels are degenerate, and one can think of the emitted light as coming also from the second sub-level. There is then a partial change in polarization because of the different magnetic quantum number. This is a very crude way of looking at what is properly regarded as a superposition of two quantum states of the atom, which at the particular field B_0 are degenerate and radiate coherently. With larger values of J and I there is more than one such level-crossing. Measurements of the critical fields lead to determinations of the hfs constants.

The width of the resonance curve (polarization change as a function of B) gives the lifetime of the excited state. This can be understood, again very crudely, by considering each of the levels in Fig. 7.15 to have a finite width proportional to $1/\tau$. The two crossing sub-levels then overlap at a finite distance each side of B_0, so that they can be considered as mixed over a finite range of B-values, proportional to $1/\tau$. This method of determining lifetimes is one of those in current use, although it is more often used in the zero-field level crossing technique otherwise known as Hanle effect. This will be further considered in Chapter 10 in the discussion on lifetime measurements.

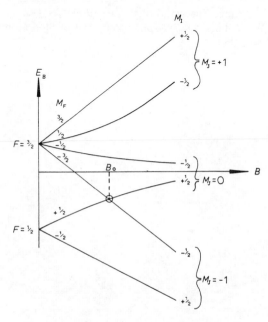

Fig. 7.15 Level crossing for $J = 1, I = \frac{1}{2}$.

The accuracy with which the position of a crossing may be measured is limited ultimately by the natural width $1/\tau$ of the resonance, and this may be a significant limitation in the case of very close components or complex structures. By using a pulsed light source for the exciting radiation and observing the re-emitted radiation after a finite delay T, where $T > \tau$, it is possible to reduce the width of the resonance below the natural width. In effect, one observes a signal only from those atoms that have survived longer than average in the excited state. A pulsed tunable dye laser is an excellent light source for this type of experiment because of its high intensity and short pulse duration.

References

1. Ingram, D. J. E. 'Spectroscopy at Radio and Microwave Frequencies', Butterworths, 1967
2. Walker, S. and Straw, H. 'Spectroscopy', Vol. 1, Chapman and Hall, 1961
3. Whiffen, D. H. 'Spectroscopy', Longmans, 1966
4. Ramsey, N. F. 'Molecular Beams', Oxford, 1956
5. Kopfermann, H. 'Nuclear Moments' (Trans. Schneider, E.E.), Academic Press, 1958
6. Kuhn, H. G. 'Atomic Spectra', Longmans, 1969
7. Series, G. W. Radiofrequency spectroscopy of excited atoms, *Rep. Prog. Phys.* **22**, 280, 1959

CHAPTER EIGHT

Width and shape of spectral lines

8.1 Line profiles

In discussing spectral line widths so far, we have been concerned only with instrumental widths – that is, the way in which a dispersing device spreads out an ideal monochromatic 'line'. No such line can in fact exist. Any atomic or molecular transition is associated with a finite spread of energy and hence of frequency. We shall now transfer the ideal behaviour from the source to the spectrograph and consider the line shape produced by a real light source and a spectrograph of infinite resolving power.

Three different processes may contribute to the finite width of a spectral line: natural broadening, Doppler broadening, and interactions with neighbouring particles. In the solid state this last may take many different forms but if the discussion is limited to free atoms and molecules the interactions with other particles may be treated under the heading of pressure broadening. Even for free atoms a full treatment of pressure broadening is beyond the scope of this book, and the survey in this chapter must necessarily be qualitative rather than quantitative.

Line shapes may be studied experimentally either in emission or in absorption. In either case, let us assume that we scan the line with a spectrometer of truly negligible instrumental width, giving an output signal I_ν proportional to the incident intensity. Then an emission line of central frequency ν_0 will have some such form as that in Fig. 8.1. Similarly, an absorption line seen against a background continuum of intensity I_0 (assumed constant over the line profile) will appear as in Fig. 8.2.

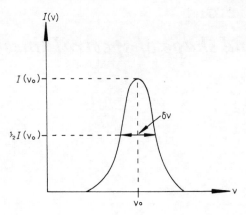

Fig. 8.1 Emission line profile. The half-value width, $\delta\nu$, is the full width at half the peak intensity.

The shapes of these curves are determined primarily by the frequency dependence of the emission and absorption coefficients of the source gas, j_ν and k_ν respectively, but they can also be affected by the absorption and re-emission of radiation from the deeper layers on its way out. In this chapter we shall be concerned only with the intrinsic line profiles represented by j_ν and k_ν. This means for the emission line assuming all the emitted radiation to reach the spectrometer without re-absorption: that is, the source is 'optically thin'. Then I_ν is simply proportional to the emissivity j_ν, and Fig. 8.1 may be regarded as a plot of j_ν. For the absorption line, k_ν is found from Fig. 8.2 as follows. From the definition of absorption coefficient as the fractional decrease in intensity per

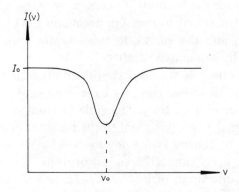

Fig. 8.2 Absorption line profile.

unit path length (Section 9.5), we have $-dI_\nu = I_0 k_\nu \, dl$. For a homogeneous layer of thickness l this integrates to give

$$I_\nu = I_0 e^{-k_\nu l} \tag{8.1}$$

k_ν is therefore found from Fig. 8.2 by calculating

$$k_\nu = 1/l \, \ln(I_0/I_\nu) \tag{8.2}$$

at each frequency. This holds good whatever the depth of the absorbing layer. However, in the case of an optically thin layer, $k_\nu l \ll 1$, and Equation 8.1 can be written as

$$I_\nu = I_0(1 - k_\nu l)$$

or

$$k_\nu = \frac{1}{l} \frac{I_0 - I_\nu}{I_0} \tag{8.3}$$

so that the profile of k_ν is identical with that of the dip in Fig. 8.2.

The ratio j_ν/k_ν is known as the source function. For a given transition in a given environment it can normally be taken as constant over the line profile. Hence with an appropriate change of ordinate, Fig. 8.1 may equally well be taken as a plot of k_ν against frequency.

The importance of any line broadening process is usually measured by its half-value width, which is the full width $\delta\nu$ of the line at half its peak intensity (see Fig. 8.1). The half-value width may of course equally well be expressed as $\delta\lambda$ or $\delta\sigma$ if j and k are plotted as functions of wavelength or wavenumber. However, the complete line shape depends on the particular broadening process or processes, and, as we shall see, the half-value width does not necessarily tell one much about the wings of the line. It should be noted that the term 'half-width', which is frequently met in the literature, is unfortunately somewhat ambiguous: while it usually has the meaning of half-value width, it is occasionally taken to be half the width of the line, or *half* the width at half intensity.

Fig. 8.3 indicates how the profiles of emission and absorption lines change when the depth of the emitting or absorbing layer is increased beyond the 'optically thin' limit – i.e., $k_\nu l \ll 1$. In absorption it is obvious that as $k_\nu l \to \infty$ a saturation point is approached at which no light is transmitted at the central frequency ν_0. In emission the light from the centre of the line profile (where the absorption coefficient is highest) is preferentially absorbed on its way out, and so the emission line peak also approaches a saturation value. These effects are discussed more fully in Sections 9.8–9.10, but they are mentioned here because of their effect on the apparent width of the line. In investigating line broadening processes in emission it is essential either to use an optically thin source or to make appropriate corrections for self-absorption; and in absorption experiments k_ν must be found from Equation 8.2 unless the absorbing layer is thin enough to justify the use of Equation 8.3.

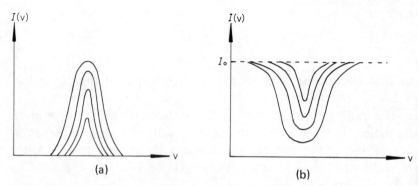

Fig. 8.3 Effect of increasing optical depth on line profile. (a) shows the effect for an emission line and (b) for an absorption line.

Whereas in normal conditions an emission line becomes broader by selective absorption as the optical depth is increased, the contrary effect occurs in a laser; the line is actually made narrower by selective amplification as it passes through the gas (Section 3.9). The contrast to broadening by self-absorption can be understood by regarding laser action as a negative absorption process, greatest at the line peak where the absorption coefficient is highest.

8.2   Natural broadening

Natural broadening may be looked at from either the classical or the quantum mechanical point of view. We take first the quantum mechanical picture, in which absorption and emission consist of transitions between two discrete energy levels. These levels cannot be infinitely narrow, because the uncertainty principle in the form

$\Delta E \cdot \Delta t \sim \hbar$ requires an energy spread $\Delta E \sim \hbar/\Delta t$ where Δt is the uncertainty in time associated with finding the atom in that particular state and is measured by the mean lifetime τ of the state. The frequency spread for the state j can thus be written

$$\delta \nu_j \approx \frac{1}{2\pi\tau_j} \qquad (8.4)$$

Whereas $\delta \nu_j$ is negligible for the ground or a metastable state $(\tau_j \to \infty)$, the upper states of allowed optical transitions have lifetimes of order 10^{-6} to 10^{-9} sec. Any spectral line starting or finishing at such a level must have a corresponding frequency spread, which is of the order of 0.1 to 100 MHz or, say, 10^{-5} to

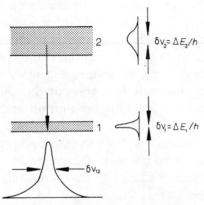

Fig. 8.4 Natural broadening of spectral line. The energy levels are smeared out according to the probability distribution shown on the right of the diagram. The half-value width of the spectral line is $\delta \nu_{12} = \delta \nu_1 + \delta \nu_2$ (see text).

10^{-2} cm^{-1}. Fig. 8.4 shows the situation when both levels are broadened. It will be seen later that the width of the line is then given by

$$\delta \nu_{12} = \delta \nu_1 + \delta \nu_2 .$$

The lifetime of an excited state, in the absence of collisions, is related to the transition probability for spontaneous emission A_{21} (see Chapter 9) by $\tau_2 = 1/A_{21}$, or, if more than one transition from level 2 is possible, by

$$\tau_2 = 1/\sum_i A_{2i}.$$

It is shown in Section 9.3 that A_{21} is proportional to ν^3. Natural line width therefore decreases rapidly in the infra-red and microwave regions, but may become appreciable in the far ultra-violet. In the visible region it is normally a couple of orders of magnitude smaller than Doppler broadening, although, because the two effects have different line shapes, it may still make a significant contribution to the wings of the line. The auto-ionizing levels mentioned in Section 2.10 constitute an important exception to the rule of small natural width: these levels have a very strong probability of making a radiationless transition into the neighbouring continuum, leading to very short lifetimes, say $\sim 10^{-13}$ sec, and correspondingly broad lines, several hundred cm^{-1} or more in width.

The shape of the broadened line depends on the shape of the energy distributions shown on the right-hand side of Fig. 8.4. This can be derived from Dirac's quantum theory of radiation, which is beyond the scope of this book. However, the classical derivation yields the same line shape by more elementary methods, and as the classical oscillator will appear again in Chapter 9 in connection with oscillator strength, absorption and dispersion it is worth considering briefly here.

The classical picture has the electron performing simple harmonic motion at a characteristic frequency but with amplitude decreasing with time because of the energy radiated away. The radiation therefore acts as the damping term in the equation of motion

$$\ddot{x} + \gamma \dot{x} + \omega_0^2 x = 0$$

where ω_0 is the characteristic angular frequency. For small damping the solution of this equation is

$$x = a_0 e^{-\gamma/2t} \cos \omega_0 t$$
$$= a(t) \cos \omega_0 t \tag{8.5}$$

But the energy of the oscillator is

$$E(t) = \frac{1}{2} m \omega_0^2 a^2(t)$$

$$\therefore \qquad E(t) = E(0)e^{-\gamma t}$$

$$-\frac{\partial E}{\partial t} = \gamma E(t) \tag{8.6a}$$

γ can be evaluated by equating the rate of loss of energy to the energy radiated by an accelerated charge according to classical electromagnetic theory:

$$-\frac{\partial E}{\partial t} = \frac{2}{3} \frac{e^2}{c^3 (4\pi\epsilon_0)} \overline{\ddot{x}^2} \qquad (8.6b)$$

From Equation 8.5

$$\ddot{x} = -\omega_0^2 a(t) \cos \omega_0 t = -\omega_0^2 x$$

Averaging over a cycle,

$$\overline{\ddot{x}^2} = \omega_0^4 \overline{x^2} = \frac{1}{2} \omega_0^4 a^2(t) = \omega_0^2 E/m \qquad (8.7)$$

$$\therefore \qquad -\frac{\partial E}{\partial t} = \frac{2}{3} \frac{e^2}{mc^3 (4\pi\epsilon_0)} \omega_0^2 E$$

Comparing this with Equation (8.6a) we get for the damping constant

$$\gamma = \frac{2}{3} \frac{e^2 \omega_0^2}{mc^3 (4\pi\epsilon_0)} = \frac{8\pi^2 e^2 \nu_0^2}{3mc^3 (4\pi\epsilon_0)} = \frac{2\pi e^2 \nu_0^2}{3\epsilon_0 mc^3} \qquad (8.8)$$

The damped oscillations have the form shown in Fig. 8.5a, with the amplitude decaying as $e^{-\gamma/2 t}$. The corresponding energy

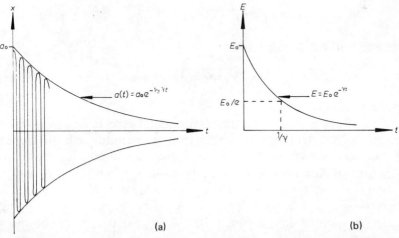

(a)   (b)

Fig. 8.5 Damped classical oscillator showing decay of (a) amplitude and (b) energy. γ is the classical damping constant. The period of the oscillations in (a) is shown much too long relative to the decay rate. In (b) the time for decay to $1/e$ of the initial value defines the classical lifetime $\tau_{class} = 1/\gamma$.

decay is shown in Fig. 8.5b. The oscillations can be thought of as having a lifetime $\tau_{class} = 1/\gamma$. Now only an infinite wave train of constant amplitude is truly monochromatic. A pulse of finite duration can be formed only by superposing waves with a spread of frequency around ω_0. This spread can be found by Fourier analysis, which effectively determines the amplitude of each frequency component $A(\omega)$ required to build up the pulse $x(t) = a_0 e^{-\gamma/2 t} \cos \omega_0 t$. $A(\omega)$ is in fact complex, and AA^* gives the intensity $I(\omega)$ as a function of frequency. The result of the Fourier transform for $t = 0$ to $t = \infty$ (see, for example, [8]) is the Lorentzian or dispersion distribution

$$I(\omega) = \frac{const}{(\omega - \omega_0)^2 + (\gamma/2)^2}$$

The accompanying term with $(\omega + \omega_0)^2$ in the denominator is negligibly small in practice. Transforming from ω to ν we have

$$I(\nu) = I_0 \frac{(\gamma/4\pi)^2}{(\nu - \nu_0)^2 + (\gamma/4\pi)^2} \qquad (8.9)$$

where I_0 is the central intensity. The form of this distribution is shown in Fig. 8.6. The frequency $\nu_{1/2}$ for half-intensity is given by

$$(\nu_{1/2} - \nu_0)^2 = (\gamma/4\pi)^2 \qquad \text{or} \qquad \nu_{1/2} - \nu_0 = \pm \gamma/4\pi$$

leading to the half-value width

$$\delta\nu_N = \gamma/2\pi = 1/2\pi\tau_{class} \qquad (8.10)$$

by definition of τ_{class}.

The effect is similar to the instrumental broadening in Fourier transform spectroscopy resulting from the finite path difference X. It was shown in Section 6.13 that a rectangular chopping function of duration X resulted in a $\sin \alpha/\alpha$ line shape of width $\delta\sigma = 1/2X$, or, taking ν and the time delay τ as the co-ordinates instead of σ and Δ, $\delta\nu = 1/2\tau_{max}$. A differently shaped chopping function resulted in a different line shape, the triangular chopping function producing a

$$\frac{\sin^2 \alpha/2}{(\alpha/2)^2}$$

line profile. An exponential chopping function applied to the same interferogram would have produced the Lorentzian line shape (Equation 8.9).

The full quantum mechanical treatment also yields a Lorentzian line profile, and if only state (2) is broadened the damping constant $\gamma_{class} = 1/\tau_{class}$ is simply replaced by $\gamma_2 = 1/\tau_2$. The corresponding frequency spread from Equation 8.10 becomes $\delta\nu_N = 1/2\pi\tau_2$, in agreement with Equation 8.4. If level (1) also has a finite lifetime τ_1, the associated frequency spread $\delta\nu_1$ will contribute to the line width. The convolution of two Lorentzians

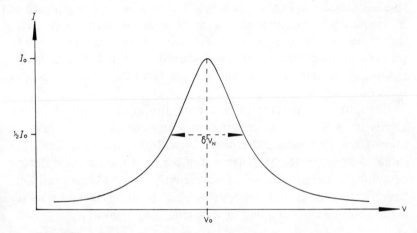

Fig. 8.6 Lorentzian line profile. The half-value width is given by

$$\delta\nu_N = \gamma/2\pi = 1/2\pi\tau_{class}$$

with damping constants γ_1 and γ_2 is just another Lorentzian with $\gamma = \gamma_1 + \gamma_2 = 1/\tau_1 + 1/\tau_2$ so the line shape remains Lorentzian but with its width increased to

$$\delta\nu_N = 1/2\pi\tau_1 + 1/2\pi\tau_2 = \delta\nu_1 + \delta\nu_2$$

One might ask finally whether the classical lifetime τ_{class} bears any relation to the quantum mechanical lifetime. Equation 8.8 implies that τ_{class} should depend only on the frequency. In the visible region it works out to about $1 \cdot 6 \times 10^{-8}$ sec. This is close to some, but by no means all, actual lifetimes. Similarly, the classical natural width $\delta\nu_N = \gamma_{class}/2\pi$, if expressed in wavelength units, becomes $\delta\lambda_N = e^2/3\epsilon_0 mc^2 = 0 \cdot 00016$ Å – constant for all wavelengths. Experimentally, the true lifetimes and half-value widths come closest to the classical values for very strong lines having *f*-values (see Chapter 9) close to unity. The transition between two states of an atom then corresponds closely with one classical oscillator. The resonance lines of the alkalis are good examples of such transitions.

8.3   Doppler broadening

Doppler broadening is a result of the well known 'Doppler effect', which is the apparent shift in wavelength of the signal from a source moving towards or away from the observer. The first of these – motion towards – causes a decrease in wavelength (or rise in frequency), and the second an increase in wavelength. With sound waves the effect, in the form of a rise and then a fall in pitch, is a familiar one in every-day life. With light waves the cosmological 'red shift' of the radiation from the receding galaxies is also well known.

Even in a stationary light source the emitting or absorbing atoms or molecules are moving, and any component of velocity away from or towards the observer will give rise to a 'red' or a 'blue' shift respectively. A large number of atoms having different velocities will emit a spread of wavelengths – i.e., a broadened line. Specifically, for an emitter approaching the observer with velocity *u* the Doppler shift in a line of wavelength λ_0 is given by

$$\lambda = \lambda_0 (1 + u/c)$$

$$\therefore \qquad - \Delta\lambda/\lambda_0 = \Delta\nu/\nu_0 = u/c \qquad (8.11)$$

If we know the proportion of atoms with a given velocity we can calculate the contribution to the spectral line at the corresponding wavelength and hence build up the line profile. If the motion is thermal – that is, the gas is in thermal equilibrium at a temperature *T* – we do know this proportion: it is given by the Maxwell distribution

$$\mathrm{d}n_u = n/(\alpha\sqrt{\pi}) \, e^{-u^2/\alpha^2} \, \mathrm{d}u \qquad (8.12)$$

where $\mathrm{d}n_u/n$ is the fraction of atoms having velocity between *u* and *u* + d*u* along any one axis (the line of sight in this case) and α is the 'most probable velocity'

$$\alpha = \sqrt{\left(\frac{2kT}{m}\right)} = \sqrt{\left(\frac{2RT}{M}\right)}$$

(m is the atomic mass, M the mass number, k Boltzmann's constant and R the universal gas constant).

Substituting u from Equation 8.11 into Equation 8.12, we have for the fraction of atoms emitting in the frequency interval v to $v + dv$

$$dn_v/n = \frac{1}{\alpha\sqrt{\pi}} e^{-c^2(\Delta v)^2/v_0^2\alpha^2} \frac{c}{v_0} dv$$

Since the intensity at v is proportional to dn_v, the line profile can be written in terms of the central intensity I_0 as

$$I_v = I_0 e^{-c^2(v_0 - v)^2/v_0^2\alpha^2} \tag{8.13}$$

This is a Gaussian distribution about the central frequency v_0 with a width determined by α. The half intensity points are the values $v_{1/2}$ for which

$$I_v = \frac{1}{2} I_0$$

i.e. $$\frac{c^2}{v_0^2\alpha^2} (v_0 - v_{1/2})^2 = \ln 2$$

The half-value width δv_D is therefore

$$\delta v_D = 2|v_0 - v_{1/2}| = 2 \frac{v_0 \alpha}{c} \sqrt{(\ln 2)}$$

or

$$\delta v_D = \frac{2v_0}{c} \sqrt{\left(\frac{2RT \ln 2}{M}\right)} = 7 \cdot 16 \times 10^{-7} v_0 \sqrt{\left(\frac{T}{M}\right)} \tag{8.14}$$

This can conveniently be expressed in the dimensionless forms:

$$\frac{\delta v_D}{v_0} = \frac{\delta \lambda_D}{\lambda_0} = \frac{\delta \sigma_D}{\sigma_0} = 7 \cdot 16 \times 10^{-7} \sqrt{\left(\frac{T}{M}\right)} \tag{8.15}$$

The Doppler distribution is often written in terms of the half-value width as

$$I = I_0 \exp(-y^2)$$

with

$$y = \frac{2(v_0 - v)}{\delta v_D} \sqrt{(\ln 2)} \tag{8.16}$$

Fig. 8.7 shows the Doppler, or Gaussian, line shape. This falls off much more rapidly in the wings than does the Lorentzian, an important point to which we shall return later.

Putting in a few figures, one finds that in the visible region at room temperature Doppler widths range from 0·06 Å (hydrogen) to 0·004 Å (mercury) – i.e., ~0·2 to 0·01 cm^{-1}. In a furnace they may be up to three times greater. In the infra-red and microwave regions the Doppler width in Å increases, but expressed in energy units (cm^{-1} or Hz) it decreases. For the purpose of measuring small energy differences, for example hyperfine structure, the longer the wavelength the better. In the visible region, where $\lambda/\delta\lambda_D$ is of order 10^5–10^6, the Doppler width is comparable to

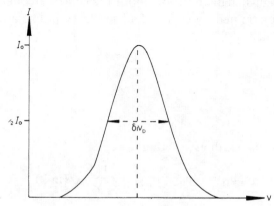

Fig. 8.7 Doppler line profile. The half-value width is given by $\delta\nu_D = 7\cdot16 \times 10^{-7}\ \nu_0\ (T/M)^{1/2}$ (see text).

the resolution limit of a large grating spectrograph. In round figures, 10^6 represents the useful limit of resolving power with a normal source, even when cooled in liquid air. The only light source that can exploit really high resolving power is the laser, in which one particular mode of oscillation of a cavity defines a narrow band within the Doppler profile, and the radiation within this narrow band is yet further concentrated in wavelength range by selective amplification.

For those branches of spectroscopy that require narrow lines – accurate measurement of wavelength, resolution of hyperfine structure or isotope shift – the limiting accuracy is usually set by the Doppler width. The best conventional general purpose source of narrow lines is the hollow cathode discharge tube (Section

3.6.4) cooled with liquid air or liquid helium and run at a current of a few mA. The Doppler width can be reduced by a further order of magnitude by using an atomic beam source, in which a system of collimated slits serves to eliminate those atoms having an appreciable velocity in the line of sight; but the greater the collimation the lower the intensity, and such sources are particularly difficult to use in emission.

An interesting new technique for velocity selection known as saturation spectroscopy (reviewed in [10]) has been made possible by the development of lasers. If a gas or vapour is irradiated by a laser operating near a resonant frequency, atoms can be excited from the ground state faster than they can return by spontaneous radiation or collisions, and the ground state becomes significantly depopulated. This is exactly the situation met in radiofrequency spectroscopy as power saturation, but in the optical region powers high enough to affect the ground state population appreciably can be attained only with lasers. If the laser is tuned to a frequency band narrow compared to the Doppler width of the resonance line and centred on ν_L, radiation can be absorbed only by those atoms whose velocity component v_x in the direction of the laser beam satisfies $(\nu_0 - \nu_L)/\nu_0 = v_x/c$, where ν_0 is the central frequency of the resonance line. The laser is said to 'burn a hole' in the population of ground state atoms centred on the velocity v_x. The width of this hole depends on the band width of the laser and on the natural and collisional broadening of the resonance line, and it can be detected by scanning through the Doppler profile with a second laser. An identical laser beam passing through the gas in the opposite direction burns a hole at velocity $-v_x$. If the lasers can be scanned over the Doppler profile, the central frequency ν_0 can be identified from the point at which one laser beam passes through the hole burned by the other, corresponding to selective absorption by atoms with $v_x = 0$. Such an experiment can be performed in practice with a single tunable dye laser together with a beam-splitter and a couple of mirrors arranged to send the two beams through the column of vapour in opposite directions. One beam burns the hole – i.e., acts as 'bleacher' –, and the other acts as probe. A detector recording the intensity of the probe registers a sharp increase in signal when $\nu_L = \nu_0$, because most of the atoms having $v_x = 0$ are removed from the ground state by the bleaching beam. If the resonance line has hyperfine structure, a signal is registered at the central frequency of each hyperfine component, even when the entire hyperfine pattern is within the Doppler profile.

Before leaving Doppler broadening it should be pointed out that observed Doppler widths are not necessarily attributable to gas kinetic motion at some temperature T. In many astrophysical and some laboratory sources there may be turbulence, or bulk motion of the gas. Usually this results in a Gaussian line profile of width greater than the thermal value, but in certain conditions the line profile may depart from the Gaussian.

8.4 Pressure broadening

8.4.1 Methods of approach

It is a well known experimental fact that as the pressure in an emitting or absorbing gas is increased the spectral lines are broadened and in many cases also shifted. Sometimes additional lines due to 'forbidden transitions' appear. These perturbations – widening, shifting and mixing of energy levels – are due to the interactions with the other particles. We shall look at them in a mainly qualitative fashion; the full treatment is beyond the scope of this book, and no exact theory is yet available that covers all relevant combinations of density, temperature and frequency.

One may classify pressure broadening effects either by the type of perturber (charged particles, neutral atoms or molecules of the same gas, neutral atoms or molecules of a different gas) or by the approximations made in treating the perturbation. This can be understood by looking at the problem in the most simple-minded way possible.

We start by assuming, as was first done by Lorentz, that the perturbation takes the form of a quenching collision, cutting off the wave train abruptly, and that the collision involves only one perturber – the so-called binary interaction approximation. There is obviously a close resemblance to natural broadening, and one would expect Fourier analysis of the truncated wave train to yield a similar sort of line profile, with a collisional damping constant γ_c replacing γ_N.

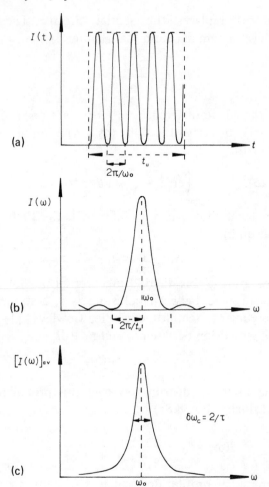

Fig. 8.8 Effect of abrupt cut-off of radiation. (a) shows the rectangular cut-off function of duration t_u, represented by the dashed line. The period of the individual oscillations of angular frequency ω_0, is much exaggerated. (b) shows the frequency spread resulting from the cut-off, of width $2\pi/t_u$. (c) shows the Lorentzian distribution obtained by averaging (b) over all possible values of t_u; τ is the mean value of t_u.

More precisely, let us suppose that the wave train from a particular atom has constant amplitude during the period t_u between collisions, when the atom is unperturbed, and zero amplitude outside this period (Fig. 8.8a). We already know the Fourier transform of this 'top hat' function: it is exactly analogous to Fraunhofer diffraction at a slit except that a temporal cut-off replaces the spatial cut-off. The intensity distribution has the familiar form (cf. Chapter 4)

$$I = I_0 \left(\frac{\sin \alpha}{\alpha} \right)^2$$

where 2α is the phase difference across the slit (as a function of diffraction angle) or across the pulse (as a function of frequency). In the present case

$$\alpha = (\omega_0 - \omega)t_u/2 \qquad \left(\text{cf. } \alpha = \frac{2\pi}{\lambda} \sin \theta \, \frac{a}{2} \text{ for a slit of width } a \right).$$

$I(\omega)$ has the form shown in Fig. 8.8b with the first minima at frequencies given by

$$\Delta \omega = \pm \, 2\pi/t_u \tag{8.17}$$

We must now average this distribution over all possible values of t_u. If the mean time between collisions is τ, a straightforward probability argument (see, for example, Cowley [1]) gives the probability of a collision between t_u and $t_u + dt_u$ as

$$dP(t_u) = 1/\tau \, e^{-t_u/\tau} \, dt_u.$$

Averaging the intensity distribution over this probability distribution we get finally (Fig. 8.8c)

$$I(\omega) = I_0 \frac{(1/\tau)^2}{(\omega_0 - \omega)^2 + (1/\tau)^2} \tag{8.18}$$

This is a Lorentzian profile identical with Equation 8.9 if γ_N is replaced by $\gamma_c \equiv 2/\tau$. The half-value width is

$$\delta \omega_c = 2/\tau \qquad \text{or} \qquad \delta \nu_c = 1/\pi\tau \tag{8.19}$$

Now this type of approach, the so-called impact approximation, can only be expected to work for discrete, separated collisions such that the duration of a perturbation t_p is small compared to the time between collisions t_u – i.e., $t_p \ll t_u$. Using Equation 8.17 the condition becomes

$$t_p \ll 1/\Delta \nu \tag{8.20}$$

where $\Delta \nu$ is a measure of the maximum frequency spread of the line. For a perturber moving in a straight line with mean velocity $\bar{v}$

and an impact parameter ρ_0 sufficiently small to define a quenching collision (we shall return to this point later) the time $\rho_0/\bar{v}$ can be taken as a measure of t_p (see Fig. 8.9). So the idea of separated collisions is tenable only over the frequency range defined by $|\nu_0 - \nu| < \bar{v}/\rho_0$ and will tend to become unrealistic in the following conditions:

(a) slow-moving perturber → small $\bar{v}$ (large t_p)
(b) high density of perturbers → small t_u (and small τ)
(c) line wings → large $\Delta\nu$
(d) long-range forces → large ρ_0 (large t_p)

From these considerations, the impact approximation may be expected to break down with increasing pressure (b), in the wings of the line first (c), for charged perturbers before neutrals (d) and for ions before electrons (a).

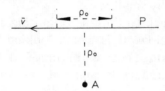

Fig. 8.9 Relation between impact parameter and duration of perturbation. A is the radiating atom and P the path of the perturber.

Let us now start from the opposite extreme, as was first done by Holtsmark, and assume the collision to be so extended that the perturbation may be taken as constant. This condition is approached if the perturbers can be regarded as almost stationary ($\bar{v} \to 0$); it is known as the quasi-static approximation (the alternative, statistical approximation, is not a good name because any theory of pressure broadening is essentially statistical). Still considering only binary interactions, we can now regard the interaction energy ΔE between a given atom/perturber pair as causing a shift $\Delta\nu_0 = \Delta E/h$ in the frequency radiated by that particular atom. A statistical average of $\Delta\nu_0$ over all such atom/perturber pairs gives the line profile from the assembly of atoms. The quasi-static approach is valid only if the shift $\Delta\nu_0$ is large compared to the broadening of the shifted line caused by the finite duration t_p of the perturbation. But we already know that a pulse of duration t is associated with a frequency spread of $1/t$ (Equation 8.17), so the validity condition becomes

$$\Delta\nu_0 \gg 1/t_p \qquad \text{or} \qquad t_p \gg 1/\Delta\nu_0 \qquad (8.21)$$

which is just the opposite of the condition for validity of the impact approximation (Equation 8.20). As before, $t_p \sim \rho/\bar{v}$. Obviously the quasi-static approximation is most likely to hold good in the wings of the line (large $\Delta\nu_0$) for slow-moving perturbers (small $\bar{v}$). At first sight it also seems to be favoured by large impact parameter ρ (distant perturber), but this is not so: as we shall see, the magnitude of the perturbation $\Delta\nu_0$ increases rapidly with *decreasing* ρ, so that the inequality

$$\rho \frac{\Delta\nu_0}{\bar{v}} \gg 1$$

is actually favoured by small ρ.

We shall come back to these two limiting approximations in a little more detail below. It is worth looking first at the possible types of perturbation, because the range of the forces involved often affects the arguments and the validity of the approximations.

8.4.2   Types of interaction

Let us assume that the perturbation shifts each energy level E_i of the atom by an amount ΔE_i which depends both on i and on the distance r between the two particles. The central frequency ν_0 of the emitted line is shifted to a frequency $\nu(r)$ given by

$$\Delta\nu(r) \equiv \nu_0 - \nu(r) = 1/h\{\Delta E_2(r) - \Delta E_1(r)\}$$

as shown in Fig. 8.10. The equation implies that kinetic energy is conserved and the 'collision' is elastic: the perturber does not induce any transitions in the emitting atom. This assumption is known as the adiabatic approximation. We shall return later to its validity.

In the quasi-static approach r is regarded as virtually fixed for any one atom-perturber pair, and the problem, as already indicated, is to calculate $\Delta E_i(r)$ and take a statistical average of $\Delta\nu(r)$ over all r. In the impact approach r is a rapidly changing function of time; ν therefore also varies with time during the

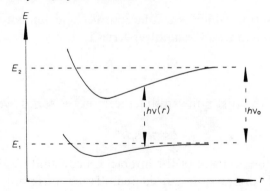

Fig. 8.10 Perturbed energy levels and shift of corresponding transition frequency.

period t_p, and at the end of this period the wave train is out of phase with the unperturbed wave train by an amount

$$\int_0^{t_p} 2\pi\Delta\nu\,dt$$

This is illustrated schematically in Fig. 8.11, where the decreased frequency during the perturbation results in a phase change of 2π. We can tie this up with the idea of an abrupt cut-off of the wave train by counting as a quenching collision any interaction that produces a total phase change large enough to destroy the coherence of the wave train.

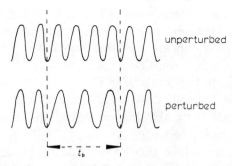

Fig. 8.11 Phase change due to frequency change during perturbation. The decreased frequency of the lower wave train during the time t_p results in a phase change of 2π.

Any interaction which is a function of r and tends to zero for large r may be expressed as a power series in $1/r$:

$$\Delta v = \sum_k C_k/r^k$$

Keeping only the first non-vanishing term in the series, we have

$$\Delta v = C_n/r^n \qquad (8.22)$$

where the value of n and of the interaction constant C_n depend on the type of interaction considered. Three different types of perturber may be distinguished.

(a) *Charged particles* An emitting atom at a distance r from an ion or electron is perturbed by an electric field $F = e/4\pi\epsilon_0 r^2$, and the interaction between the atom and the field is described by the Stark effect. For this reason, broadening by charged particles is often known as Stark broadening. It was seen in Section 2.9 that a perturbation proportional to F exists only in the case of the hydrogen atom; for all other atoms the first non-vanishing interaction is the quadratic Stark effect, proportional to F^2 and hence to $1/r^4$. For hydrogen and H-like ions, then, we have $n = 2$ and a constant C_2 determined by the linear Stark coefficients, which are readily calculable from the hydrogen wave functions. For all other atoms we have $n = 4$; the constant C_4 can in principle be calculated from the relevant wave functions but is more likely to be found from the experimental determination of the quadratic Stark coefficients in a static external electric field.

The linear Stark effect splits the energy levels symmetrically, resulting in a symmetric line pattern. The statistical averaging process effectively smears out this line pattern to give a symmetrically broadened, unshifted line. The quadratic effect, on the other hand, splits the levels asymmetrically and also shifts their centre of gravity, usually downward (see Fig. 2.15). Since the shift is usually greater for the higher states, the frequency of the transitions is usually reduced ($\Delta v < 0$; $C_4 < 0$). A line broadened by the quadratic effect therefore tends to be asymmetric and shifted to longer wavelengths. It should be noted that because of the F^2 dependence the direction of the shift is the same whether the charged particles are ions or electrons.

(b) *Resonance* Resonance interactions occur only between identical atoms and are confined to states combining with the ground state. They take the form of a dipole–dipole interaction, for which n in Equation 8.22 is 3. Since an atom in a stationary state has no permanent electric dipole moment, this form of interaction might seem at first sight strange. To understand how it comes about, consider the quasi-molecule formed by an excited atom A and a nearby identical atom B in the ground state. Because of the identity of the atoms this system is degenerate with the system A (ground-state) + B (excited). To form a non-degenerate wave function one takes a linear combination of the two degenerate functions, implying that both A and B are partly in the ground state and partly excited. (There is a close analogy to the sharing of excitation energy between the two electrons in a helium atom and also to the sharing of an electron between the two protons in the H_2^+ molecule). Now the electric dipole matrix elements evaluated between two states connected by an allowed transition do not vanish; indeed, their squares are proportional to the oscillator strength f of the transition (Section 9.6). If the perturbation $\Delta v = C_3/r^3$ is envisaged as an interaction between two dipoles of magnitude proportional to $\sqrt{f}$, it is seen that C_3 should be proportional to f and should depend on the orientation of the dipoles with respect to the inter-atomic axis – in other words the m-values – positive and negative values being equally likely. Resonance interactions should therefore give symmetrically broadened, unshifted lines. It should be emphasized that they occur strictly between individual pairs of atoms: the extension of the theory to include multiple interactions is not at all simple.

(c) *Neutral particles (van der Waals)* Van der Waals forces act between any two atoms or molecules but have a smaller range than either of the other two types we have considered: $n = 6$ in Equation 8.22. They take the form of a fixed dipole-induced dipole interaction. Classically the fixed dipole is regarded as a statistical fluctuation; quantum mechanically it is related to the expectation value of the square of the electron co-ordinate. We write it as $\sqrt{\bar{p}^2}$ in order to emphasize its root-mean-square character. The electric field F' that it produces in the neighbourhood of the perturber is approximately $\sqrt{(\bar{p}^2)}/4\pi\epsilon_0 r^3$. If α is the polarizability of the perturber, this field induces in it a dipole moment $\alpha F' \approx \alpha \sqrt{\bar{p}^2}/4\pi\epsilon_0 r^3$. The interaction energy between the induced dipole and the field is $-\alpha F'$. $F' = -\alpha \bar{p}^2/(4\pi\epsilon_0)^2 r^6$; this is always negative in sign, corresponding to an attractive force. It is generally greatest for the most highly excited levels and for the heaviest atoms, which is understandable if one takes the expectation value of the square of the atomic radius a^2 as a measure of $\bar{p}^2$. If indeed ΔE_2 is greater than ΔE_1, $\Delta\nu$ is negative (Fig. 8.10), and a red shift is to be expected. If several perturbers have to be considered, their potentials add scalarly; the van der Waals $1/r^6$ potential is the only perturbation for which this is true.

Shorter range   Obviously other terms due to higher multipoles (dipole-quadrupole, quadrupole-quadrupole, etc.) can be included, but as they involve yet higher powers of r they are often neglected by comparison with the leading term. However, the van der Waals forces have themselves such short ranges that by the time they become important, particularly if C_6 is small (as with helium, which has a very low polarizability), the atoms may be close enough for repulsive forces to matter. These are bound to become important when r is small enough for the two charge distributions to overlap. The interaction potential as a function of r must therefore go through some sort of minimum as in Fig. 8.10. The repulsion is often taken into account by using the Lennard-Jones potential $V = C_6/r^6 + C_{12}/r^{12}$ where $C_6 < 0$ and $C_{12} > 0$. The choice of the 12th power for the repulsive part is for mathematical convenience rather than physical significance. In any case, many calculations are not very sensitive to the exact power chosen. The repulsive forces can usually be taken as negligible at separations greater than two atomic radii, say 3 to 5 Å. For comparison, the average distance $\bar{d}$ between particles at N.T.P. is of order $1/N_L^{1/3}$ (where N_L is Loschmidt's number), which is about 30 Å.

It is worth making an order of magnitude calculation of the strengths of the interactions (a), (b) and (c) so as to have some idea of the effective ranges of the forces and the validity of the binary approximation in each case. For this purpose we shall arbitrarily define the effective range by $\Delta E \geqslant 10^{-24}$ joule, because this corresponds to $\Delta\sigma \geqslant 0.05$ cm^{-1}, a typical value for Doppler broadening.

Starting with the shortest range, case (c), we have

$$C_6 \sim \alpha \bar{p}^2 / (4\pi\epsilon_0)^2$$

Atomic polarizabilities in the visible region are of order 2×10^{-29} ϵ_0 m^3 (it follows from Section 9.6 that the polarizability is related to the refractive index and number density of the gas by $n - 1 = N\alpha/2\epsilon_0$). As an order of magnitude for p we can take the product of e and a typical atomic radius for a moderately excited state, say $3a_0$. Then $C_6 \sim 10^{-77}$ joule m^6, and $r_{max} \sim 15$ Å. Experimental values for C_6 are actually in the range 10^{-78} to 10^{-76} J m^6. Because of the sixth power dependence r_{max} is very insensitive to the value of C_6 or ΔE.

The next shortest range interaction is the quadratic Stark effect, for which we should expect an interaction proportional to the polarizability of the atom and the square of the field strength:

$$\Delta E \sim \frac{\alpha e^2}{(4\pi\epsilon_0)^2 r^4} \quad \text{or} \quad C_4 \sim \frac{\alpha e^2}{(4\pi\epsilon_0)^2} \sim 5 \times 10^{-58} \text{ J m}^4$$

from which one finds $r_{max} \sim 50$ Å. For the resonance interaction we expect something of the order of a dipole-dipole energy, $p^2/4\pi\epsilon_0 r^3$. Putting $p \sim ea_0$, we get $C_3 \sim 5 \times 10^{-49}$ and $r_{max} \sim 80$ Å. The linear Stark effect will obviously have a yet longer range; indeed, one can often take as the cut-off for this interaction the Debye radius, which, as will be seen in Section 11.3, is a measure of the distance at which any charged particle is effectively screened by charged particles of the opposite sign.

It is evident that the binary approximation cannot be expected to hold, except perhaps for the van der Waals interaction, unless the perturber density is considerably below the N.T.P. value of 3×10^{25} m^{-3}, corresponding to $\bar{d} \sim 30$ Å. The pressure should be restricted to about ~ 1 torr for resonance broadening and the number density of charged particles to say 10^{22} m^{-3} (10^{16} cm^{-3}) for Stark broadening. To include more than one collision makes the mathematics exceedingly complicated. In any case it is often justifiable to treat the extreme wings of the line in terms of the binary approximation, because these are attributable to a very strong interaction, or a very close perturber, compared to which the effect of more distant perturbers is negligible.

8.4.3 Impact theory

To carry a little further the impact theory outlined in Section 8.4.1, one makes three more assumptions, known as the binary, classical path, and adiabatic approximations. The first of these has already been discussed. The second assumes the perturber to move along a classical path – i.e., a straight line, unless both emitter and perturber are charged. The third implies that the perturbers do not induce transitions between, or mix up, different close-lying states of the emitting atom. Collisions are assumed to cause phase changes in the radiated wave train, but not to cut it short by knocking the radiating atom out of its excited state. One would expect that both processes might actually occur, but it turns out that for neutral perturbers the adiabatic approximation is usually good. When electrons are present, however, the inelastic collisions do contribute appreciably to the line width because of the high electron velocities, as we shall see shortly.

The replacement of the billiard ball type of collision by an 'optical collision', causing merely a change of phase in the emitted radiation, represents the next degree of sophistication in the impact theory. The total phase change η caused by the perturbation $\Delta v(r)$ over the duration of the collision is

$$\int_0^{t_p} 2\pi \Delta v(r) \, dt$$

We may regard η as a function of the impact parameter ρ for that particular collision and write

$$\eta(\rho) = 2\pi \int_{-\infty}^{\infty} \Delta v(r) \, dt \qquad (8.23a)$$

where the change in the limits of integration alters nothing because Δv is zero outside the perturbation period t_p. It is often convenient to use the approximate expression

$$\eta(\rho) \simeq 2\pi \Delta v(\rho) \cdot \rho/\bar{v} \qquad (8.23b)$$

which results from taking η as the product of the frequency shift at closest approach, $\Delta v(\rho)$, and the collision duration $\rho/\bar{v}$.

In the Weisskopf theory a collision producing a phase change of unity is taken to destroy the coherence of the wave train and thus to play the part of a quenching collision in the Lorentz theory. The corresponding value of the impact parameter, ρ_0, is known as the Weisskopf radius. Writing the interaction in the form of Equation 8.22)

$$\Delta\nu(r) = C_n/r^n$$

and putting

$$dt = \frac{dx}{\bar{v}} = \frac{1}{\bar{v}} \rho \sec^2 \theta \, d\theta, \quad r = \rho \sec \theta$$

(see Fig. 8.12), we get

$$\eta(\rho) = 2\pi \int_{-\pi/2}^{\pi/2} \frac{C_n \cos^{n-2} \theta}{\rho^{n-1} \bar{v}} \, d\theta = \frac{2\pi a_n C_n}{\rho^{n-1} \bar{v}}$$

where a_n is a numerical constant of order unity, depending on the power n. So for $\eta(\rho_0) \approx 1$ the appropriate impact parameter is

$$\rho_0 \approx (2\pi C_n/\bar{v})^{1/n-1} \tag{8.24}$$

(an expression that could also have been obtained by using the approximate form (Equation 8.23b)). The number of collisions per second for which $\rho \leqslant \rho_0$ is $\pi\rho_0^2 \bar{v} N$, where N is the number

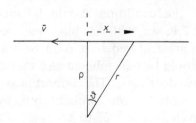

Fig. 8.12 Geometry for calculating phase change.

density, and this is set equal to $1/\tau$ in the Lorentzian (Equation 8.18). The Weisskopf theory thus leads to a Lorentzian line profile of half-value width

$$w = \delta\nu_c = 1/\pi\tau \approx \left(\frac{2\pi C_n}{\bar{v}}\right)^{2/n-1} \bar{v} N \tag{8.25}$$

with no shift of central frequency.

This simple model is not very satisfactory for broadening by electrons, but it predicts a number of features that have been verified experimentally for neutral atoms. For example, it gives the right order of magnitude for ρ_0: with the approximate values of C_3 and C_6 derived in the last section (divided by h for use in Equation 8.22) we find $\rho_0 \sim 25$ Å for resonance and $\rho_0 \sim 10$ Å for van der Waals broadening. It also gives the right temperature dependence. For $n = 3$, w is independent of $\bar{v}$ and hence of T, while for $n = 6$, $w \propto \bar{v}^{3/5} \propto T^{3/10}$.

The Weisskopf radius ρ_0 is the parameter already mentioned in assessing the validity of the impact treatment: $\Delta\nu \ll 1/t_{\mathrm{p}} \sim \bar{v}/\rho_0$. There are really two points at issue: first, whether the impact theory is valid for the central part of the line, $\Delta\nu \leqslant w/2$, and, secondly, how far into the line wings it may be used. The first requirement, putting

$$w = 1/\pi\tau = \rho_0^2 \bar{v} N \qquad (8.26)$$

amounts to

$$\tfrac{1}{2}\rho_0^2 \bar{v} N \ll \bar{v}/\rho_0$$

or

$$\rho_0 \ll N^{-1/3} \qquad (8.27)$$

The Weisskopf radius must be much smaller than the average distance $\bar{d}$ between particles. From the estimates of ρ_0 above one can deduce that the condition should be satisfied for neutral perturbers at N.T.P. ($\bar{d} \approx 30$ Å) but that considerably lower particle densities are required if it is to be valid for the longer range forces. It should be remembered also that ρ_0 increases as the perturber velocity decreases. The importance of the perturber velocity can be more clearly brought out by eliminating ρ_0 between Equations 8.26 and 8.27 to give

$$w \ll \bar{v} N^{1/3} \qquad (8.28)$$

Let us take as an example a hydrogen plasma at 25 000 K with a density of both electrons and ions of 10^{22} m^{-3}; then $N^{1/3} \sim 2 \times 10^7$ m, and $\bar{v}$ is about 6×10^5 m sec^{-1} for the electrons and 1.5×10^4 m sec^{-1} for the ions. The criterion (Equation 8.28) expressed in wave numbers, $w_\sigma = w/c$, becomes $w_\sigma \ll 400$ cm^{-1} for the electrons, which is realistic, and $w_\sigma \ll 10$ cm^{-1} for the ions, which is not. The impact approximation is obviously much more generally useful for electrons than for ions.

Assuming the conditions for the central region of the line to be satisfied, we come to the second point – how far out in the line wings may the impact theory be used? This is best approached from the quasi-static end, for which we require that the static frequency shift $\Delta\nu$ from a perturber at distance ρ should be greater than the frequency spread associated with the finite duration t_p of the perturbation, which is of order $\bar{v}/\rho$. Using Equation 8.22, the quasi-static condition becomes

$$C_n/\rho^n > \bar{v}/\rho.$$

which can obviously be satisfied by sufficiently small ρ because n is greater than one. The value of ρ determining the transition region between the two regimes is given by $\Delta\nu\,\rho/\bar{v} \sim 1$. But from Equation 8.23b this relation defines the Weisskopf radius ρ_0 to within a factor 2π. The frequency shift that would be produced by a *static* perturber at a distance ρ_0 is therefore a measure of the impact theory limit. Any part of the spectral line appreciably beyond about C_n/ρ_0^n sec^{-1} from the centre must be attributed to perturbers closer than ρ_0 which are to be treated as effectively stationary.

It remains to consider collisions with $\rho > \rho_0$, which are ignored in the Weisskopf theory. This theory, as we have seen, predicts a broadened but unshifted line, whereas experimentally many pressure-broadened lines with a perfectly good Lorentzian profile are found to be shifted, usually towards the red. The phase shift idea was extended to include small as well as large phase shifts, first by Lindholm and later, with the inclusion of inelastic collisions, by Foley and by Anderson (1945–47). A Lorentzian with shift d and width w

$$I_\nu = I_0 \frac{1}{(\nu - \nu_0 - d)^2 + (w/2)^2} \tag{8.29}$$

is obtained by applying a Fourier transform to a damped oscillation when the damping constant has an imaginary part (responsible for d) as well as a real part (responsible for w). In the phase shift model d and w are related quantities: they are both associated with the phase shift $\eta(\rho)$ through the real and imaginary parts of $(1 - e^{i\eta})$. As in the Weisskopf theory, $\eta(\rho)$ is the total phase change produced by a complete collision at impact parameter ρ. For $\rho \geqslant \rho_0$, as we have seen, the collisions are well separated in time, so a single complete collision is a valid concept. It is now necessary to average over all impact parameters. Fortunately, the assumption of separated collisions allows one to replace $\overline{(1 - e^{i\eta})}$ by $(1 - e^{i\overline{\eta}})$, essentially because the perturbations from weak collisions can be added linearly, while a strong collision can be assumed to be an isolated event. The probability of a phase shift $\eta(\rho)$ is given by

$$d\eta(\rho) = 2\pi\rho \, d\rho N\overline{v}$$

The final expressions for the shift and width then come out as

$$d = 2\pi N\overline{v} \int_0^\infty \rho \, \sin \eta(\rho) \, d\rho$$

$$w/2 = 2\pi N\overline{v} \int_0^\infty \rho \, \{1 - \cos \eta(\rho)\} \, d\rho$$

(8.30)

Equations 8.30 allow d and w to be evaluated if the interaction (Equation 8.22) is known. Although the lower limit of integration is written as zero, there must in fact be an arbitrary cut-off below ρ_0.

Looking at these expressions qualitatively, one can deduce that the weak collisions (large ρ, small η) must be responsible for the line shift and the strong collisions (small ρ, large η) for most of the broadening. This is because η has the same sign for all perturbers, depending on the sign of C_n, so for $0 < |\eta| < \pi$, $\sin \eta$ has a non-zero average. For $C_n < 0$ there is an average retardation of phase, corresponding to a slightly longer average wavelength – i.e., a red shift. The very distant collisions will scarcely contribute to w because $\cos \eta \sim 1$. But for strong collisions, $|\eta| \sim \pi$, $\sin \eta$ has both positive and negative values, and its average contribution to d will be small, while most of the contribution to the w integral comes from the region $|\eta| \to \pi$. To summarize, then, collisions for which $\rho \sim \rho_0$ are responsible for most of the line broadening, while collisions for which $\rho \gg \rho_0$ cause most of the shift.

The width/shift ratio depends on the interaction potential C_n/r^n. For $n = 6$ (van der Waals), $w/d = 2\cdot8$, while for $n = 4$ (quadratic Stark) $w/d = 1/\sqrt{3} \simeq 0\cdot6$. The experimental ratio in the second case usually turns out to be larger than that predicted by the phase shift model. This is because of the inelastic collisions, which stop the wave train by inducing transitions and so contribute to the width but not to the shift. We return to this point below.

When applying Equations 8.30 to long range forces (electron perturbers) the upper limit of integration must be looked at critically. Collisions at large impact parameters may still produce appreciable effects, and if ρ is made large enough the collision duration $\rho/\bar{v}$ will also become uncomfortably large. However, an upper limit to ρ is set by the Debye shielding radius, ρ_D. The emitting atom is effectively shielded from the effects of all charged particles further away than ρ_D. The upper limit of t_p is therefore $\rho_D/\bar{v}$, and the reciprocal of this is just the plasma frequency ω_p (see Section 11.4). The impact criterion can then be expressed as $\Delta\nu \ll \nu_p$ where

$$\nu_p = \frac{\omega_p}{2\pi} = \left(\frac{N_e e^2}{4\pi^2 \epsilon_0 m}\right)^{1/2}$$

For the central region of the line this allows electron densities up to 10^{24}–10^{26} m^{-3} or 10^{18}–10^{20} cm^{-3}.

We end this section by returning to the question of inelastic collisions, which may have a significant effect on the line width in the case of electron perturbers. The modern theory, which is largely due to Baranger and may be found in his article [5] and in Griem's book [4], is beyond the scope of this book; however, a simple qualitative argument enables one to see the conditions in which these collisions are likely to be important – that is, the adiabatic approximation is no longer valid. For a perturbation lasting a time t_p the uncertainty principle predicts a corresponding uncertainty in the energy of $\Delta E \sim \hbar/t_p$. If the emitting atom has an energy level E' near its initial level E_2 of the same parity, it will not be possible to determine which of the two levels it is actually in if $|E_2 - E'| \lesssim \hbar/t_p$. So the criterion for the breakdown of the adiabatic approximation can be expressed as

$$t_p \lesssim \frac{\hbar}{|E_2 - E'|}.$$

Writing, as usual, $t_p \approx \rho/\bar{v}$, it is clear that this inequality is far more likely to hold for electrons than for ions or neutrals. Obviously, by choosing collisions with small ρ we can make t_p as small as we like for any given $\bar{v}$, but as soon as ρ is reduced to a value approaching ρ_0 the *elastic* collisions will produce broadening. So inelastic collisions will matter much only if they are important for $\rho > \rho_0$, i.e., if

$$\rho_0/\bar{v} \lesssim \hbar/|E_2 - E'|$$

From Equation 8.24 ρ_0 itself depends on $\bar{v}$ according to

$$\rho_0 \sim \left(\frac{C_n}{\bar{v}}\right)^{1/n-1}$$

Our condition therefore becomes

$$(\bar{v})^{n/n-1} \geqslant \frac{C_n^{1/n-1}}{\hbar} \cdot |E_2 - E'|$$

The critical velocity is approximately proportional to the energy interval for quadratic Stark broadening and to its square root for linear Stark effect. If $|E_2 - E'|$ is of the order of the fine structure splitting, one finds that, for a plasma at a few thousand degrees Kelvin, the electrons have velocities above the critical value but the ions do not. Inelastic collisions can be ignored altogether for neutral perturbers for most practical purposes. Hydrogen is something of a special case because of the degeneracy of the fine structure levels, but it turns out from the quantum mechanical treatment that the conditions on critical velocity are similar to the non-degenerate cases.

8.4.4 Quasi-static theory

This theory, as we have seen, starts from the assumption that the perturbers are almost stationary and the perturbation is nearly constant over the whole time that the emitting atom is radiating. There are two steps to be taken: first, to calculate the effect of a single perturber on the emitter and secondly to perform a statistical average over all perturbers.

As in the impact approximation one considers a perturbation $\Delta v(r) = C_n/r^n$ but this is now treated simply as a shift of the energy levels. In general, the statistical averaging of this shift over the probability distribution of perturbers at all distances r results in an asymmetric line with a shift of central frequency. If, as is usually the case, C_n is negative and the upper level is shifted more than the lower (Fig. 8.10), the shift is to the red. The simplest model, that of Kuhn, takes a single perturber for each emitting atom, situated somewhere at random within a sphere of radius R corresponding to an arbitrary cut-off of the interaction. The intensity corresponding to a shift $\Delta v(r)$ is just the probability of finding a perturber at r – i.e.,

$$I(v) \propto 4\pi r^2 \, dr$$

Substituting for r from $r^n = C_n/\Delta v$ gives

$$I(v) \propto (\Delta v)^{-(n+3)/n}$$

For $n = 6$ (van der Waals interaction) this becomes $I(v) \propto (\Delta v)^{-3/2}$. The assumption of a single perturber is justified only if R is small compared to the average inter-particle distance $\bar{d}$ – in other words, if we confine ourselves to small values of r. But a small r implies a large Δv, so that this model is most useful in the extreme wing of the line. Margenau extended the range of application by considering several perturbers whose effects could be added linearly, as for the van der Waals' case, for which he obtained the expression

$$I(v) \propto (\Delta v)^{-3/2} \exp\left(-\frac{4\pi^3}{9} N^2 C_6 \Delta v\right)$$

This modification extends the range of validity somewhat towards the line centre, but it must be remembered that as we go towards small Δv the static approximation eventually ceases to be valid. When $r \to \rho_0$, it is no longer realistic to think of the perturbers as stationary. Various attempts have been made to smear out the static broadening by including the motion of the perturbers (Jablonski, Anderson, and others), but the procedure is too complicated to be discussed here. We note again that the transition region is defined by $\Delta v \sim C_n/\rho_0^n$.

The situation for charged particle broadening is a little different, because a plasma contains both ionic and electronic perturbers. As we saw in the last section, the impact approximation is usually valid over most of the line for the electrons, whereas the quasi-static approximation usually hold good for the ions. The problem may be tackled by calculating the line shape from the ion fields (as was first done by Holtsmark) and then considering each element of the line to be broadened and shifted

by the electron impacts. The asymmetry of the quadratic Stark
effect gives rise to asymmetric lines. However, hydrogen lines
(linear Stark effect) are symmetric and much wider.

It may also be necessary to include in the treatment forbidden
transitions induced by the ion fields. It was pointed out in Section
2.9 that an external electric field effectively mixes configurations
of different parity, so that transitions such as *s-d* and *s-s* appear.
The ion micro-fields can have the same effect, and any such
'forbidden' neighbours of the 'allowed' line have to be included in
the calculation of the line profile of the latter.

8.4.5   *Experimental aspects*

The experimental work on broadening by neutrals and by charged
particles has been reviewed in various places – for example [6] for
neutrals and [4], [7] for plasmas. A few very general comments
follow.

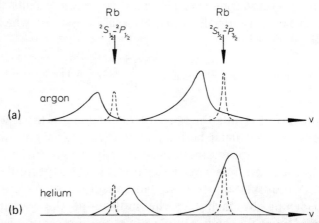

Fig. 8.13 Rubidium lines broadened by (a) argon and (b) helium. In each
case the dotted curve represents the rubidium principal series doublet at low
pressure and the full curves represent it at 12 atm. inert gas pressure.

There are considerable experimental difficulties associated with
obtaining a true pressure-broadened line profile. It is necessary
either to use an instrument of sufficiently high resolving power
that the instrumental width is negligible or to disentangle the
latter, using perhaps a very narrow spectral line to obtain the true
instrumental profile. Doppler broadening may have to be taken
into account. In emission lines, the profile may be further

distorted by self-absorption. These problems are, of course, much more severe for van der Waals broadening, which is a couple of orders of magnitude less than charged particle broadening for the same density of perturbers.

Most of the experimental work on broadening by neutral atoms has been done on alkali metal lines broadened by inert gases. With the heavier gases the Weisskopf radius is of order 5–10 Å. For experiments carried out in conditions to which the theory can be expected to apply, the line shape, the shift/width ratio and the temperature and density dependences agree very well with the predictions. In most cases, however, experimental values of C_6 are

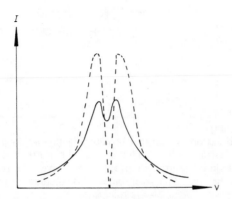

Fig. 8.14 Broadening of H_β line showing contributions of ions and electrons. The dashed curve is the Holtsmark profile (for ion broadening only). The full curve includes the effects of electron impacts.

larger than would be expected from the theoretical estimates. With helium as broadener the C_6 is so small that the repulsive terms become significant, and there is often a violet instead of a red shift. Fig. 8.13 shows an example of rubidium lines broadened by argon and helium. Empirical combinations of several different interaction potentials have been used successfully in some cases to reproduce the experimental results – for example, a C_8 term added to the C_6-C_{12} Lennard-Jones potential.

Resonance broadening has been studied in resonance lines themselves and in lines ending on the upper level of a resonance transition. ρ_0 is an order of magnitude greater than for van der Waals broadening. The predictions of zero shift, independence of temperature, and linear dependence on density have been con-

firmed. Moreover, the absolute values of the broadening constants
agree with theory to within 5- 10%.

   In the case of broadening by charged particles, an immense
amount of theoretical and experimental work has been done in the
course of plasma spectroscopy. With a number of exceptions,
particularly when forbidden components are involved, the modern
theory is usually capable of calculating profiles that fit the

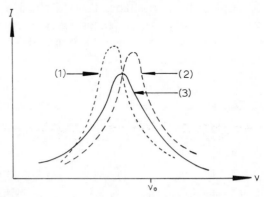

Fig. 8.15 Modification of impact-broadening profile by inclusion of quasi-
static and inelastic effects. The curve (1) represents electron impact
broadening, adiabatic approximation. The curve (2) includes the effect
of electron collisions. The full curve (3) includes also quasi-static ion
broadening.

experimental ones. Much of the work has been concentrated on
hydrogen and the hydrogen-like ions, and the agreement is now
very good. Fig. 8.14 shows how the profile of H_β for ion
broadening only (Holtsmark profile) is modified when electron
impacts and correlation effects are included. Fig. 8.15 shows how
an adiabatic electron-impact profile is modified by the inclusion of
quasi-static ion broadening and inelastic collisions.

8.5   Combination of Gaussian and Lorentzian line profiles

In the frequent cases when a pressure-broadened line can be
represented by a Lorentzian, the combination of pressure and
natural broadening is again a Lorentzian, with damping constant
$\gamma = \gamma_{nat} + \gamma_{coll}$ where in practice usually $\gamma_{nat} \ll \gamma_{coll}$. Since
Doppler broadening produces a Gaussian profile, very many

spectral lines are combinations in some proportion of a Gaussian and a Lorentzian profile.

The actual line shape is obtained by 'folding' the two profiles together. One can think of each velocity-shifted element as having a Lorentzian profile. The net intensity at any point is then the sum of the contribution from the Lorentzians drawn round every Doppler element. In terms of the parameter y of Equation 8.16 the intensity becomes

$$I(x) = \text{const.} \int_{-\infty}^{\infty} \frac{e^{-y^2} \, dy}{(x - y)^2 + a^2} \tag{8.31}$$

where a, the damping ratio, is defined by

$$a = \sqrt{(\ln 2)} \, \delta\nu_L / \delta\nu_D \tag{8.32}$$

The intensity distribution (Equation 8.31) is known as the Voigt profile and is of great practical importance. The integration cannot be done analytically, but it has been computed and tabulated (see, for example, [2, 9]). Without going further into the analysis, there are a couple of points worth making. First, the folding of any number of Gaussian and Lorentzian distributions produces a Voigt profile, which can be unfolded into an equivalent Gaussian and an equivalent Lorentzian, with half-value widths given by $(\delta\nu_D)^2 = (\delta\nu_{D1})^2 + (\delta\nu_{D2})^2 + \cdots$ and $\delta\nu_L = \delta\nu_{L1} + \delta\nu_{L2} + \cdots$ respectively. The instrumental profile may sometimes be approximated by a Lorentzian in order to use the Voigt analysis for separating instrumental width from true spectral line width when the latter is dominated by Doppler broadening. Secondly, as remarked earlier, the Doppler profile has a fairly compact form, while the Lorentzian spreads out a long way in the wings. At two full half widths from the centre a pure Doppler profile has dropped to ~0·2% of the peak intensity, whereas a pure Lorentzian profile has dropped only to ~6%. If, as is usually the case, the damping constant a is fairly small, the intensity distribution in the central part of a Voigt profile is determined by the Doppler component, but the intensity in the wings of the line may come almost entirely from the Lorentzian component. This is an important point when using integrated absorption or total emission methods to measure transition probabilities (Chapter 10), and we shall also come back to it in discussing curves of growth (Chapter 9).

References

Introductory treatment

1. Cowley, C. R. 'The Theory of Stellar Spectra', Gordon and Breach, 1970
2. Kuhn, H. G. 'Atomic Spectra', Longmans, 1969

Detailed treatment

3. Breene, R. G. 'Shift and Shape of Spectral Lines', Pergamon, 1961
4. Griem, H. R. 'Plasma Spectroscopy', McGraw-Hill, 1964
5. Baranger, M. Spectral Line Broadening in Plasmas in 'Atomic and Molecular Processes' (ed. Bates), Academic Press, 1962

Recent reviews

6. Hindmarsh, W. R. and Farr, J. M. Collision Broadening of Spectral Lines by Neutral Atoms, *Prog. in Quantum Electronics* **2**, 143, 1972.
7. Burgess, D. D. Spectroscopy of Laboratory Plasmas, *Space Science Reviews* **13**, 493, 1972

Miscellaneous references

8. Ditchburn, R. W. 'Light', Blackie, 1963
9. Mitchell, A. C. G. and Zemansky, M. S. 'Resonance Radiation and Excited Atoms', Cambridge University Press, 1934
10. Demtroder, W. High Resolution Spectroscopy with Lasers, *Physics Reports* **7c**, 223, 1973 (covering saturation spectroscopy)

CHAPTER NINE

Emission and absorption of line radiation

9.1 Introduction

This chapter is concerned with the total energy emitted or absorbed in a radiative transition between discrete energy levels rather than with its frequency spread. The intensity of a spectral line in emission is determined by the population of the upper level of the transition and the transition probability. Similarly, the strength of an absorption line is determined by the population of the lower level, the transition probability, and the density of radiation from the background source. These quantities have appeared in various forms in earlier chapters. The population of excited states was discussed briefly in Section 3.3; the transition probabilities were introduced in Section 3.4 in the form of the Einstein coefficients and earlier still in section 2.8 in the form of electric dipole strength. Emission and absorption coefficients j_ν and k_ν appeared in Chapter 8. Transition probabilities can also be expressed in terms of the lifetime of the excited state, which also appeared in Chapter 8, and of yet another quantity, the oscillator strength of the line.

The first purpose of this chapter is to derive the relations between all these quantities. The second purpose is to discuss the effect on the observed radiation of increasing the number of emitting or absorbing atoms in the line of sight – that is, the optical depth. Fig. 8.3 showed qualitatively the dependence of line shape on optical depth; this chapter will consider the effect on the total energy emitted or absorbed and its representation by a curve of growth, a device that is widely used in astrophysics.

9.2   Boltzmann distribution of population and state sum

Boltzmann's formula for the ratio of the populations of two energy levels E_1 and E_2 in thermal equilibrium at temperature T has already been quoted in Equation 3.6:

$$N_2/N_1 = g_2/g_1 \; e^{-(E_2 - E_1)/kT}$$

where g is the statistical weight or degeneracy of a level. The N's are usually taken to represent number densities – that is, number per unit volume. It is often necessary to express the population of a given state N_j in terms of N, the total number density of atoms of that particular species. If N_0 is the population of the ground state, N_1 that of the first excited level, and so on, then

$$\begin{aligned}
N &= N_0 + N_1 + N_2 + \cdots \\
&= \frac{N_0}{g_0} (g_0 + g_1 e^{-E_1/kT} + g_2 e^{-E_2/kT} + \cdots) \\
&= \frac{N_0}{g_0} \sum_{j=0}^{\infty} g_j e^{-Ej/kT} = \frac{N_0}{g_0} U(T)
\end{aligned}$$

where $U(T)$ is defined by

$$U(T) = \sum_{j=0}^{\infty} g_j e^{-Ej/kT} \tag{9.1}$$

Therefore $N_0 = g_0 N/U(T)$, and the population of the jth level is given by

$$N_j = \frac{N}{U(T)} g_j e^{-Ej/kT} \tag{9.2}$$

$U(T)$, or sometimes $Q(T)$, is known as the partition function or state sum, the first term being the more usual and the second the more descriptive.

At moderate temperatures such that the first excited level has a fairly high energy compared to kT ($E_1 > kT$) the partition function converges rapidly, and only a few terms have to be included in the sum. Simple atoms, for example, usually have their resonance lines in the visible or near ultra-violet region, corresponding to $E_1 \sim 25\,000 \text{ cm}^{-1}$. Since $k = 0 \cdot 7 \text{ cm}^{-1}$ per degree, T has to be of order 30 000 K for $E_1 \sim kT$, and the populations of the higher levels are very small indeed. In complex atoms, however, there may be a number of low-lying states that are appreciably populated even at fairly low temperatures. In molecules one can often ignore the excited electronic states, but a large number of vibrational and rotational levels must usually be included in the state sum. Luckily this does not have to be done level by level, because the energies can be expressed in terms of the vibrational and rotational constants and the state sum obtained analytically (see, for example, [6]).

9.3 Einstein coefficients and line strength

The Einstein coefficients for spontaneous emission, A_{21}, induced emission, B_{21}, and absorption, B_{12}, were defined in Section 3.4. The relations between the coefficients were found to be (Equation 3.8)

$$\left. \begin{array}{c} g_1 B_{12} = g_2 B_{21} \\[2mm] A_{21} = \dfrac{8\pi h\nu^3}{c^3} B_{21} \end{array} \right\} \tag{9.3}$$

Since these coefficients are intrinsic properties of the atom or molecule, they can in principle be calculated if the wave functions are known. To calculate A directly it is necessary to quantize the electromagnetic field as well as the atomic energy states. However, the B coefficients can be calculated from simple wave mechanics by treating the oscillating electric and magnetic fields as a time-dependent perturbation (see, for example, [1] and [2]). A can then be found from B by Equation 9.3. The results obtained in this way are identical with those of quantum electrodynamics. For electric dipole transitions

$$B_{21} = \frac{8\pi^3}{3h^2(4\pi\epsilon_0)} |R_{12}(er)|^2 = \frac{2\pi^2}{3\epsilon_0 h^2} \{|R_{12}(ex)|^2 + |R_{12}(ey)|^2 + |R_{12}(ez)|^2\}$$

$$\tag{9.4}$$

R_{12} is the electric dipole transition moment, and the matrix elements $R_{12}(ex)$, etc., are defined by

$$R_{12}(ex) = \int \psi_1 ex \psi_2 \, d\tau$$

etc., as in Equation 2.12. Since

$$\int \psi_1 x \psi_2 \, d\tau = \int \psi_2 x \psi_1 \, d\tau,$$

the order of the subscripts is immaterial. These matrix elements vanish identically unless certain 'selection rules' specifying the relation between ψ_1 and ψ_2 are satisfied, as explained in Section 2.8.

Using the notation

$$\int \psi_1 e x \psi_2 \, d\tau \equiv e|x_{12}|$$

etc., Equation 9.4 can be written as

$$B_{21} = \frac{2\pi^2}{3\epsilon_0 h^2} e^2 \{|x_{12}|^2 + |y_{12}|^2 + |z_{12}|^2\} \qquad (9.4a)$$

From Equation 9.3,

$$A_{21} = \frac{16\pi^3 \nu^3}{3\epsilon_0 h c^3} e^2 \{|x_{12}|^2 + |y_{12}|^2 + |z_{12}|^2\} \qquad (9.5)$$

The electric dipole *line strength* S of the transition is defined by

$$S_{12} = |R_{12}|^2 = e^2 \{|x_{12}|^2 + |y_{12}|^2 + |z_{12}|^2\}$$

Equations 9.4a and 9.5 expressed in terms of S become

$$B_{21} = \frac{2\pi^2}{3\epsilon_0 h^2} S_{12}$$

$$A_{21} = \frac{16\pi^3 \nu^3}{3\epsilon_0 h c^3} S_{12}$$

No account has been taken so far of the degeneracy of either level. If level 1 has degeneracy g_1, the total transition probability $2 \to 1$ is found by summing over all values of m_1, the quantum number specifying the g_1 different sub-levels.

$$\therefore \quad A_{21} = \frac{16\pi^3 \nu^3}{3\epsilon_0 h c^3} e^2 \sum_{m_1} \{|x_{12}|^2 + |y_{12}|^2 + |z_{12}|^2\}$$

If level 2 is also degenerate, the total transition probability must include transitions from all sub-levels m_2 – that is, the matrix elements must be summed over m_2 as well as m_1. On the other hand, the population of level 2 is now divided among the g_2 sub-levels, so that there are only $1/g_2$ as many atoms in each sub-level. The total transition probability therefore becomes

$$A_{21} = \frac{1}{g_2} \frac{16\pi^3 \nu^3}{3\epsilon_0 h c^3} e^2 \sum_{m_1} \sum_{m_2} \{|x_{12}|^2 + |y_{12}|^2 + |z_{12}|^2\}$$

In view of the symmetry of the individual matrix elements ($|x_{12}| = |x_{21}|$, etc.) it is convenient to make the line strength symmetrical also. This is done by defining

$$S = S_{12} = S_{21} = e^2 \sum_{m_1} \sum_{m_2} \{|x_{12}|^2 + |y_{12}|^2 + |z_{12}|^2\} \quad (9.6)$$

and leads finally to

$$\left. \begin{aligned} A_{21} &= \frac{16\pi^3 \nu^3}{3\epsilon_0 hc^3} \frac{1}{g_2} S \\[2mm] B_{21} &= \frac{2\pi^2}{3\epsilon_0 h^2} \frac{1}{g_2} S \\[2mm] B_{12} &= \frac{2\pi^2}{3\epsilon_0 h^2} \frac{1}{g_1} S \end{aligned} \right\} \quad (9.7)$$

It is worth reverting for comparison to the model of the classical oscillator used to explain natural line broadening in Section 8.2. Let us assume that the quantum mechanical system corresponds to a classical dipole of moment $P_x = ex$ oscillating with frequency ν_0, so that at any instant

$$P_x = ex_0 \cos 2\pi\nu_0 t = P_{0x} \cos 2\pi\nu_0 t$$

This dipole radiates energy at the rate given by Equation 8.6b:

$$-\frac{\partial E}{\partial t} = \frac{2}{3} \frac{e^2}{c^3 (4\pi\epsilon_0)} \overline{\ddot{x}^2}$$

But

$$e\ddot{x} = -(2\pi\nu_0)^2 ex_0 \cos 2\pi\nu_0 t$$

$$\therefore \quad e^2 \overline{\ddot{x}^2} = (2\pi\nu_0)^4 P_{0x}^2 \overline{\cos^2 2\pi\nu_0 t} = 16\pi^4 \nu_0^4 P_{0x}^2 \tfrac{1}{2}$$

giving

$$-\frac{\partial E}{\partial t} = \frac{4\pi^3 \nu_0^4}{3\epsilon_0 c^3} P_{0x}^2$$

If this rate of radiation is equated to the energy emitted per second per atom,

$$-\frac{\partial E}{\partial t} = A_{21} h\nu_0$$

we have

$$A_{21} = \frac{4\pi^3 \nu_0^3}{3\epsilon_0 c^3 h} P_{0x}^2$$

Comparison with Equations 9.6 and 9.7 shows that the two expressions are equivalent if the x-component of the classical oscillator $P_{0x} = ex_0$ is replaced by $2e|x_{12}|$. The factor 2 comes in because both the matrix elements $|x_{12}|$ and $|x_{21}|$ contribute to the equivalent classical dipole. When the y and z components are included, the equivalent classical dipole is just twice the quantum mechanical transition moment:

$$P_0 = 2R_{12}(er)$$

Transitions for which electric dipole radiation is forbidden may occur by magnetic dipole or electric quadrupole radiation. Section 2.8 made the point that the corresponding transition probabilities are much smaller, being second order terms in the time-dependent perturbation. The relevant line strengths are determined essentially by the atomic unit of magnetic moment, the Bohr magneton, and the atomic unit of electric quadrupole moment respectively – i.e., $S_{m.d.} \sim (\mu_B)^2$ and $S_{e.q.} \sim (e a_0^2)^2$, as compared with $S_{e.d.} \sim (e a_0)^2$. The relations between A_{21} and S analogous to Equation 9.7 for the magnetic dipole and electric quadrupole radiation are given in Table 9.3. In Table 9.4 the orders of magnitude are brought out more clearly by relating A_{21} to the line strength expressed in the above atomic units, which in each case is typically of order unity. Evidently $A_{m.d.}/A_{e.d.} \sim 10^{-5}$, while $A_{e.q.}/A_{e.d.} \sim 1/\lambda^2$, where λ is in Ångströms, giving 10^{-7}–10^{-8} in the visible region. Furthermore, for those magnetic dipole transitions discussed in Chapter 7 that involve *nuclear* spin magnetic moment rather than electronic magnetic moment the line strength is smaller still by a factor of order $(\mu_N/\mu_B)^2 \sim 10^{-6}$.

9.4　Lifetimes

If induced emission is neglected, the lifetime τ of an excited state is easily defined in terms of the Einstein A-coefficient. If only one transition from level 2 is possible, then

$$\tau_2 = 1/A_{21}$$

More generally, if level 2 can combine with several distinct lower levels,

$$\tau_2 = 1/\sum_j A_{2j} \tag{9.8}$$

In this case we must know the branching ratios, or relative A-values, in order to determine any one A absolutely from τ. However, the lifetime is often determined predominantly by the transition probability of one line, either because that line has an intrinsically greater line strength or because its wavelength is much shorter, so that the ν^3 term increases its relative importance.

The lifetimes of most excited states which can decay by electric dipole radiation are in the range 10^{-6} to 10^{-9} sec. In Section 8.2 the lifetime of a classical oscillator τ_{class} was found from the classical damping constant to be

$$\tau_{class} = 1/\gamma_{class} = \frac{3\epsilon_0 mc^3}{2\pi e^2 \nu_0^2}$$

which is about 10^{-8} sec for transitions in the visible region. The classical damping constant therefore gives the right order of magnitude in many cases for the natural line width, but the correct value of γ to be used in Equation 8.9 to find the natural line width for a particular transition is

$$\gamma_{12} = \gamma_1 + \gamma_2 = 1/\tau_1 + 1/\tau_2$$

If level 1 is the ground state or is metastable,

$$\gamma_{12} = 1/\tau_2 = \sum_j A_{2j}$$

9.5 Absorption coefficient

When light passes through an absorbing medium, the energy absorbed in a distance δx is proportional to both δx and the incident flux, the proportionality constant being in general a function of frequency. The equation

$$-\delta I_\nu(x) = k_\nu I_\nu(x)\delta x \tag{9.9}$$

defines the absorption coefficient k_ν, usually expressed in cm^{-1} or m^{-1}. For a thickness l of absorbing medium, the relation between

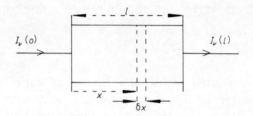

Fig. 9.1 Absorption by a column of gas.

the incident and emerging intensities (see Fig. 9.1) is found by integrating Equation 9.9:

$$I_\nu(l) = I_\nu(0)e^{-\int_0^l k_\nu \mathrm{d}x}$$

$\int_0^l k_\nu \, \mathrm{d}x$ is the optical depth of the medium. If this is homogeneous, so that k_ν is independent of x, the optical depth is $k_\nu l$, and we have

$$I_\nu(l) = I_\nu(0)e^{-k_\nu l} \tag{9.10}$$

The absorption from an isolated spectral line is spread over a finite frequency range by the various broadening mechanisms, and a plot of $I_\nu(l)$ against ν has the general form of Fig. 8.2, reproduced here as Fig. 9.2. This plot provides the experimental data from which k_ν is obtained by means of Equation 9.10:

$$k_\nu = \frac{1}{l} \ln \frac{I(0)}{I_\nu(l)}$$

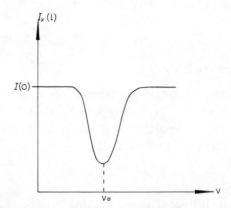

Fig. 9.2 Absorption line profile. $I_\nu(0)$ is assumed to be independent of ν – i.e., it has the constant value $I(0)$.

Fig. 9.3 shows a typical plot of k_ν against ν. It is important to remember that, appearances notwithstanding, Fig. 9.3 is *not* just Fig. 9.2 upside-down, *unless* $k_\nu l \ll 1$ so that Equation 9.10 can be written as

$$I_\nu(l) = I(0)(1 - k_\nu l) \qquad (9.11)$$

In this case the dip $I(0) - I_\nu(l)$ is indeed proportional to k_ν. But if terms of order $k_\nu^2 l^2$ cannot be neglected in the expansion of the exponential the two curves differ in shape. As the optical depth $k_\nu l$ is increased, Fig. 9.3 stays the same shape, but Fig. 9.2 changes in the way illustrated in Fig. 8.3.

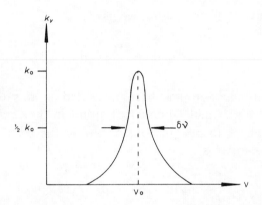

Fig. 9.3 Absorption coefficient as a function of frequency. For an optically thin absorbing column, this profile is identical with that of the dip in Fig. 9.2.

While the shape of the k_ν curve depends on the broadening mechanisms involved, the total area under it must depend on the total light absorbed – i.e., on B_{12}. The relation between k_ν and B_{12} may be found by considering the energy absorbed in a slab of gas of unit cross-section and thickness δx exposed to radiation of specific intensity I_ν as in Fig. 9.1. The energy incident per unit time on this slab in the frequency band $d\nu$ and the solid angle $\delta\Omega$ is $I_\nu \delta\Omega \, d\nu$. The energy absorbed per unit time is, by Equation 9.9, $I_\nu k_\nu \, \delta\Omega \, d\nu \, \delta x$. If I_ν is assumed not to vary appreciably over the line profile, the total energy absorbed per second is

$$I_\nu \delta\Omega \delta x \int_{\text{line}} k_\nu \, d\nu$$

The energy density is related to the specific intensity by $I_\nu = c\rho(\nu)/\delta\Omega$.

$$\therefore \qquad -\frac{dE}{dt} = c\rho(\nu)\delta x \int_{\text{line}} k_\nu \, d\nu$$

But the number of transitions $1 \to 2$ in this same slab per second is $\rho(\nu) B_{12} N_1 \delta x$ where N_1 is the number density of absorbing atoms. Each absorption removes energy $h\nu_0$, neglecting the small variation of ν over the line width.

$$\therefore \qquad -\frac{dE}{dt} = \rho(\nu) B_{12} N_1 \, \delta x \, h\nu_0$$

Equating these two expressions, we have

$$\int_{\text{line}} k_\nu \, d\nu = \frac{h\nu_0}{c} N_1 B_{12} \qquad (9.12)$$

Stimulated emission has been neglected in this equation. If there are an appreciable number of atoms in the excited level 2, the effect will be to put photons back into the beam so that the net energy removed is

$$-\frac{dE}{dt} = \rho(\nu)\{B_{12} N_1 - B_{21} N_2\} \delta x \, h\nu_0$$

Using Equation 9.3 to relate the B's, we have

$$\int_{\text{line}} k_\nu \, d\nu = \frac{h\nu_0}{c} N_1 B_{12}\left(1 - \frac{g_1 N_2}{g_2 N_1}\right) \qquad (9.13)$$

Absorption is sometimes expressed in terms of the *atomic* absorption coefficient, usually written as α_ν. This is related to k_ν by $\alpha_\nu = k_\nu/N_1$. Since it has the dimensions of area, it is often referred to as the atomic cross-section.

9.6   Oscillator strength

The quantum mechanical and classical models for the absorption of radiation can be related to one another by equating the light energy absorbed at frequency ν_0 in the transition $1 \to 2$ by N_1 atoms to that absorbed at the same frequency by $\mathcal{N}$ classical oscillators of resonant frequency ν_0. An 'oscillator strength' f for each atom can then be defined from the relation $\mathcal{N} = N_1 f$. Since there is also a change proportional to $\mathcal{N}$ in the phase velocity of the light near a resonance, and hence a change in refractive index, f is sometimes referred to as the number of dispersion electrons per atom. f is evidently related to the line strength S and the Einstein coefficients, and the form of the relation is found by calculating the absorption coefficient for light passing through an assembly of damped oscillators.

In the classical model the electrons are set into damped forced oscillation by the driving force due to the oscillating electric vector of the incident light wave. For simplicity, we shall consider oscillations in the x-direction only, produced by a plane polarized light beam travelling in the z-direction, through a medium of permittivity ϵ. The solution to Maxwell's equation for the electric vector F is

$$F_{t,z} = F_{0,0}\, e^{i\omega(t - z/v)}$$

where the phase velocity v is given by

$$v = 1/\sqrt{(\epsilon\mu_0)} = \frac{c}{\sqrt{(\epsilon/\epsilon_0)}} \equiv \frac{c}{n'} \tag{9.14}$$

n' is the complex refractive index, which can be written as

$$n' = n + i\kappa$$

Then

$$\frac{z}{v} = \frac{zn'}{c} = \frac{zn}{c} - i\frac{\kappa z}{c}$$

and

$$F_{t,z} = F_{0,0}\, e^{-\omega\kappa z/c}\, e^{i\omega(t - zn/c)} \tag{9.15}$$

Thus κ determines the absorption and n the phase velocity. κ can be related to the more familiar absorption coefficient k very simply. The intensity of the light is proportional to the square of the electric vector, so $I_z = I_0 e^{-2\omega\kappa z/c}$. Comparing this with Equation 9.10, with z replacing l,

$$2\omega\kappa/c = k \tag{9.16}$$

Both κ and n are obtained by solving the equation for damped forced oscillations of the bound electrons in the medium. At a fixed value of z Equation 9.15 may be written as $F_{t,z} = F_{0,z}\, e^{i\omega t}$. The equation of motion of the electrons subject to a driving force $-e\, F_{t,z}$ is

$$\ddot{x} + \gamma\dot{x} + \omega_0^2 x = -\,(e/m)F_{0,z}e^{i\omega t}$$

ω_0 is the resonant angular frequency, and γ covers all forms of damping, including radiation and collision damping. The solution to this equation is well known to be

$$x_{t,z} = x_{0,z}e^{i\omega t}$$

with

$$x_{0,z} = \frac{-\,(e/m)F_{0,z}}{(\omega_0^2 - \omega^2) + i\omega\gamma} \tag{9.17}$$

Now the displacement x of the electron from its equilibrium position determines the polarizability of the medium and hence its dielectric constant. The instantaneous value of the dipole due to one electron displaced a distance $x_{t,z}$ is

$$ex_{t,z} = ex_{0,z}e^{i\omega t}$$

If there are $\mathcal{N}$ electron oscillators per unit volume the polarization is given by

$$P_{t,z} = -\mathcal{N}ex_{0,z}e^{i\omega t}$$

The susceptibility, or polarization per unit field is given by $\chi = P/F$.

Using Equation 9.17,
$$\chi = \frac{\mathcal{N}e^2/m}{(\omega_0^2 - \omega^2) + i\omega\gamma}$$

This is related to the permittivity by

$$\epsilon = \epsilon_0 + \chi$$

The expression for the dielectric constant is therefore

$$\frac{\epsilon}{\epsilon_0} = 1 + \frac{\mathcal{N}e^2}{\epsilon_0 m} \frac{1}{(\omega_0^2 - \omega^2) + i\omega\gamma} \tag{9.18}$$

and the complex refractive index is found from Equation 9.14, $n' = \sqrt{(\epsilon/\epsilon_0)}$.

Equation 9.18 can be considerably simplified for the case of a gas ($n \sim 1$) and for the spectral region near the absorption line ($\omega \sim \omega_0$). If $|\omega_0 - \omega| \ll \omega_0$ then

$$(n')^2 = \frac{\epsilon}{\epsilon_0} = 1 + \frac{\mathcal{N}e^2}{2\epsilon_0 m \omega_0} \frac{1}{(\omega_0 - \omega) + i\gamma/2}$$

Converting to frequency units, this becomes

$$(n')^2 = 1 + \frac{\mathcal{N}e^2}{8\pi^2 \epsilon_0 m \nu_0} \frac{1}{(\nu_0 - \nu) + i\gamma/4\pi}$$

$$\therefore \quad (n - i\kappa)^2 = n^2 - \kappa^2 - 2in\kappa = 1 + \frac{\mathcal{N}e^2}{8\pi^2 \epsilon_0 m \nu_0}$$

$$\times \left\{ \frac{\nu_0 - \nu}{(\nu_0 - \nu)^2 + (\gamma/4\pi)^2} - i\frac{\gamma/4\pi}{(\nu_0 - \nu)^2 + (\gamma/4\pi)^2} \right\} \quad (9.19)$$

Making the approximation $n \approx 1$ (for a gas at N.T.P. $n - 1$ is of order 10^{-3} to 10^{-4}), we can find κ by equating the imaginary parts of this equation:

$$\kappa = \frac{\mathcal{N}e^2}{16\pi^2 \epsilon_0 m \nu_0} \frac{\gamma/4\pi}{(\nu_0 - \nu)^2 + (\gamma/4\pi)^2}$$

Putting $n = 1 + \delta$, we have for the real parts

$$1 + 2\delta - \kappa^2 = 1 + \frac{\mathcal{N}e^2}{8\pi^2 \epsilon_0 m \nu_0} \frac{\nu_0 - \nu}{(\nu_0 - \nu)^2 + (\gamma/4\pi)^2}$$

where the second order term δ^2 has been neglected. If $\delta \ll 1$, then $\kappa \ll 1$ also, because the second and third terms of Equation 9.19 differ only in that $\nu_0 - \nu$ replaces $\gamma/4\pi$ in the numerator. As long as we are 'outside' the line, $|\nu_0 - \nu| > \gamma/4\pi$ and $\kappa < \delta$. κ^2 is therefore also a second order term, and we have

$$\delta = \frac{\mathcal{N}e^2}{16\pi^2 \epsilon_0 m \nu_0} \frac{\nu_0 - \nu}{(\nu_0 - \nu)^2 + (\gamma/4\pi)^2}$$

The final result for a gas near an absorption line ν_0 is:

$$n - 1 = \frac{\mathcal{N}e^2}{16\pi^2\epsilon_0 m\nu_0} \frac{\nu_0 - \nu}{(\nu_0 - \nu)^2 + (\gamma/4\pi)^2} \qquad (9.20)$$

$$k = \frac{4\pi\nu_0\kappa}{c} = \frac{\mathcal{N}e^2}{4\pi\epsilon_0 mc} \frac{\gamma/4\pi}{(\nu_0 - \nu)^2 + (\gamma/4\pi)^2} \qquad (9.21)$$

(using Equation 9.16). These curves are sketched in Fig. 9.4.

Equations 9.20 and 9.21 can be rewritten in terms of the oscillator strength f_j for a particular transition of frequency ν_j by setting the number of classical oscillators $\mathcal{N}$ equal to $N_j f_j$, where N_j is the number density of atoms or molecules in the lower state of the transition considered. f_j can be interpreted as the effective number of electrons per atom for that particular transition.

$$n_\nu - 1 = \frac{e^2}{16\pi^2\epsilon_0 m} \frac{N_j f_j}{\nu_j} \frac{\nu_j - \nu}{(\nu_j - \nu)^2 + (\gamma/4\pi)^2} \qquad (9.22)$$

$$k_\nu = \frac{e^2}{4\pi\epsilon_0 mc} N_j f_j \frac{\gamma/4\pi}{(\nu_j - \nu)^2 + (\gamma/4\pi)^2} \qquad (9.23)$$

A useful relation between k and f is found by integrating Equation 9.23 to give the total area under the absorption curve in Fig. 9.4.

$$\int_0^\infty k_\nu \, d\nu = -\frac{e^2 N_j f_j}{4\pi\epsilon_0 mc} \left[\tan^{-1} \frac{\nu_j - \nu}{\gamma/4\pi} \right]_0^\infty = \frac{e^2 N_j f_j}{4\pi\epsilon_0 mc} \left\{ \frac{\pi}{2} + \tan^{-1} \frac{4\pi\nu_j}{\gamma} \right\}$$

Since ν_j is much greater than the line width $\gamma/2\pi$ the second term is also $\pi/2$.

$$\therefore \qquad \int k_\nu \, d\nu = \frac{e^2}{4\epsilon_0 mc} N_j f_j \qquad (9.24)$$

Equations 9.22 to 9.24 are expressed in c.g.s. units by replacing $e^2/4\pi\epsilon_0$ by e^2:

$$n - 1 = \frac{e^2}{4\pi m} \frac{N_j f_j}{\nu_j} \frac{\nu_j - \nu}{(\nu_j - \nu)^2 + (\gamma/4\pi)^2},$$

$$k = \frac{e^2}{mc} N_j f_j \frac{\gamma/4\pi}{(\nu_j - \nu)^2 + (\gamma/4\pi)^2}, \quad \int k_\nu \, d\nu = \frac{\pi e^2}{mc} N_j f_j$$

Equation 9.24 expresses the important result that $\int k_\nu \, d\nu$ is independent of the line shape. Although the particular expression (Equation 9.23) was derived for a classical oscillator and represents a Lorentz profile, Equation 9.24 is also valid for Doppler broadening or for any combination of damping and Doppler broadening and does not depend on classical assumptions. It was

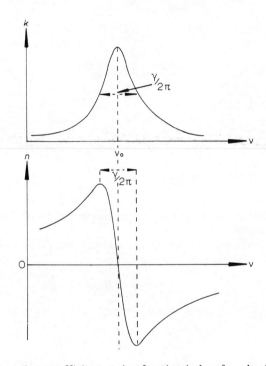

Fig. 9.4 Absorption coefficient and refractive index for classical oscillator. The absorption coefficient k and the refractive index n are plotted on the same frequency scale.

in fact implicitly assumed in the derivation of Equation 9.12 that $\int k_\nu \, d\nu$ had no line shape dependence.

The two expressions for $\int k_\nu \, d\nu$, Equations 9.12 and 9.24, complete the inter-linkage of the intensity parameters by relating f to the Einstein coefficients:

$$f_j = f_{12} = \frac{4mh\epsilon_0}{e^2} \, \nu_{12} B_{12} \qquad (9.25)$$

Table 9.1   Conversion factors for intensity parameters

		A_{21}	B_{12}	B_{21}	f_{12}	$S_{\text{el.dip.}}$
A_{21}	=	1	$\dfrac{g_1}{g_2}\dfrac{8\pi h\nu^3}{c^3}$	$\dfrac{8\pi h\nu^3}{c^3}$	$\dfrac{g_1}{g_2}\dfrac{2\pi e^2\nu^2}{\epsilon_0 mc^3}$	$\dfrac{1}{g_2}\dfrac{16\pi^3\nu^3}{3\epsilon_0 hc^3}$
B_{12}	=	$\dfrac{g_2}{g_1}\dfrac{c^3}{8\pi h\nu^3}$	1	$\dfrac{g_2}{g_1}$	$\dfrac{e^2}{4\epsilon_0 mh\nu}$	$\dfrac{1}{g_1}\dfrac{2\pi^2}{3\epsilon_0 h^2}$
B_{21}	=	$\dfrac{c^3}{8\pi h\nu^3}$	$\dfrac{g_1}{g_2}$	1	$\dfrac{g_1}{g_2}\dfrac{e^2}{4\epsilon_0 mh\nu}$	$\dfrac{1}{g_2}\dfrac{2\pi^2}{3\epsilon_0 h^2}$
f_{12}	=	$\dfrac{g_2}{g_1}\dfrac{\epsilon_0 mc^3}{2\pi e^2\nu^2}$	$\dfrac{4\epsilon_0 mh\nu}{e^2}$	$\dfrac{g_2}{g_1}\dfrac{4\epsilon_0 mh\nu}{e^2}$	1	$\dfrac{1}{g_1}\dfrac{8\pi^2 m\nu}{3e^2 h}$
$S_{\text{el.dip.}}$	=	$g_2\dfrac{3\epsilon_0 hc^3}{16\pi^3\nu^3}$	$g_1\dfrac{3\epsilon_0 h^2}{2\pi^2}$	$g_2\dfrac{3\epsilon_0 h^2}{2\pi^2}$	$g_1\dfrac{3e^2 h}{8\pi^2 m\nu}$	1

Read: $A_{21} = \dfrac{g_1}{g_2}\dfrac{8\pi h\nu^3}{c^3} B_{12}$ etc.

Notes: (1) The B's are defined in terms of *energy density* per *unit frequency interval*.

(a) The energy density per *unit wavelength interval* is c/ν^2 times that per unit frequency interval, i.e.

$$A_{21} = \frac{g_1}{g_2}\frac{8\pi hc}{\lambda^5} B_{12}, \text{ etc.}$$

(b) The *specific intensity* of radiation is $c/4\pi$ times the density, i.e.

$$A_{21} = \frac{g_1}{g_2}\frac{2h\nu^3}{c^2} B_{12}^I, \text{ etc.}$$

(2) The formulae can be converted to c.g.s. units by replacing ϵ_0 by $1/4\pi$.

(3) The integrated absorption coefficient is related to the f-value by

$$\int_{\text{line}} k_\nu \, d\nu = \frac{e^2}{4\epsilon_0 mc} N_1 f$$

where N_1 = number density in lower state.

or in c.g.s. units

$$f_{12} = \frac{mh}{\pi e^2}\nu_{12} B_{12}$$

f may then be related to A_{21} and S by Equations 9.3 and 9.7. Table 9.1 summarizes the inter-relationships of the five quantities

Table 9.2   Numerical conversion factors for intensity parameters
A is in sec^{-1}, λ in Å, f dimensionless and S in $coulomb^2$ m^2

		A_{21}	f_{12}	$S_{e.d.}$
A_{21}	$=$	1	$\dfrac{6{\cdot}670 \times 10^{15}\, g_1}{g_2 \lambda^2}$	$\dfrac{2{\cdot}819 \times 10^{76}}{g_2 \lambda^3}$
f_{12}	$=$	$\dfrac{1{\cdot}499 \times 10^{-16}\, g_2 \lambda^2}{g_1}$	1	$\dfrac{4{\cdot}226 \times 10^{60}}{g_1 \lambda}$
$S_{e.d.}$	$=$	$3{\cdot}548 \times 10^{-77}\, g_2 \lambda^3$	$2{\cdot}366 \times 10^{-61}\, g_1 \lambda$	1

Other relations

(1) $B_{12} = 60{\cdot}1 \dfrac{g_2 \lambda^3}{g_1} A_{21}$, $B_{21} = 60{\cdot}1\,\lambda^3 A_{21}$

where B is in sec^{-1} per (joule m^{-3} sec)

(2) $\displaystyle \int_{\text{line}} k_\nu\, d\nu = 2{\cdot}65 \times 10^{-6}\, N_1 f$

where k_ν is in m^{-1}, ν in sec^{-1} and N_1 in m^{-3}.

A_{21}, B_{21}, B_{12}, f and S. We have taken f to represent the absorption oscillator strength f_{12}, which is its normal meaning, but one does occasionally come across the emission f-value, f_{21}, which is related to f_{12} by

$$f_{21} = - (g_1/g_2)f_{12} \qquad (9.26)$$

The reason for the minus sign should become apparent in the next section. Table 9.2 gives the numerical conversion factors between

Table 9.3   Magnetic dipole and electric quadrupole radiation

Magnetic dipole	$A_{21} = \dfrac{1}{g_2}\dfrac{16\pi^3 \mu_0 \nu^3}{3hc^3}S_{m.d.} = \dfrac{3{\cdot}136 \times 10^{59}}{g_2 \lambda^3}S_{m.d.}$
Electric quadrupole	$A_{21} = \dfrac{1}{g_2}\dfrac{8\pi^5 \nu^5}{5\epsilon_0 hc^5}S_{e.q.} = \dfrac{8{\cdot}346 \times 10^{96}}{g_2 \lambda^5}S_{e.q.}$

The relations between A, B, and f are as in Tables 9.1 and 9.2.
In the numerical relations, A is in sec^{-1} and λ .n Å. For magnetic dipole radiation S is in units of (amp $m^2)^2$ or (J per (weber $m^{-2}))^2$. For electric quadrupole radiation S is in units of (coulomb $m^2)^2$.

Table 9.4 Conversion factors for atomic units

	Atomic unit	A_{21}	f
Elec. dip.	$7 \cdot 187 \times 10^{-59}$ (coulomb m)^2	$\dfrac{2 \cdot 026 \times 10^{18}}{g_2 \lambda^3} S^{\mathrm{a}}_{\mathrm{e.d.}}$	$\dfrac{303 \cdot 7}{g_1 \lambda} S^{\mathrm{a}}_{\mathrm{e.d.}}$
Mag. dip.	$8 \cdot 599 \times 10^{-47}$ (amp m^2)^2	$\dfrac{2 \cdot 697 \times 10^{13}}{g_2 \lambda^3} S^{\mathrm{a}}_{\mathrm{m.d.}}$	$\dfrac{4 \cdot 043 \times 10^{-3}}{g_1 \lambda} S^{\mathrm{a}}_{\mathrm{m.d.}}$
Elec. quad.	$2 \cdot 021 \times 10^{-79}$ (coulomb m^2)^2	$\dfrac{1 \cdot 680 \times 10^{18}}{g_2 \lambda^5} S^{\mathrm{a}}_{\mathrm{e.q.}}$	$\dfrac{251 \cdot 8}{g_1 \lambda^3} S^{\mathrm{a}}_{\mathrm{e.q.}}$

The atomic units of line strength are $(a_0 e)^2$, (Bohr magneton)^2 and $(a_0^2 e)^2$ for electric dipole, magnetic dipole and electric quadrupole radiation respectively. S^{a} is the line strength in atomic units, so $S_{\mathrm{e.d.}} = (e a_0)^2 S^{\mathrm{a}}_{\mathrm{e.d.}}$, etc.

A, f and S. $\int k_\nu \, \mathrm{d}\nu$ is also included in both tables. Tables 9.3 and 9.4 give conversion factors for magnetic dipole and electric quadrupole radiation and for atomic units.

The relationships between f and S for molecular lines and bands are considerably more complex. It is not apparent at first sight, for example, what, if anything, is meant by the oscillator strength of a band. Molecular transitions are therefore treated in a special post-script to this chapter.

9.7 The f-sum rule

From the interpretation of f as the optically effective number of electrons per atom, one would expect some connection between the value of f and the number of valence electrons – that is, f should be about 1 for hydrogen and the alkalis, 2 for the alkaline earths, etc. Actually, of course, each electron can take part in several different transitions, and the total oscillator strength is accordingly split between several lines. f thus represents the fraction of the available electrons participating in a particular transition and has values in the range 1 to 0·01 for reasonably strong lines. Because of its convenient size it is probably the most widely used of the line strength parameters.

The Thomas-Kuhn-Reiche sum rule states that the sum of *all* transitions from a given state should equal the number of 'optical' or valence electrons, z.

i.e.,
$$\sum_u f_{ju} + \sum_l f_{jl} + \int_0^\infty f_{j\epsilon}\, d\epsilon = z$$

where ju refers to transitions upwards from a particular state j and jl refers to transitions downwards (Fig. 9.5), while $j\epsilon$ refers to transitions to the continuum. The downward transitions represent

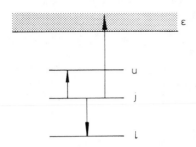

Fig. 9.5 Transitions involved in f-sum rule. Upward transitions from a given level j may go to another bound level u or to the continuum ϵ; downward transitions go to a lower bound level l.

induced emission rather than absorption. This is the reason for the minus sign in Equation 9.26: if the downward transitions are expressed in terms of absorption oscillator strengths f_{ju} they have to be subtracted from the sum. In terms of absorption oscillator strengths only,

$$\sum_{u > j} f_{ju} - (g_l/g_j) \sum_{l < j} f_{lj} + \int_0^\infty f_{j\epsilon}\, d\epsilon = z$$

This rule can occasionally be used to put relative oscillator strengths on an absolute scale, but it is less useful than might at first sight appear. Even if all the absorption f-values to higher bound states are known, the cross-section for continuous absorption has to be either known or estimated; still worse, the virtual transitions to lower levels (which cannot actually take place, because these levels are already full) have to be included in the sum. Partial f-sum rules, relating the f-values for multiplets from a given configuration, are usually more useful in practice.

9.8   Optical depth and equivalent width

The optical depth of a medium was defined in Section 9.5 as $\int_0^l k_\nu \, dx$ which, for a homogeneous medium, is $k_\nu l$. In what follows k_ν is assumed to be independent of depth, but it must be remembered that in many cases, particularly in astrophysical applications, this assumption is not justified.

A slab of gas is defined as optically thin if $k_\nu l \ll 1$ and optically thick if $k_\nu l \gg 1$. Because of the large variations of k_ν with ν, the medium may be optically thick for one spectral line and thin for another, or thick at the centre of a line but not in the wings. In

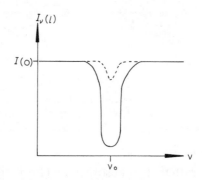

Fig. 9.6 Transmitted intensity for increasing optical depth. The dotted curve is for small optical depth ($k_0 l \ll 1$) and the full curve for large optical depth ($k_0 l \to \infty$).

absorption an optically thin path may be regarded as one in which the absorption shows no signs of saturation – doubling the path length halves the transmitted intensity. In emission, it is one in which there is no self-absorption – a photon emitted anywhere in the gas has an appreciable chance of getting out without being absorbed on the way.

Let us start with the case where the medium is optically thin over the whole profile of an absorption line and examine the effect of increasing the optical depth, by increasing l or the gas pressure. Figs. 9.6 and 9.7 show respectively the transmitted intensity $I_\nu(l)$ and the optical depth $k_\nu l$ as a function of ν. $k_\nu l$ is computed from $I_\nu(l)$ by means of Equation 9.10,

$I_\nu(l) = I(0)\,e^{-k_\nu l}$. The dotted curves represent the optically thin case, for which Equation 9.11 is valid:

$$k_\nu l = \frac{I(0) - I_\nu(l)}{I(0)}.$$

Obviously the optically thin approximation is least likely to hold at the line centre. Since it amounts to neglecting the third and higher terms in the expansion

$$I_{\nu_0}(l) = I(0)\{1 - k_0 l + \tfrac{1}{2}k_0^2 l^2 - \cdots\}$$

it is valid to within, say, 1% provided $k_0 l \leqslant 0\cdot14$ – i.e., 14% peak absorption.

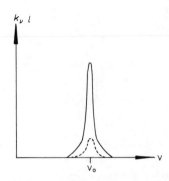

Fig. 9.7 Absorption $k_\nu l$ for increasing optical depth. The dotted and full curves in this figure correspond to those in Fig. 9.6.

In the case of the resonance line ($f \sim 1$) of a gas or vapour at a pressure of 1 milli-torr, l has to be of the order of micrometres to satisfy this condition.

If now the optical depth is increased, the $k_\nu l$ curve stays the same shape (all the ordinates are simply increased proportionately), but the I_ν curve changes shape, as shown by the solid curves in Figs. 9.6 and 9.7. Since the dip in the I_ν curve cannot go down below the base line corresponding to zero transmitted intensity, the absorption must saturate, first at the centre of the line and then outwards towards the wings.

The area of the dip in Fig. 9.6, $\int_{\text{line}} \{I(0) - I_\nu(l)\}\,d\nu$, is equal to the rate at which energy is absorbed from the incident beam in a particular spectral line. This quantity, or rather the fractional energy loss, is used to define the *equivalent width W* of the line:

$$W_\nu = \int_{\text{line}} \frac{I(0) - I_\nu(l)}{I(0)} \, d\nu \qquad (9.27)$$

The equivalent width is so called because it is the width of the rectangle shown in Fig. 9.8 that has the same area as the actual dip. To a large extent the equivalent width is independent of instrumental resolution; increasing the slit width, for example, makes the dip broader and shallower, but leaves the total area

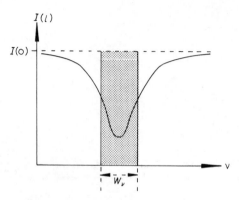

Fig. 9.8 Illustration of equivalent width. The shaded area is equal to the area between the curve and the $I(0)$ base-line.

almost unchanged. This property makes it a particularly useful quantity in astrophysics, where low light levels often preclude high resolution. As defined in Equation 9.27, W_ν is evidently measured in frequency units. It does, however, masquerade in several slightly different forms: astrophysicists tend to work with wavelengths rather than frequencies and so to measure the equivalent width W_λ in Ångström units, and either W_ν or W_λ is sometimes multiplied by 2π and called the 'total absorption'.

Equation 9.27 may be re-written in terms of the absorption coefficient as

$$W_\nu = \int_{\text{line}} (1 - e^{-k_\nu l}) \, d\nu \qquad (9.28)$$

Since k_ν is proportional to $N_j f_j$, W_ν is some function of $N_j f_j l$. For the optically thin case,

$$W_\nu = \int k_\nu l \, d\nu = l \int k_\nu \, d\nu = \frac{e^2}{4\epsilon_0 mc} N_j f_j l \qquad (9.29)$$

and the area of the dotted dip in Fig. 9.6 is equal to the area under the dotted line in Fig. 9.7. This is indeed obvious if one remembers that the two curves are in this case identical. When the medium is *not* optically thin, the area of the dip is less than the area under the $k_\nu l$ curve because of the saturation near the line peak. W_ν therefore increases less rapidly than Nfl as the optical depth is increased. This behaviour can conveniently be described by a 'curve of growth'.

9.9 Curve of growth

The curve of growth for a given spectral line describes the behaviour of its equivalent width as the number of absorbing atoms in the line of sight is increased. It usually takes the form of a plot of $\log W_\nu$ (or $\log W_\lambda$) against $\log Nfl$. The exact form of the curve depends on the width and shape of the particular line, but the width dependence can be taken care of by plotting as ordinate the dimensionless quantity $W_\nu/\delta\nu_D$ ($\equiv W_\lambda/\delta\lambda_D$) as shown in Fig. 9.9. This figure follows the reproduction in [3], [4], [5] of the curves of growth originally computed by van der Held in 1931.

Whatever the line shape, Equation 9.29 must hold as long as the absorption path is optically thin. If both sides of the equation are divided by $\delta\nu_D$ it becomes

$$\frac{W_\nu}{\delta\nu_D} = \frac{e^2}{4\epsilon_0 mc} \frac{Nfl}{\delta\nu_D} = 2\cdot 65 \times 10^{-6} \frac{Nfl}{\delta\nu_D}$$

for N in m^{-3}, l in m and $\delta\nu_D$ in sec^{-1}.

$$\therefore \qquad \log(W_\nu/\delta\nu_D) = -5\cdot 58 + \log(Nfl/\delta\nu_D) \qquad (9.30)$$

The curve of growth must therefore start as a straight line of unit gradient. The absorption in this linear part, marked A on Fig. 9.9, is shown in Fig. 9.10a.

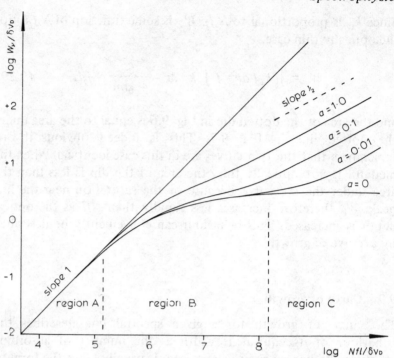

Fig. 9.9 Curves of growth. The numerical values of the abscissae correspond to N in m^{-3}, l in m and $\delta\nu_D$ in sec^{-1}. If N and l are in cm^{-3} and cm respectively, each number should be reduced by 4. The damping ratio a is defined by $a = \sqrt{(\ln 2)} \, \delta\nu_L/\delta\nu_D$.

When the centre of the line is no longer optically thin, W increases less rapidly than Nfl, and the curve starts to flatten out (region B). The linear limit and the behaviour of the curve thereafter depend on the spectral line shape. The reason for this is that a Lorentzian profile has much more prominent wings than a

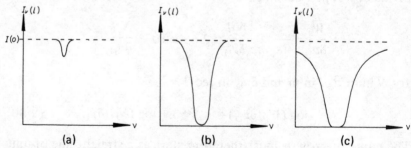

Fig. 9.10 Absorption line profile corresponding to different sections of curve of growth. Figures 9.10(a), (b), and (c) refer to the regions A, B and C respectively of Figure 9.9.

Doppler profile (Section 8.5) and consequently a lower peak height for the same total area, as can be seen in Fig. 9.11. For 1% accuracy the linear limit is reached at $W_\nu \approx 0.31 \, \delta\nu_L$ for a pure Lorentzian profile of half-value width $\delta\nu_L$, as compared with $W_\nu \approx 0.15 \, \delta\nu_D$ for a pure Doppler profile; these figures correspond to about 20 and 14% peak absorption respectively, or $k_0 l \approx 0.15$. The limiting value of W is typically 0.01 to 0.001 cm^{-1}, or a few milli-ångströms. In a great many cases of practical interest the line shape can be represented by a Voigt profile (Equation 8.31). The relative importance of the contributions from the Lorentzian and Doppler components is measured

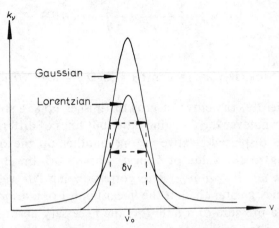

Fig. 9.11 Comparison of Lorentzian and Doppler line profiles. The two curves have the same total area (i.e. the same equivalent width) and the same half-value width $\delta\nu$.

by the ratio of their half-value widths – that is, by the damping ratio a defined in Equation 8.32:

$$a = \sqrt{(\ln 2)} \; \delta\nu_L/\delta\nu_D \approx \delta\nu_L/\delta_D$$

The curves in Fig. 9.9 are labelled according to their a-values. For the relatively small a-values usually met in practice, the central part or core of a Voigt profile is of almost pure Doppler shape, out to perhaps 3 $\delta\nu_D$ each side of the centre, while the wings are of almost pure Lorentzian shape [5]. As a result all the curves in region B remain fairly flat while absorption is mainly confined to the Doppler core where saturation is approached.

If *Nfl* is increased still further, region C, the absorption in the Lorentzian wings becomes significant (Fig. 9.10c), and *W* starts to increase much faster. In this region it is again possible to find an analytical expression for *W*. The absorption in the wings is very closely represented by Equation 9.23 with the further approximation $|v - v_j| \gg \delta v_L$, where $\delta v_L \equiv \gamma/2\pi$, giving

$$k_v = \frac{e^2 Nf}{8\pi\epsilon_0 mc} \frac{\delta v_L}{(v - v_j)^2}$$

With this expression for k_v, $e^{-k_v l}$ can be integrated to give

$$W_v = \left(\frac{e^2}{2\epsilon_0 mc} \delta v_L\right)^{1/2} (Nfl)^{1/2} \qquad (9.31a)$$

or

$$\log (W_v/\delta v_D) = \text{const} + \tfrac{1}{2} \log a + \tfrac{1}{2} \log (Nfl/\delta v_D) \quad (9.31b)$$

Consequently, the curve of growth finishes up as a straight line of slope $\tfrac{1}{2}$ whatever the damping ratio, but lines of different damping ratio are displaced relative to one another up the ordinate axis according to the value of $\tfrac{1}{2} \log a$. For $a = 0 \cdot 1$ the B-C transition occurs when $W \approx 4 \, \delta v_D$. The corresponding *Nfl* value is about 1000 times greater than at the linear limit. For a resonance line of a gas at a pressure of a milli-torr the necessary absorption path is still only a few millimetres.

Curves of growth have been considerably used in the determination of *f*-values. They also play an important part in astrophysics. A curve of growth calculated for a stellar atmosphere assuming plausible values of temperature, pressure, electron density and thickness – that is, a particular model atmosphere – can be compared with measured equivalent widths of lines from a given element. Special calculations are necessary because the curves of Fig. 9.9 apply only to a homogeneous absorbing layer. Then if Boltzmann's formula is valid the abscissa $\log N_j f_j l$ can be written as $\log Nl + \log g_j f_j - E_j/kT$, where N is the abundance of the element in question. The partition function, which is calculable and not very sensitive to temperature, can be incorporated in the constant. With an appropriate choice of T, the points corresponding to measurements on lines of known $g_j f_j$ from different initial levels E_j should fall on a single curve that can be brought into coincidence with the theoretical curve by sliding it along the abscissa axis. The distance slid gives $\log N \, l$, or, since abundances are usually given relative to hydrogen, N/N_H. Failure to match the curves, or a bad scatter in the experimental points, may indicate that the model atmosphere should be improved or that the laboratory f-values are suspect. Unreliable f-values are indeed an important source of uncertainty in work of this kind. For a fuller account of the use of curves of growth in astrophysical problems a reference such as [3] or [4] should be consulted.

9.10 Emission lines, source function and radiative transfer

In the last section it was tacitly assumed that the layer of gas under study absorbed line radiation of frequency ν_j without re-emission. This assumption is justified if the absorbing gas is much cooler than the background light source, as is the case for light from the interior of a star passing through the stellar atmosphere. In a more general treatment of the transfer of radiation through an absorbing medium, however, both emission and absorption have to be taken into account. A full treatment of radiative transfer may be found in a reference such as Cowley [3]; in this section we shall consider the relatively simple case of line emission from a homogeneous slab of hot gas.

As a starting point, the line emission from an optically thin layer of hot gas is easily found. The number of spontaneous transitions per second in a slab of unit cross-section and thickness δx is $A_{21} N_2 \, \delta x$, where N_2 is the number density of the upper state. With each of these is associated energy $h\nu_{12}$. The frequency ν_{12} can be assumed to vary little over the line profile and set equal to the central frequency ν_0. In the visible and ultraviolet regions of the spectrum induced emission can usually be neglected; for simplicity it is omitted here, although it can perfectly well be included in the analysis if necessary. The spontaneous radiation is emitted in all directions, and the energy emitted per unit solid angle per second by unit area is therefore

$$B(\nu_0) = \frac{1}{4\pi} A_{21} N_2 \, \delta x \, h\nu_0 \qquad (9.32)$$

It is convenient in radiative transfer problems to work with the emission coefficient j_ν. This is defined as the energy emitted per unit solid angle per second per unit volume per unit frequency interval – that is, the brightness per unit thickness. The integral of $j_\nu \delta x$ over the emission line,

$$\int_{\text{line}} \delta x j_\nu \, d\nu,$$

must be equal to $B(\nu_0)$.

$$\therefore \quad \int_{\text{line}} j_\nu \, d\nu = \frac{h\nu_0}{4\pi} N_2 A_{21} \qquad (9.33)$$

$B(\nu_0)$ is the brightness of the line source provided that nearly all the emitted photons escape without being absorbed – in other words, the layer of gas is optically thin.

Equation 9.33 is evidently analogous to Equation 9.12 connecting $\int k_\nu \, d\nu$ with B_{12}:

$$\int_{\text{line}} k_\nu \, d\nu = h\nu_0/c \, N_1 B_{12}$$

Taking these two equations together, we have

$$\frac{\int j_\nu \, d\nu}{\int k_\nu \, d\nu} = \frac{c}{4\pi} \frac{N_2}{N_1} \frac{A_{21}}{B_{12}}$$

The right-hand side of this equation is a constant in the sense that it is independent of the line profile. In normal conditions the various line-broadening processes for a given transition can be taken as identical in emission and absorption, so the ratio of the areas on the left-hand side can be replaced by the ratio of the ordinates, j_ν/k_ν. Using Equation 9.3 for the ratio $A_{21}B_{12}$,

$$\frac{j_\nu}{k_\nu} = \frac{2h\nu_0^3}{c^3} \frac{N_2}{N_1} \frac{g_1}{g_2}$$

j_ν/k_ν is known as the source function, S_ν.

If the levels N_2 and N_1 are thermally populated so that the ratio N_2/N_1 is given by Boltzmann's formula (Equation 9.2) this equation becomes

$$\frac{j_\nu}{k_\nu} = \frac{2h\nu_0^3}{c^2} e^{-h\nu_0/kT}$$

The right-hand side is Wien's approximation to the black body function, Equation 11.13c – that is, Equation 3.2b with $h\nu \gg kT$ (induced emission neglected).

$$\therefore \quad S_\nu = j_\nu/k_\nu = B_0^{\text{Wien}}(\nu_0, T) \tag{9.34a}$$

This result is scarcely surprising, for it is simply an application of Kirchhoff's Law relating coefficients of emission and absorption in general to the particular case of line radiation. The relation holds true even if induced emission is not negligible, provided that for k_ν is substituted k'_ν, representing absorption *minus* induced emission. Then

$$S_\nu = j_\nu/k'_\nu = B_0(\nu_0, T) \tag{9.34b}$$

where $k'_\nu = k_\nu(1 - e^{-h\nu/kT})$ and B_0 is the complete black body function (Equation 3.2b). The expression for k'_ν is obtained from

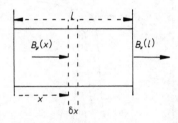

Fig. 9.12 Radiation from a column of hot gas.

Equation 9.13 together with the Boltzmann formula. Both Equations 9.34a and 9.34b, depend on the population following a Boltzmann distribution at temperature T. Alternative expressions for the source function S_ν can be found if different assumptions are made [3].

We can now investigate the line emission from a slab of hot gas as the optical depth is increased beyond the optically thin limit. Let the surface brightness of a column of homogeneous hot gas of length x be $B_\nu(x)$. In an additional layer of thickness δx (see Fig. 9.12) the energy per unit of area, solid angle, frequency and time is increased by an amount $j_\nu \, \delta x$ and decreased by $B_\nu(x) \, k_\nu \, \delta x$.

$$\therefore \quad dB_\nu = (j_\nu - B_\nu(x)k_\nu) \, dx = (S_\nu - B_\nu(x))k_\nu \, dx$$

For a homogeneous column, this equation can be integrated over the length l to give

$$[\ln(B_\nu - S_\nu)]_0^l = -k_\nu l$$

$$\frac{B_\nu(l) - S_\nu}{-S_\nu} = e^{-k_\nu l}$$

$$B_\nu(l) = S_\nu(1 - e^{-k_\nu l}) \qquad (9.35a)$$

When Equation 9.34 for the source function is valid, this becomes

$$B_\nu(l) = B_0(\nu, T)(1 - e^{-k_\nu l}) \qquad (9.35b)$$

As $l \to \infty$, $B_\nu \to B_0(\nu, T)$, the black body function for the appropriate ν and T. This is what one would expect from a qualitative argument: at large optical depth each photon must be

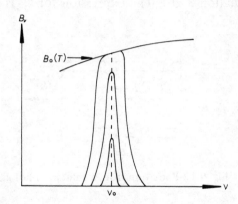

Fig. 9.13 Effect of increasing optical depth on surface brightness for line radiation. The curves represent increasing optical depth of a homogeneous column of gas at temperature T. $B_0(T)$ is the black body curve for T.

absorbed and re-emitted many times on its way out, and as this is just what happens to radiation in equilibrium in an enclosure the surface brightness must reach a saturation value equal to that of a black body. This behaviour is illustrated in Fig. 9.13 (and also Fig. 8.3). Saturation is reached first at the centre of the line and then spreads out towards the wings. At the opposite extreme, as $k_\nu l \to 0$, $B_\nu(l) \to S_\nu k_\nu l = j_\nu l$ which, integrated over the line, agrees with Equations 9.32 and 9.33.

The saturation of emission of radiation in a gas at constant temperature is not to be confused with the self-absorption and self-reversal which occur if the edges are cooler than the middle. Whereas in the homogeneous case each layer of gas contributes to the emission as well as the absorption, in the case of a temperature gradient there will be fewer excited atoms at the edges, and therefore more absorption than emission. The line peak will be first flattened and then reversed. When emission in the cooler layers is almost negligible, we get back to the case of a pure absorption line discussed in the last section.

9.11 *f*-values in diatomic molecules

This section forms a rather lengthy post-script, which may be omitted without loss of continuity by any reader not concerned with molecular *f*-values. The reasons for treating the topic at apparently disproportionate length here are, first, that it is not immediately obvious what is meant by, for example, a band oscillator strength; and, secondly, the information is not readily available in the standard textbooks. It will be seen, too, that there is ample room for confusion over the statistical weights that enter into the relations between the quantities in Table 9.1. The convention adopted here is that advocated by Tatum [7].

If we start with the nuclei frozen in position, we should have a line strength S given as in Equation 9.6 by

$$S = e^2 \{|x_{12}|^2 + |y_{12}|^2 + |z_{12}|^2\}$$

which we may write in terms of the transition moment $R_{12}(r)$ as $S = |R_{12}|^2$. The result of a slow vibrational or rotational motion of the nuclei is not to change significantly the value of this transition moment, but rather to distribute the intensity associated with it over a large number of lines. To be specific, let us consider absorption; the same arguments can be applied to emission by interchanging the single and double primes throughout. A given rotational level J'' belonging to the vibrational state v'' of the lower electronic state can combine with several different upper vibrational states v', and for each v' there are several possible J' levels corresponding to the various branches (in the simplest case these are the P, Q and R branches corresponding to $\Delta J = 1, 0, -1$). Ideally we should like to be able to write the strength of one such rotational line in the form

$$S_{v'v''J'J''} = R_e^2 \alpha_{v'v''} \beta_{J'J''}$$

where R_e is the electronic transition moment for stationary nuclei and the factors $\alpha_{v'v''}$ and $\beta_{J'J''}$ give the fractions appropriate to the particular v' and J' upper levels. If this factorization is in fact valid, the band strength $R_e^2 \alpha_{v'v''}$ is obtained by summing over all branches starting from a given J'' level (since $\sum_{J'} \beta_{J'J''} = 1$) and the electronic transition moment by summing all the band strengths with a common v'' (since $\sum_{v'} \alpha_{v'v''} = 1$). The distribution of intensity, or in this case absorption coefficient, over different *lower* vibrational and rotational levels comes about because the absorbing molecules are distributed among these levels.

Let us now look at the justification for the factorization. Using the Born–Oppenheimer approximation, which regards the effects of nuclear motion as perturbations to the electronic motion (Section 2.14), one can write the total wave function as the product $\psi_{el}(r)\psi_{vib}(r)\psi_{rot}$. The first two factors are functions of the inter-nuclear distance r, but the last has only angular dependence. The transition moment can then be written as the product of a radial part

$$\int \psi_{v'} R_e(r) \psi_{v''} \, dr$$

(where $\psi_{v'}$ and $\psi_{v''}$ are the vibrational wave functions of the two states) and an angular part which depends on the rotational wave functions only. This last corresponds to our factor $\beta_{J'J''}$. It is actually written in the form $\mathscr{S}_{J'J''}/(2J'' + 1)$ where $\mathscr{S}_{J'J''}$ is the so-called Hönl–London factor, which can be calculated reliably for almost all types of transition in diatomic molecules. According to the rules advocated by Tatum, the Hönl–London factors should be normalized so that $\sum_{J'} \mathscr{S}_{J'J''} = 2J'' + 1$, the statistical weight of the lower level J''. $\mathscr{S}_{J'J''}/(2J'' + 1)$ summed over all branches is therefore unity. If, however, there is an electronic degeneracy g'' in the lower state due to electron spin and/or Λ-doubling (see [6] and [7]), there will be g'' sub-levels of given J'', and $\sum_{J'} \mathscr{S}_{J'J''}$ summed over all of these becomes $g'' (2J'' + 1)$. It is then necessary to incorporate the g'' in the denominator of the relative line strength in order to preserve the sum rule:

i.e., $$\sum_{\text{sub-levels}} \sum_{J'} \frac{\mathscr{S}_{J'J''}}{g''(2J'' + 1)} = \sum_{\text{sub-levels}} \frac{1}{g''} = 1 \qquad (9.36)$$

We now return to the radial part of the transition moment, the square of which gives the absolute band strength, denoted by either $S_{v'v''}$ or $p_{v'v''}$:

$$S_{v'v''} = p_{v'v''} = |\int \psi_{v'} R_e(r) \psi_{v''} \, dr|^2 \qquad (9.37)$$

Since both the vibrational and the electronic wave functions (which are included in R_e) are functions of r, the integral cannot be exactly separated into the factors $R_e^2 \alpha_{v'v''}$ as we should like. Moreover, it cannot be easily or reliably calculated, because electronic wave functions are not known with anything like so much certainty for molecules as for atoms. If, however, R_e is assumed to vary only slowly with r over the relevant range of r, one can make the approximation

$$p_{v'v''} \simeq R_e^2 |\int \psi_{v'} \psi_{v''} \, dr|^2 \equiv R_e^2 q_{v'v''} \qquad (9.38)$$

$\int \psi_{v'} \psi_{v''} dr$ is known as the overlap integral, and its square $q_{v'v''}$ is the Franck–Condon factor. If the inter-nuclear potential can be expressed in a standard form – the Morse potential is the best known such – the vibrational wave functions and hence the Franck–Condon factors can be computed. The Franck–Condon factors correspond to our original factors $\alpha_{v'v''}$ and give the *relative* band strengths. They are normalized so that

$$\sum_{v'} q_{v'v''} = \sum_{v''} q_{v'v''} = 1.$$

(Vibrational levels have unit statistical weight.)

To summarize thus far: by making the two approximations

$$\psi_{el,vib,rot} = \psi_{el} \, \psi_{vib} \, \psi_{rot}$$

and R_e almost independent of r (the first being better justified than the second), one can express the relative strength of any line in any band in the form

$$q_{v'v''} \frac{\mathscr{S}_{J'J''}}{g''(2J'' + 1)}$$

The quantity R_e^2 puts these strengths on an absolute scale. The first of the approximations allows us to express the band strength $S_{v'v''}$ as the sum of all the line strengths starting from a given J''

level (or set of sub-levels) in v'' and finishing on various J' levels in v':

$$S_{v'v''J'J''} \equiv \frac{\mathscr{S}_{J'J''}}{g''(2J''+1)} S_{v'v''} \qquad (9.39a)$$

$\therefore$ by Equation 9.36 $\sum_{\text{sub-levels}} \sum_{J'} S_{v'v''J'J''} = S_{v'v''}$ (9.39b)

The second of the approximations, rather less meaningfully, allows us to express the strength of the band *system* by the relation:

$$\sum_{v'} S_{v'v''} \simeq R_e^2 \sum q_{v'v''} = R_e^2 \qquad (9.40)$$

Let us now consider the line and band *oscillator* strengths. The general relation between f and S (see Table 9.1) is

$$f_{12} = \frac{1}{g_1} \frac{8\pi^2 m\nu}{3he^2} S \equiv C_1 \frac{\nu}{g_1} S$$

where $C_1 \equiv 8\pi^2 m/3he^2$ and g_1 is the statistical weight of the lower level. The potential confusion comes from the value to be given to g_1. With the rotational line strength defined as in Equation (9.39a), the statistical weight of the lower level $2J''+1$ is already incorporated in $S_{v'v''J'J''}$. Therefore for a single line from a single sub-level:

$$f_{v'v''J'J''} = C_1 \nu S_{v'v''J'J''} = C_1 \nu \frac{\mathscr{S}_{J'J''}}{g''(2J''+1)} S_{v'v''} \qquad (9.41)$$

Can we now define a band oscillator strength? Summing f over all branches from *one* of the J'' sub-levels, by Equation 9.36:

$$\sum_{J'} f_{v'v''J'J''} = C_1 \sum_{J'} \nu S_{v'v''J'J''} = \frac{C_1 \nu}{g''} S_{v'v''}$$

since ν is almost constant for different branches with the same J''. If, furthermore, ν varies little over the whole band it is possible to define a *band oscillator strength* by

$$f_{v'v''} = \sum_{J'} f_{v'v''J'J''} = \frac{C_1 \nu_0}{g''} S_{v'v''} \qquad (9.42)$$

where ν_0 is usually taken to be the frequency of the band origin. Note that the electronic statistical weight enters into this relation.

Thus the band f-value is obtained from the line f-values by summing over all transitions from any *one* of the J'' sub-levels; and conversely the line f-value is obtained from the band f-value by means of the Hönl–London factor and the relation

$$f_{v'v''J'J''} = \frac{\mathscr{S}_{J'J''}}{2J'' + 1} f_{v'v''}$$

Let us take a little more space to check that we have arrived at a meaningful definition. Suppose there are $N_{v''}$ molecules in the lower vibrational state, distributed among the various rotational levels so that $\sum_{J''} N_{J''} = N_{v''}$. The total absorption W in the band should be proportional to the total 'Nf' value. If the band structure is quite unresolved, this is simply $W \propto N_{v''} f_{v'v''}$. But if the lines are resolved we have to sum over all transitions from all sub-levels of given J'' and then sum over all values of J'', remembering that, in the first of these sums, there are only $N_{J''}/g''$ molecules in each of the g'' sub-levels.

$$W \propto \sum_{J''} \frac{N_{J''}}{g''} \sum_{\text{sub-levels }J''} \sum f_{v'v''J'J''} = \sum_{J''} \frac{N_{J''}}{g''} \sum_{\text{sub-levels}} f_{v'v''} = \sum_{J''} N_{J''} f_{v'v''}$$

$$= f_{v'v''} \sum_{J'} N_{J''} = f_{v'v''} N_{v''} \qquad (9.43)$$

If the frequency spread over the band is not small enough to be ignored, $f_{v'v''}$ as expressed in Equation 9.42 will depend on ν and hence on J''; the sum in Equation 9.43 will then depend on the distribution of particles over the different J'' levels, and the band oscillator strength will become temperature-dependent. An objection of this type applies to any attempt to define an oscillator strength for a *band system*, as we shall now see.

The electronic oscillator strength, f_{el}, is still met in the literature on molecular intensities, despite attempts to discourage its use. It is defined as the sum of the band oscillator strengths for a given progression:

$$f_{\text{el}} = \sum_{v'} f_{v'v''} = C_1/g'' \sum_{v'} \nu S_{v'v''}$$

One can get a unique value for f_{el} only by assuming that ν is effectively constant for the whole band *system*, so that it may be brought outside the sum with some sort of mean value $\bar{\nu}$. Since a band system may extend over many hundreds of Ångströms, ν may vary by 25% or more (compared to 1 or 2% within a single band), and f_{el} is not a very meaningful quantity. In practice it is usual to make the further approximation that R_e is independent of r and hence of vibrational state so that

$$S_{v'v''}(\equiv p_{v'v''}) = q_{v'v''}R_e^2 \tag{9.44}$$

Together these two approximations yield

$$f_{el} = \frac{C_1}{g''}\,\bar{\nu}\sum_{v'} S_{v'v''} = \frac{C_1}{g''}\,\bar{\nu}R_e^2 \sum_{v'} q_{v'v''} = \frac{C_1}{g''}\,\bar{\nu}R_e^2$$

Substituting

$$R_e^2 = \frac{S_{v'v''}}{q_{v'v''}}$$

and using Equation 9.42 one can also write:

$$f_{el} = f_{v'v''}/q_{v'v''}$$

References

General

1. Kuhn, H. G. 'Atomic Spectra', Longmans, 1969
2. Woodgate, G. K. 'Elementary Atomic Structure', McGraw-Hill, 1970
3. Cowley, C. R. 'The Theory of Stellar Spectra', Gordon and Breach, 1970
4. Aller, L. H. 'The Atmospheres of the Sun and Stars', Ronald Press Co., 1963
5. Mitchell, A. C. G. and Zemansky, M. S. 'Resonance Radiation and Excited Atoms', Cambridge University Press, 1934

Diatomic Molecules

6. Herzberg, G. 'Spectra of Diatomic Molecules', Van Nostrand, 1950
7. Tatum, J. B. Interpretation of Intensities in Diatomic Molecular Spectra, *Ap. J. Supp.* **14**, 21, 1967
8. Nicholls, R. W. and Stewart, A. L. Allowed Transitions in 'Atomic and Molecular Processes', p. 247 (ed. Bates), Academic Press, 1962

CHAPTER TEN

Experimental determination of transition probabilities and radiative lifetimes

10.1 General remarks

The experimental methods described in this chapter are of rather wider application than the title would indicate. Although the determination of transition probabilities is itself important, particularly in astrophysics, the same techniques are widely used in the spectroscopy of plasmas, both laboratory and astrophysical. For instance, measurements of emission intensities and absorption coefficients yield essentially the product of a population density and an oscillator strength, and if the latter is known the former can be deduced. This aspect will be followed up in the next chapter.

Apart from the experimental determinations, calculations of transition probabilities have been made for a large number of lines. Large discrepancies between experimental and calculated values are not uncommon, and attempts to reduce these have done much towards improving both experimental and theoretical techniques. The existing data for the elements at the beginning of the periodic table (up to calcium, to date) have been collected in the National Bureau of Standards 'Atomic Transition Probabilities' [1], and these volumes also include brief accounts of, and references to, experimental and theoretical methods. For the rest of the periodic table the information exists in the form of a bibliography, also published by the Bureau of Standards [2]. No comparable data exist for molecular transition probabilities.

This chapter gives the most important methods in current use, with brief indications of their relative advantages and limitations.

It is important to realize at the outset that the one thing that all methods have in common is their inherent uncertainty. Whereas it is commonplace to measure wavelengths, and hence energy levels, to 1 part in 10^5 or 10^6, the transition probabilities between the levels are known to about 2% in a few very favourable cases (the resonance lines of sodium, for example) and to 20% or worse – considerably worse – in most of the others. Moreover, it is only too likely that results obtained by two different methods, each with a quoted accuracy of 20%, may differ by a factor of 2 or more. The N.B.S. compilation classifies an accuracy of 10–20% as 'very good' and excludes only those values for which the error seems likely to be greater than 50%. In the case of calculated values it is very difficult even to estimate the probable errors, except for the hydrogen and hydrogen-like lines and a few helium lines, which should be good to better than 1%.

The situation for transition probabilities in molecular spectra is even less satisfactory. While relative rotational line strengths, and often relative vibrational band strengths, of simple molecules can be calculated with good accuracy, the absolute transition probabilities require a knowledge of the radial part of the electronic wave function, and molecular wave functions are far less reliable than atomic. An additional complication arises in that there is often sufficient interaction between any or all of the electronic, vibrational and rotational functions to make their separation according to the Born-Oppenheimer approximation unjustifiable. There is also a considerable dearth of experimental data on molecules, presumably because of the experimental complications associated with the band structure. In recent years, however, some of the lifetime methods have been applied to the more important transitions in a number of common molecules.

10.2   Remarks on the calculation of transition probabilities

A summary of the various methods of calculation is given in the N.B.S. 'Atomic Transition Probabilities' with further references in the N.B.S. bibliography [1], [2]. The calculation of allowed atomic transition probabilities is discussed in two recent review articles [8], [9], and forbidden transitions have also been reviewed [10].

Apart from transitions in the hydrogen-like, and to some extent the helium-like, spectra, the results obtained from most methods have an accuracy between about 10 and 40%, the lower figure being very difficult to achieve. In certain cases – when both states of the optical electron may be treated as almost independent of the core – the Coulomb approximation gives satisfactory results. Variational methods based on Hartree-Fock wave functions (the self-consistent field approach with allowance made for electron exchange) are the most generally applicable, but may go badly wrong if there is appreciable configuration interaction. Iso-electronic sequences may be treated by the Z-expansion method, which is essentially a perturbation method using the parameter $1/Z$, Z being the atomic number. This again is affected by configuration interaction.

The calculation of transition probabilities is a much harder task than the calculation of energy levels, because the former are more sensitive to inaccurate wave functions. The results are particularly unreliable if there is a large degree of cancellation in the integrals $|\int \psi_1 ex \psi_2 \, d\tau|^2$ that determine the line strength (Section 9.3). Moreover, wave functions that are good for small r and bad for large r will tend to give good energy values and bad transition moments because of the r-weighting in the latter.

The calculation of relative vibrational and rotational transition probabilities in molecular spectra is also covered by review articles [11], [12], but there is no comparable review available for electronic transitions.

10.3 Survey of experimental methods

The experimental determinations of transition probabilities can be grouped under three broad headings: emission measurements, absorption and refractive index measurements, and lifetime determinations. The long-established methods are fully discussed in Mitchell and Zemansky's book [3]. A number of review articles cover more recent methods, in addition to the brief account in the introduction to 'Atomic Transition Probabilities [1]. A very full review was published by Foster in 1964 [6]. The principal developments since then have been in lifetime determinations, with some improvement in arc techniques, and these have been reviewed by Weise [7]. One of the new lifetime techniques is that of beam foil spectroscopy, and the report of the 1968 conference

[5] on this subject also includes reviews of other methods. The references listed at the end of the chapter are mostly general reviews, but a few of them review one specific method.

All the methods discussed below, except beam foil, are in principle applicable to molecular as well as atomic spectra, but in the molecular case difficulties frequently arise in resolving individual rotational lines. Most of the molecular data, therefore, refer to complete bands, and as the bands often overlap one another the reduction of the data is a good deal more complicated than for atomic spectra.

10.4   Emission measurements

The emission method is the most generally applicable, but by no means necessarily the most accurate. The basis of determinations from emission measurements is that the total line energy radiated from unit cross section of a layer of gas of thickness l is $N_2 A_{21} l h\nu_0$, so that if the layer is optically thin the brightness of the line source is $1/4\pi\, N_2 A_{21} l h\nu_0$ (Equation 9.32). The difficulties associated with this apparently simple method are easily summarized.

First, absolute intensity measurements are notoriously difficult to make with great accuracy. The normal practice is to compare the source with a standard source (Section 3.2), with due precautions over solid angles, equivalent light paths, apertures, etc. High temperature sources have necessarily to be compared with a source of very different brightness temperature.

Secondly, the condition for optical thinness may be difficult to satisfy. Sources other than shock tubes usually have cylindrical symmetry and a radial temperature gradient viewed end-on and tend to be optically thick; viewed side on, the radiation comes from an inhomogeneous layer. The temperature gradient has to be reliably known if it is to be allowed for properly. Small corrections for self-absorption in a uniform layer can be made [3] but tend to be unreliable unless they are indeed small.

Thirdly, there may be a certain amount of continuous emission from the source, and the base line must be very carefully drawn so as not to include either this or any scattered light of different wavelength. Furthermore, unless the line has a pure Doppler profile the wings extend a very long way out beyond the half width. As emission measurements have to be made at rather high pressures (atmospheric or above) in order to establish population equilibrium – see below – there is almost bound to be an appreciable Lorentzian component in the line profile, and there may easily be an error of ~7% in cutting off as far as 5 full half-widths from the line centre. This makes accurate measurement for close lines extremely difficult, and adds to the problems of establishing a true base line.

Fourthly, and probably most importantly, comes the determination of N_2, the population density of the upper state. With gases and volatile compounds one needs to know the distribution among the various excited states, and with solids there is the additional problem of the total amount of material present in the discharge. Assuming this last quantity to be known, the distribution among excited states can be calculated from Boltzmann's relation (Equation 9.2) provided there is population equilibrium at some known temperature T. Complete thermodynamic equilibrium cannot in fact exist in a source such as an arc, because energy is being constantly fed in in electrical form and let out as radiation; but in the situation known as local thermodynamic equilibrium, or LTE, roughly speaking everything *except* the radiation is in equilibrium (Sections 11.8 and 11.9). Collisions are then principally responsible for population and depopulation of excited states, and the Boltzmann and Saha relations are valid. The actual value of the temperature (and of the electron density where relevant), has to be found by one of the methods described in Sections 11.16 and 11.17. Considerable errors have been caused by de-mixing effects in arc work; the various components of the arc plasma tend to separate out, with the more highly excited atoms apparently congregating in the hotter bits of the arc. The necessary corrections for this and other effects are discussed in the more detailed references already given.

Furnaces may be used for emission work up to about 3000 K, but although they avoid the temperature and equilibrium uncertainties they are not hot enough for most purposes and may well not be optically thin. Shock tubes would seem to be suitable sources, but have been very little used in practice. In principle the temperature can be calculated from the shock conditions, but in practice this is often unreliable, and reversal temperature measurements are more reliable. The temperature range is say

2000–20 000 K, but there are liable to be problems arising from cooler boundary layers. Nearly all measurements have in fact been made with arcs of various kinds. Much of the work done with free-burning arcs is now considered unreliable because of de-mixing, inhomogeneity and deviations from LTE due to the rather low electron density ($<10^{15}$ cm^{-3}). A number of different ways of stabilizing higher current arcs have been tried – rotating wall, gas vortex, liquid vortex – and are discussed in [6]. The 'standard' source at present is the wall-stabilized arc, in which an arc column some 10 cm long and a few mm in diameter runs through holes in a series of water-cooled copper washers. The current may be as high as 100 amp, giving axial temperatures up to 13 000– 15 000 K and electron densities greater than 10^{16} cm^{-3}.

The temperature is usually obtained by plotting log $B_\nu/(g_2 A_{21}\nu)$ against ν for lines of known *relative* transition probabilities. If there is true LTE the plot is a straight line of slope h/kT.

The accuracy of relative f-value measurements is, of course, far higher than that of absolute f-values, partly because relative intensity measurements are much easier to make than absolute and partly because the absolute number density of atoms no longer enters into the calculation. Of course the problems of absolute temperature determination and temperature inhomogeneities still remain unless the lines all have the same upper state.

10.5 Absorption and dispersion measurements

Near an absorption line both the real part of the refractive index, which gives rise to dispersion, and the imaginary part, which gives rise to absorption, are proportional to $N_1 f$, where N_1 is the number density of the lower state of the transition and f is the corresponding oscillator strength (Equations 9.22 and 9.23). To obtain absolute f-values from either of these relations it is necessary to know N_1. As with N_2 in the case of emission measurements, this is very often the largest source of error. Even in the case of lines starting from the ground state, the vapour pressure is likely to be subject to uncertainty. For absorption from an excited state it is also necessary to establish the existence of local thermodynamic equilibrium at a known temperature in order to calculate the Boltzmann factor and the partition function. Since the quantity actually measured is $N_1 fl$ it is further necessary to have a well-defined path length *l* or to make suitable allowances for inhomogeneity along the path. Apart from these difficulties, which are common to both emission and absorption methods, it must be remembered that in LTE it is difficult to populate excited states, other than low-lying or metastable states, sufficiently to obtain measurable absorption from them.

This section describes two absorption methods, integrated absorption coefficient and equivalent width, and one dispersion method, the hook technique. These three are the most widely used methods, but there are a number of variations on both themes [3], some of them extremely ingenious – for example, line absorption and Faraday rotation.

10.5.1 Integrated absorption

This method uses the relations

$$\int k_\nu \, d\nu = \frac{e^2}{4\epsilon_0 mc} N_1 f \quad \text{or} \quad \int k_\nu d\nu = N_1 B_{12} \, h\nu_0/c$$

(Equations 9.24 and 9.12). The absorption coefficient is not measured directly, but has to be deduced from

$$k_\nu = \frac{1}{l} \ln \frac{I_0}{I_\nu}.$$

The integrated absorption is then found from the area under the curve when k_ν is plotted against ν. The principal difficulty associated with this method is that of using high enough resolution to obtain a true line profile. Ideally the instrumental width, or resolution limit, should be much less than the line width. If this condition is not fulfilled, the folding together of the instrumental and true line profiles flattens the curve out and gives too low a value for the area. This is basically because one is going over to measuring the equivalent width, which is no longer proportional to Nfl once one is off the linear part of the curve of growth (Section 9.9). The greater the absorption, the worse is the error; for example, for 2/3 peak absorption ($k_0 l \sim 2$) the error in $\int k_\nu d\nu$ is 10% for a pure Doppler profile and 5% for a pure Lorentzian, and this error increases rapidly with increasing peak absorption. Since the Doppler width of a moderately heavy atom at room temperature is comparable with the resolution limit of a large grating, accurate measurements of integrated absorption require considerable care. One common device is to pressure-broaden the line. There are two more possible sources of error to be guarded against in both this and the equivalent width method. One is the presence of scattered light, which reduces the apparent absorption. The other is the absorption in the far wings of the line, which is difficult to measure accurately. Both errors result in too low a value for the oscillator strength.

10.5.2 *Equivalent width*

The equivalent width W_ν or W_λ, defined in Equation 9.27, measures the total energy extracted from the beam and is independent of the resolution over a wide range. It is in fact the analogue of the absolute emission method, and, like this, requires an optically thin gas layer if it is to be used directly in the form $W_\nu = e^2/(4\epsilon_0 mc)\, Nfl$ (Equation 9.29). This of course corresponds to the linear part of the curve of growth, Fig. 9.9. It is, however, possible to use the curve of growth to extend the range of application of the equivalent width method. This is done as follows. Theoretical curves of growth are plotted for various values of the damping constant a (these are obtained by numerical integration of the Voigt profile, Equation 8.31). W is then measured as a function of Nfl, either by increasing the path length or pressure for one line or by using several lines with the same N but known relative f-values (an atomic multiplet or the rotational lines of a band). If the experimental plot of log W against log Nfl is then slid horizontally over the theoretical curves it should eventually fit over one of them. One can then read off both the experimental value of a and the additive constant to log Nfl, i.e., the absolute value of Nf. In practice, the experimental scatter makes this difficult to apply very accurately, particularly if the points fall on the flattish part of the curve, where a small error in W leads to a very large error in Nfl. Very weak lines fall on the linear part, where the equivalent width and integrated absorption are identical, and very strong lines go up to the square root part where

$$W \propto \sqrt{(\delta\nu_L)}\sqrt{(Nfl)}$$

If pressure broadening is negligible, so that the Lorentzian contribution to the line is determined entirely by the natural width, the experimentally determined value of

$$\delta \nu_L = \frac{a \, \delta \nu_D}{\ln 2}$$

then gives an independent estimate of τ, the lifetime of the upper state.

The fitting of experimental equivalent widths to a theoretical curve of growth is the same procedure as is used to determine abundances in astrophysics (Section 9.9), only there lines of known f were used to find N, and here known values of N are required to determine f.

10.5.3   Hook technique

This method of determining the oscillator strength of a line from the anomalous dispersion at its edges was first devised by Roshdestwensky; it is a particularly convenient and accurate way of using the relationship between n and f given by Equation 9.22. If one is concerned with the region 'outside' the line ν_0 – that is,

$$|\nu_0 - \nu| \gg \gamma/4\pi$$

this equation reduces to

$$n - 1 = \frac{e^2}{16\pi^2 \epsilon_0 m \nu_0} \frac{Nf}{\nu_0 - \nu} = \frac{e^2}{16\pi^2 \epsilon_0 mc^2} \frac{Nf\lambda_0^3}{\lambda - \lambda_0} \quad (10.1)$$

The absorbing gas is contained in one arm of a Jamin-type two beam interferometer (a Mach–Zehnder interferometer is generally used in practice, as shown in Fig. 10.1) illuminated by a continuous source and arranged to give horizontal fringes focused on the slit of a stigmatic spectrograph. Maxima occur for $by = p\lambda$, where y is distance up the slit, as described in Section 6.21. The fringes extend almost horizontally across the spectrum when there is zero path difference between the beams (Fig. 10.2a). Near an absorption line, however, the rapid change in refractive index and so in optical path nl in the top beam curves the fringes into a hyperbolic form, as shown in Fig. 10.2b. The fringes in fact trace out the $n - 1$ curve of Fig. 9.4. In principle, the f-value can be

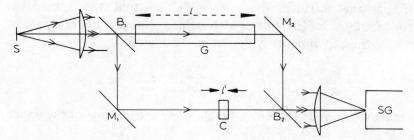

Fig. 10.1 Mach–Zehnder interferometer used for 'hook' measurements. B_1 and B_2 are the beam splitters and M_1 and M_2 are the mirrors. G is the gas cell, C the 'compensating plate' and SG the spectrograph. The source S is restricted in area but is not a point source.

found by tracing the behaviour of a given fringe, whose equation is $by + (n - 1)l = p\lambda$ (Equation 6.35). In the hook method a 'compensating plate' of thickness l' and refractive index n' is inserted in the lower beam, modifying the fringe equation to

$$by + (n - 1)l - (n' - 1)l' = p\lambda,$$

where p is now a large integer. Owing to the large path difference, the fringes away from the absorption line now slope diagonally across the spectrum (Fig. 10.3a). The variation of both n and n' with λ can be ignored in any small region clear of absorption lines, and the number of fringes Δp crossed in wavelength interval $\Delta\lambda$ at constant height on the slit is given by

$$p\Delta\lambda = -\lambda\Delta p \qquad \text{or} \qquad \lambda\,\Delta p/\Delta\lambda = -p = K, \text{say}$$

Near the absorption line, however, the path increase in the upper beam from the rapidly rising refractive index must eventually overtake the constant path increment in the lower beam, reversing the slope of the fringes and thus forming hooks (Fig. 10.3b). The

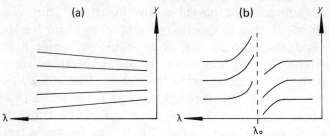

Fig. 10.2 Interference fringes (a), far from, and (b), close to, an absorption line λ_0. Wavelength is shown increasing to the left so as to conform with Fig. 9.4 showing the variation of refractive index with wavelength.

hook wavelength is defined by $(dy/d\lambda)_{const.p} = 0$ which, from Equation 10.2 with $dn'/d\lambda \sim 0$, gives

$$-l \frac{e^2}{16\pi^2\epsilon_0 mc^2} \frac{Nf\lambda_0^3}{(\lambda_H - \lambda_0)^2} = p = -K$$

where λ_H is the hook wavelength. The two roots of this equation determine the hooks on either side of the line. The distance Δ between hooks is given by

$$\Delta = 2|\lambda_H - \lambda_0|, \quad \text{or} \quad \frac{e^2}{4\pi^2\epsilon_0 mc^2} \lambda_0^3 Nfl = K\Delta^2 \quad (10.3)$$

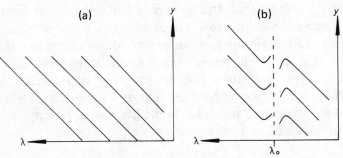

Fig. 10.3 High order interference fringes (a), far from, and (b), close to, an absorption line λ_0.

If several close lines are contributing to the refractive index, as in a multiplet, Equation 10.1 has to be summed over all relevant lines:

$$n - 1 = \frac{e^2}{16\pi^2\epsilon_0 mc^2} \sum_r \frac{N_r f_r \lambda_r^3}{\lambda - \lambda_r},$$

and the hook condition becomes

$$\frac{e^2}{16\pi^2\epsilon_0 mc^2} \sum_r \frac{N_r f_r \lambda_r^3}{(\lambda_H - \lambda_r)^2} = \frac{K}{l} \quad (10.4)$$

In general there are $2r$ such equations, corresponding to two hooks for each of r lines, and the r unknowns $N_r f_r$ can therefore be obtained from the solution of $2r$ simultaneous equations. Since the hook near any one line is mainly determined by the Nf of the nearest line, with the other lines contributing corrections, the solutions can usually be obtained quite quickly by successive approximations. The same method can be applied to the rotational lines in a molecular band, provided that the rotational structure is sufficiently open for hooks to be measured between the lines, as in a hydride. An alternative approach for close rotational structure if the band has a well-defined head is to measure the single hook near the band head; if the *relative* rotational line strengths are known, this one measurement then determines the band oscillator strength.

The hook method has two great advantages over the absorption methods and one disadvantage; the disadvantage is the sensitivity: the magnitude of Nfl required for a measurable hook is at least a factor of 10 greater than that required for respectable absorption. The advantages are first that the accuracy is not dependent on resolution, optical thinness, or any assumptions about line shape, and secondly that it is nearly always easier to measure distances accurately than intensities, and the problems of scattered light, absorption from trace impurities, and the inclusion of the far wings of the line are all irrelevant in the hook case. Reference [13] gives a full description of the method.

10.6   Lifetime measurements

There are three rather different ways of measuring lifetimes of excited states: delay time, beam foil spectroscopy, and Hanle effect. A more detailed account than is possible here is to be found in a recent review article [14]. None of these methods is universally applicable. Some work only for states that can be excited directly from the ground state, others are no good if the lifetime is too short (strong line) or the intensity too low (weak line) for accurate measurement. Amongst them, however, they cover a very wide range of transitions, both atomic and molecular, and they have in common the one immensely important advantage that there is no need to know the population of the initial level. All the methods we have so far discussed determine essentially Nf, and the discrepancies in the absolute f-values are probably in almost all cases primarily due to the uncertainties in N. It must be remembered that a lifetime does not yield directly an oscillator strength, unless there is only one allowed transition from the excited state. However, if relative values of A_{2j} can be obtained from emission, they can be put on an absolute scale by a measurement of τ. Moreover since $A \propto \nu^3$, it may be possible to ignore any very long wavelength transitions.

In the description of the delay methods that follows (and also in that of the Hanle effect) reference is made to excitation by resonance radiation. It is now becoming possible to use the tunable dye laser as an exciting source. Its high intensity makes it ideal for selective population of excited levels, and its short pulse duration allows good time resolution. New ways of exploiting these advantages are constantly being devised at the present time, and the following sections should be read with this development in mind.

10.6.1 Delay methods

The *delayed coincidence method* is shown schematically in Fig. 10.4. The gas is excited by a pulsed electron beam (or occasionally by a pulse of resonance radiation), which also triggers off the time-to-pulse-height converter. The stop signal for this is provided

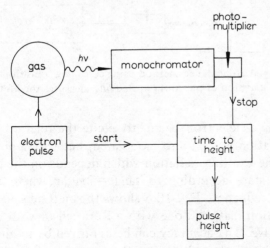

Fig. 10.4 Schematic arrangement for measurement of lifetimes by delayed coincidence. For explanation, see text.

by the emitted radiation, via a monochromator and photomultiplier, so that the pulse height is proportional to the time between the excitation and decay of the relevant excited level. Building up statistics for dN/dt as a function of t should result in an exponential of time constant τ. A variant of this is to use a coincidence counter in place of the time-to-pulse-height converter and to record coincidences between the exciting electron pulse and the emitted radiation as a function of a variable delay time t introduced to the signal from the electron pulse. In yet another variant, shown schematically in Fig. 10.5, the light pulses from transitions into and out of a given level are fed into a coincidence counter and the number of coincidences recorded as a function of the delay time introduced into the signal from the first transition. In practice it is usually difficult to find a suitable pair of transitions to which to apply this variation.

The phase shift method uses modulated instead of pulsed excitation, usually in the form of resonance radiation but

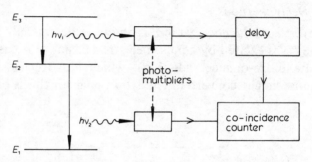

Fig. 10.5 Delayed coincidence method using cascading transitions. The delay between the emission of photons $h\nu_1$ and $h\nu_2$ measures the lifetime of E_2.

sometimes as an electron beam, to excite the relevant level. The emitted radiation is then also modulated, but with a phase shift θ relative to the exciting radiation which depends on the lifetime of the excited state according to $\tan \theta = 2\pi\nu_m \tau$ where ν_m is the modulating frequency. Fig. 10.6 shows the method schematically. The modulation may be done with a Kerr cell or with a standing ultrasonic wave. The accuracy can be improved by modulating the voltage on the photomultiplier cathode so that θ is measured at the beat frequency of a few kilocycles rather than at the modulation frequency, which has to be in the megacycle range.

The delay methods can be used for lifetimes from a few microseconds to a few nanoseconds. Apart from difficulties associated with scattered light and insufficient resolution (they are usually used with broad band-pass monochromators or filters to obtain enough intensity in the fluorescent light), they have two principal potential sources of error. The first is cascading: unless the relevant level is excited by resonance radiation or by a tuned dye laser, a number of higher levels may be excited at the same time and subsequently decay to the level under investigation. The latter is therefore re-populated from above, so that its apparent lifetime is prolonged. In principle, even with electron excitation it should be possible to excite only the required level, but in practice the cross-section near the threshold energy is so small that considerably higher electron energies have to be used. The second

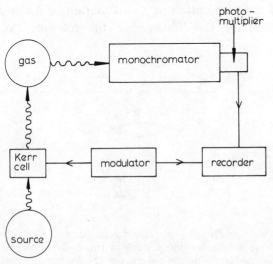

Fig. 10.6 Schematic arrangement for measurement of lifetimes by phase shift. For explanation, see text.

principal difficulty affects all levels emitting to the ground or a metastable state and is usually known as imprisonment of radiation. If the gas is not optically thin, a number of photons will be absorbed and re-emitted one or more times before they eventually get out, and the effect is, again, to prolong the apparent lifetime of the excited state. The effect can be checked for by measuring apparent lifetime as a function of pressure. A further source of error that may affect the longer-lived states is collisional depopulation; this of course shortens the apparent lifetime. Both these last troubles may be avoided by going to sufficiently low pressures (atomic beams have been used for this reason), but one may then run into difficulties with the low light intensities. The accuracy claimed for lifetime measurements is usually around 10%, is often better, but may in fact be considerably worse if systematic errors have not been eliminated.

10.6.2   Beam foil method

The beam foil light source was described in Section 3.7.5. Its application to lifetime measurements is reviewed in [15]. A beam of ions of various degrees of ionization emerges from the thin foil in different states of excitation. The excited states decay as the beam travels downstream from the foil, and the rate of decrease of intensity of any particular line as a function of distance from the foil gives directly the lifetime of the relevant excited state.

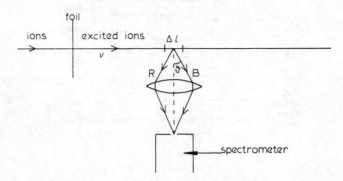

Fig. 10.7 Reason for Doppler broadening in beam foil spectroscopy. The excited ions have a velocity component in the R-direction of $v \sin \theta$ away from the spectrometer, giving a red-shift. For the B-direction there is an equal blue shift.

Historically, this method is the successor to the experiments of Wien in the 1920s with canal rays, in which the lifetime was measured from the decay of emitted radiation as the excited ions in a discharge tube travelled beyond the cathode. In practice, in the beam foil method, the spectrometer is kept fixed, and the intensity of the required lines is measured while the foil is moved upstream. It is necessary to monitor the constancy of the beam while this is going on, either by measuring the total charge collected at its end or by using a second photomultiplier.

The principal difficulties of the method are cascading from higher excited states, low light intensity, and large Doppler broadening. The reason for this last is illustrated in Fig. 10.7. The velocity of the beam is of order 5×10^8 cm sec^{-1}, and because of the finite acceptance angle of the spectrometer the direction of observation is not strictly perpendicular to the beam direction. For light travelling along ray B the ions have a component of velocity along B of $v \sin \theta$, resulting in a blue shift, whereas for ray R the velocity is $v \sin \theta$ away from the spectrometer, giving a red shift. Since all rays in the cone RB contribute to the image, the result is a broadened line. The low intensity makes a wide aperture essential, and the Doppler broadening may be as high as

$$\frac{\delta\lambda}{\lambda} = \frac{v \sin \theta}{c} \approx 0.01$$

On the other hand, the density in the beam is so low that there are no difficulties with imprisonment of radiation or collisional depopulation. More importantly, this is the only method other than that of emission line intensity which can be applied to ionic spectra. It is best suited for lifetimes of about 10^{-8} sec, during which time the ions travel a few cm. With shorter lifetimes the decay is too fast to measure accurately, bearing in mind that it is usually necessary to accept light from an appreciable length of beam Δl to obtain sufficient intensity, so that the space resolution is rather poor. If the lifetime is very long, so few ions decay in the path Δl that inadequate intensity again becomes a problem.

10.6.3   Hanle effect

This technique again is a resurrection from the 1920s, when the effect, investigated by Hanle, was known as magnetic depolarization of resonance radiation. It is now also known as zero-field level crossing, being a special case of the level-crossing experiments described in Section 7.13 for the measurement of hyperfine structure. For a very brief qualitative description of the effect it is simplest to use a semi-classical model, but when there is hyperfine structure or close fine structure in the excited level a proper quantum-mechanical treatment is required. A typical experimental lay-out is shown in Fig. 10.8. Light from the source of resonance radiation travelling in the y-direction is plane-polarized in the x-direction before entering the resonance vessel where it is absorbed and re-emitted. The emitted radiation has the same polarization as the exciting radiation, and a photo-multiplier on the x-axis therefore records no signal. If now a magnetic field is

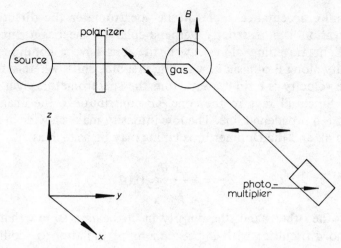

Fig. 10.8 Schematic arrangement for Hanle effect. The photomultiplier in the x-direction registers a signal when field B is applied in the z-direction.

applied to the resonance vessel in the z-direction, the resonance radiation is partly de-polarized, and a signal is registered, rising at strong field to half the intensity in the absence of the polarizer. This is because the direction of oscillation of the electrons, originally parallel to the x-axis, now precesses about the z-axis. If the precession is sufficiently rapid, half the oscillations may be considered as in the y-direction, thus radiating in the x direction. The lifetime enters into the story because if the precession is not very fast some of the atoms will have decayed before the precessional cycle is completed, so that the intensity component

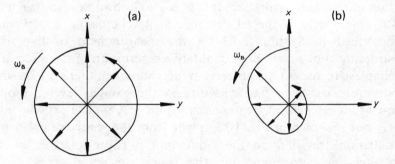

Fig. 10.9 Classical interpretation of Hanle effect. (a) represents small damping ($\omega_B > 1/\tau$) and (b) represents large damping ($\omega_B \sim 1/\tau$). The length of each spoke in these figures is proportional to the intensity of the radiation at the corresponding direction of polarization.

in the y-direction is less than the original x-component. The signal thus depends on the relative magnitudes of the precessional and decay times, as shown schematically in Fig. 10.9: (a) represents the situation when the precessional frequency ω_B is large compared to the damping constant $\gamma = 1/\tau$ and (b) shows them comparable. Simply by treating the system as an oscillator precessing with angular frequency ω_B and decaying amplitude $\exp(-\frac{1}{2}\gamma/t)$ it can be shown (see, for example, [16]) that the signal is given by

$$I = \text{const.} \left\{ 1 - \frac{\gamma^2}{\gamma^2 + 4\omega_B^2} \right\}$$

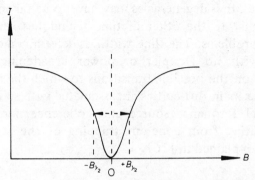

Fig. 10.10 Signal as a function of magnetic field in Hanle effect. The curve is an inverted Lorentzian, and the field at the half-value points, $B_{1/2}$, is given by $B_{1/2} = \hbar/2\mu_B \; 1/g_J\tau$.

where the Larmor precessional frequency $\omega_B = g_J\mu_B B/\hbar$ (μ_B is the Bohr magneton and g_J the Lande g-factor). This expression for I has the inverted Lorentzian form shown in Fig. 10.10. The signal has half its maximum value when

$$\frac{\gamma^2}{\gamma^2 + 4\omega_B^2} = \frac{1}{2} \qquad \text{or} \qquad \omega_B = \frac{\gamma}{2} = \frac{1}{2\tau}$$

$$\therefore \quad g_J\tau = \frac{\hbar}{2\mu_B B_{1/2}}$$

τ can therefore be determined from the half-width of the signal-versus-field curve if g_J is known.

Looked at from the quantum-mechanical point of view, all the magnetic sub-levels of the upper state are populated by the incident radiation in zero field, since they are then degenerate. The emitted radiation from these sub-levels is coherent and interferes destructively in the x-direction. To lift the degeneracy it is necessary to separate the sub-levels by an amount ΔE_B greater than their natural width $\Delta E_{nat} \sim \hbar/\tau$ leading to $g_J \mu_B B > \hbar/\tau$; this is fulfilled beyond the half-width point. However, any hyperfine structure or close fine structure within this energy range must also be brought into account. Further degeneracies, with consequent changes of polarization, occur whenever the field is such that the energies of two hyperfine components overlap, as already seen in level-crossing spectroscopy. The Hanle effect is concerned primarily with the zero-field degeneracies, but because the field must be raised to $10^{-3} - 10^{-2}$ Wb m^{-2} in order to establish the width these other degeneracies may have to be taken into account.

In contrast to the other lifetime techniques, this one has no cascading problems. The 'line width' is determined entirely by the lifetime, with no Doppler or power broadening. Apart from limitations on the possible transitions to which the method can be applied, the main difficulties are associated with self-absorption in the original resonance source and coherence narrowing of the 'line', resulting from coherent trapping of the radiation. These effects are explained in [16].

References

General

1. Wiese, W. L., Smith, M. W. and Glennon, B. M. 'Atomic Transition Probabilities', N.S.R.D.S. N.B.S. 4 vol. I H–Ne, 1966; N.B.S.22 vol. II Na–Ca, 1969
2. Miles, B. M. and Wiese, W. L. 'Bibliography on Atomic Transition Probabilities', N.B.S. Spec. Pub. 320 (1916-1969), 1970; and supplement (1969-1971), 1971
3. Mitchell, A. C. G. and Zemansky, M. S. 'Resonance Radiation and Excited Atoms', Cambridge University Press, 1934
4. Allen, C. W. 'Astrophysical Quantities', Athlone Press, 1963
5. Bashkin, S. (ed.) 'Beam Foil Spectroscopy', Gordon and Breach, 1968
6. Foster, E. W. Measurement of Oscillator Strengths, *Rep. Prog. Phys.* 27, 469, 1964
7. Wiese, W. L. (a) Transition Probabilities, *Methods of Exptl. Phys.* (ed. Bederson and Fite) 7A, 117, 1968; (b) Electric Arcs, *Methods of Exptl. Phys.* (ed. Bederson and Fite), 7B, 307, 1968

Theoretical

8. Crossley, R. J. S. Calculation of Atomic Transition Probabilities, *Adv. in Atomic and Molecular Phys.* **5**, 237, 1969
9. Layzer, D. and Garstang, R. H. Theoretical Atomic Transition Probabilities, *Ann. Rev. of Astron. and Astrophys.* **6**, 449, 1968
10. Garstang, R. H. Forbidden Transitions, 'Atomic and Molec. Processes' (ed. Bates), Academic Press, 1962
11. Nicholls, R. W. and Stewart, A. L. Allowed Transitions, 'Atomic and Molec. Processes' (ed. Bates), Academic Press, 1962
12. Tatum, J. B. Interpretation of Intensities in Diatomic Molecular Spectra, *Ap. J. Supp.* **14**, 21, 1967

Particular methods

13. (Hooks) Marlow, W. C. Hakenmethode, *Appl. Optics* **6**, 1715, 1967
14. (Lifetimes) Corney, A. The Measurement of Lifetimes of Free Atoms, Molecules and Ions, *Adv. in Electronics and Electron Phys.* **29**, 115, 1970
15. (Beam foil) Bickel, W. S. Mean Life Measurements using Beam Foil Light Source *Appl. Optics* **7**, 2367, 1968
16. (Hanle) De Zafra, R. L. and Kirk, W. Measurement of Atomic Lifetimes by the Hanle Effect, *Amer. J. Phys.* **35**, 573, 1967

CHAPTER ELEVEN

Elementary plasma spectroscopy

11.1 Introduction

It has been seen in the preceding chapters that the widths, shapes and intensities of spectral lines depend on the temperature, pressure and electron density as well as on the intrinsic atomic or molecular properties. If the broadening and other physical processes are properly understood and the necessary atomic parameters are known, analysis of the spectral lines can give information about the physical conditions in the emitting or absorbing gas. In laboratory conditions spectroscopic techniques have the advantage over some other methods – probes, for example – that they can be applied without in any way interfering with the plasma, and in the early days of laboratory plasma physics spectroscopy was a most important diagnostic tool. In this role it has now been to a large extent replaced by other optical techniques made possible by the development of lasers and masers. Nevertheless, spectroscopy still plays a major part in determining the physical processes going on in the plasma, and it is important to test the validity of theories of line broadening, emission intensities, and so on in known conditions in order to apply them to astrophysical plasmas where there is no alternative to spectroscopic diagnosis.

In this chapter a brief account of plasma characteristics is followed by the introduction of Saha's equation for ionization equilibrium and a discussion of the concepts of temperature and equilibrium in general. Continuous emission and absorption is

342

treated next, and the remainder of the chapter is concerned with the evaluation of temperatures and number densities.

The word 'plasma' is sometimes used loosely (and technically incorrectly), both in this book and elsewhere, to describe a mass of hot gas regardless of its degree of ionization. The proper definition of a plasma is quite unambiguous, as will be seen in the next two sections, but the word is a convenient shorthand to describe any vapour or gas hot enough to be partly ionized.

11.1.1   Note on units

In the S.I. system the unit of number density is m^{-3}, but in the references cited in this chapter and in nearly all current work on plasma spectroscopy number densities are given in cm^{-3}. I have used cm^{-3} in this chapter, except when the number density is incorporated in a formula, because the advantage of familiarity with the actual numerical values seems to me to outweigh the disadvantage of having to multiply occasionally by 10^6 when using the number density in a formula.

11.2   Properties of plasmas

A plasma is an assembly of atoms, molecules, ions and electrons that may be treated statistically in the same way as a solid, liquid or neutral gas. The plasma as a whole is electrically neutral, the total ionic charge being equal to the total number of electrons. It can easily be shown, in fact, that the charge neutrality must hold even when quite small volumes are considered: if N_i and N_e are the number of densities of ions (assumed singly charged) and electrons respectively, a sphere of radius r contains a charge of $4/3 \, \pi r^3 (N_i - N_e)e$ and has a potential of $r^2 (N_i - N_e)e/3\epsilon_0$. A typical value of N_e is $10^{16} \, cm^{-3}$ (i.e., $10^{22} \, m^{-3}$), and the potential required to maintain a 1% difference between N_i and N_e within a sphere of radius 1 mm then works out to nearly a million volts. Such a potential cannot be maintained in a gas, where the particles can move freely to neutralize the excess charge. It is therefore quite safe to take the average particle densities as given by $\overline{N}_i = \overline{N}_e$ until one gets down to distances so small that the out-of-balance potential is comparable to the thermal energy. The order of magnitude of such a distance is given by the Debye screening radius ρ_D.

11.3   Debye radius

A plasma is distinguished from a hot gas by the importance of collective effects. The long range of the electrostatic forces between charged particles means that each particle interacts with an appreciable number of other particles at the same time, and the motions of the particles must be correlated. The Debye shielding radius ρ_D is a criterion of the importance of such effects: it measures the distance to which the electric field of an individual electron or ion extends before it is effectively shielded by the oppositely charged particles. Individual interactions between particles are important only over distances less than ρ_D; for distances greater than ρ_D the collective effects dominate. The criterion for the existence of a plasma, or the predominance of collective effects, therefore amounts to requiring that ρ_D be much smaller than the dimensions of the plasma.

The expression normally used for the Debye radius is that originally derived by Debye for an electrolyte. Suppose a particle of charge q is injected into an electrically neutral plasma, for which the mean potential is zero and $\overline{N}_i = \overline{N}_e$. In the neighbourhood of q the potential has some value V, and there is a local charge imbalance giving a net charge density $\rho = e(N_i - N_e)$. If the particle density is sufficiently high for V to vary smoothly, unaffected by individual ions or electrons, Poisson's equation must be satisfied:

$$\nabla^2 V = - \rho/\epsilon_0 = - e(N_i - N_e)/\epsilon_0$$

An electron at potential V has energy $- eV$. In equilibrium the number of electrons with this energy is given by the Boltzmann distribution (Equation 3.6) in the form

$$N_e = \overline{N}_e \exp(eV/kT)$$

since $N_e = \overline{N}_e$ when $V = 0$. Similarly, for the ions,

$$N_i = \overline{N}_i \exp(- eV/kT) = \overline{N}_e \exp(- eV/kT)$$

The differential equation determining V becomes

$$\nabla^2 V = - \frac{e}{\epsilon_0} \overline{N}_e \{\exp(-eV/kT) - \exp(eV/kT)\}$$

Assuming that $eV \ll kT$, which will be justified below,

$$\nabla^2 V \approx \frac{e}{\epsilon_0} \bar{N}_e \frac{2eV}{kT} = \frac{2e^2 \bar{N}_e}{\epsilon_0 kT} V$$

For the notional point charge disturbance, V must have spherical symmetry, so that

$$\nabla^2 V \to \frac{1}{r^2} \frac{\partial}{\partial r} \left(r^2 \frac{\partial V}{\partial r} \right)$$

For the boundary conditions $V \to 0$ as $r \to \infty$ and $V \to q/4\pi\epsilon_0 r$ as $r \to 0$, this equation has the solution (easily verified by substitution)

$$V = \frac{q}{4\pi\epsilon_0 r} e^{-r/\lambda} \qquad \text{where} \qquad \lambda^2 = \frac{\epsilon_0 kT}{2e^2 N_e}$$

λ determines the 'decay distance' of the potential distribution produced by the charge, and this is defined as the Debye shielding distance – i.e.,

$$\rho_D = \left(\frac{\epsilon_0 kT}{2e^2 N_e} \right)^{1/2} \tag{11.1a}$$

An expression of the same order of magnitude is obtained if one considers the penetration of an external field from, say, the boundary of the plasma, in which case ρ_D may be regarded as a sheath thickness.

If one is concerned with transient plasmas or very rapid disturbances, the ions may not be able to move fast enough to conform to the instantaneous equilibrium charge distribution. Then only the electrons can be effective in the shielding, and ρ_D^2 must be increased by a factor of two:

$$\rho_D = \left(\frac{\epsilon_0 kT}{e^2 N_e} \right)^{1/2} \tag{11.1b}$$

Putting the numerical constants into Equation 11.1a,

$$\rho_D \approx 50 \sqrt{\left(\frac{T}{N_e} \right)}$$

where T is in K, N_e is in m^{-3}, and ρ_D is in m. For an arc having, say, $T \sim 10^4$ K, $N_e \sim 10^{22}$ m^{-3} (i.e., 10^{16} cm^{-3}), ρ_D comes to

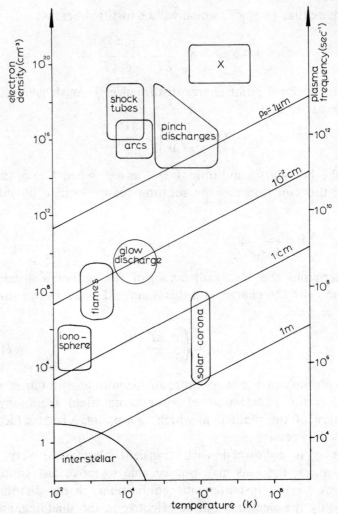

Fig. 11.1 Debye radius and plasma frequency for different types of plasma. The diagram shows the approximate ranges of temperature and electron density to be found in different types of plasma; *X* corresponds approximately to that covered by the vacuum spark, plasma focus, and laser-generated plasmas. The corresponding plasma frequencies are shown on the right. The sloping lines are lines of constant Debye radius.

about 5×10^{-8} m, which is obviously very much smaller than the dimensions of the arc column. The high temperature pinches ($T \sim 10^6$ K) of relatively low density ($N_e \sim 10^{14}$ cm^{-3}), for which $\rho_D \sim 10^{-5}$ m, still have a couple of orders of magnitude in hand. In the solar corona, with $T \sim 10^6$ K and $N_e \sim 10^6$ cm^{-3}, ρ_D

is of the order of several cm, but of course the dimensions are very many orders of magnitude greater. Fig. 11.1 shows values of ρ_D for various types of plasma.

The derivation of ρ_D was subject to the restrictions that V be a smoothly varying function and that $eV \ll kT$. In fact these two conditions amount to one and the same, at least so long as the perturbing potential is attributable to an isolated ion or electron. The first requires that the Debye sphere contain a large number of particles:

$$4/3\pi\rho_D^3 N_e \gg 1$$

But the mean electrostatic energy $e\overline{V}$ of the electrons surrounding a charge e is of order $e^2/4\pi\epsilon_0\rho_D$. Eliminating N_e between the expressions

$$\rho_D^2 \approx \frac{\epsilon_0 kT}{e^2 N_e} \quad \text{and} \quad \rho_D^3 \gg \frac{3}{4\pi N_e}$$

gives the inequality

$$\rho_D \gg \frac{3e^2}{4\pi\epsilon_0 kT},$$

leading directly to the second condition

$$e\overline{V} \approx \frac{e^2}{4\pi\epsilon_0 \rho_D} \ll kT$$

11.4 Plasma oscillations

Just as the Debye radius represents a typical length for collective action, so the plasma frequency – or rather, its reciprocal – is a typical time. The plasma frequency is the resonance frequency for collective oscillations of the electrons about their equilibrium positions. It can be derived as follows.

Consider all the electrons in a slab of plasma of unit cross-section and length l to be displaced simultaneously from their equilibrium positions a distance x (where $x \ll l$). The amount of displaced charge is $eN_e x$. The slab now has surface charges $+\sigma$ at one end and $-\sigma$ at the other, where $\sigma = eN_e x$ (see Fig. 11.2). The electric field produced within the slab by this charge is σ/ϵ_0. Each electron therefore experiences a force $e\sigma/\epsilon_0$, or $e^2 N_e x/\epsilon_0$, and its equation of motion, if one neglects collisions and thermal energy, is

$$m\ddot{x} + \frac{e^2 N_e}{\epsilon_0} x = 0$$

This is the equation of undamped simple harmonic motion and has the solution

$$x = a \sin \omega_p t$$

where the angular frequency ω_p is given by

$$\omega_p^2 = \frac{e^2 N_e}{\epsilon_0 m} \tag{11.2}$$

The plasma frequency in $\sec^{-1}$ or Hz is therefore

$$\nu_p = \frac{\omega_p}{2\pi} = \left(\frac{e^2 N_e}{4\pi^2 \epsilon_0 m}\right)^{1/2} = 9\, N_e^{1/2}\ \sec^{-1} \tag{11.3}$$

where N_e is in m^{-3}. (When N_e is in cm^{-3}, $\nu_p = 9 \times 10^3 N_e^{1/2}$.) For the laboratory plasmas discussed in the last section, with N_e in the range $10^{20}-10^{24} m^{-3}$, or $10^{14}-10^{18}\ cm^{-3}$, we have ν_p between 10^{11} and $10^{13}\ \sec^{-1}$, which puts the plasma frequency in the far

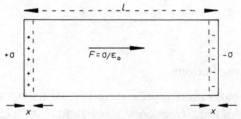

Fig. 11.2 Plasma oscillations. For explanation, see text

infra-red, between 3 mm and 30 μm in wavelength. Values of ν_p for various plasmas are shown on the right-hand side of Fig. 11.1.

It should be emphasized that plasma oscillation is not just a resonance oscillation of one electron bound to a nucleus, of the type dealt with in classical absorption/dispersion theory: it is a collective motion of *free* electrons, and the restoring force exists only because all the electrons are displaced together. This collective behaviour tends to disappear if the electron motions are randomized by collisions; a necessary condition for its existence is that the time between collisions exceed the oscillation period, or the collision frequency ν_c be much smaller than ν_p.

Equation 11.3 suggests a way of determining N_e provided that ν_p can be measured. One obvious possibility is to investigate the propagation of radiation through a plasma, since one would expect some sort of change when the frequency of the radiation coincides with the plasma frequency. This change takes the form of a cut-off, as can be understood qualitatively by considering the distances moved by the electrons. Eliminating the electron density between Equations 11.1b and 11.2, we have $\rho_D \omega_p = (kT/m)^{1/2}$. The mean electron velocity is approximately $(kT/m)^{1/2}$, so that in the time taken for one plasma oscillation the electron can travel about one Debye length. This is just what it needs to do in order to cancel out the disturbance from an external field such as an electromagnetic wave. When $\omega > \omega_p$ the electrons cannot travel fast enough to annul the waves, but when $\omega < \omega_p$ they can adjust their positions in time to prevent the wave propagating.

The same result can be reached rather more rigorously by finding the refractive index of the plasma. Equation 9.18 for the dielectric constant due to the oscillations of bound electrons is also valid for free electrons if one puts the resonance frequency $\omega_0 = 0$, since there is no restoring force for individual electrons in a plasma. The complex dielectric constant for free electrons is then

$$\frac{\epsilon}{\epsilon_0} = n_{el}^2 = 1 - \frac{N_e e^2}{\epsilon_0 m} \frac{1}{\omega^2 - i\omega\gamma}$$

For negligible damping ($\gamma \sim 0$) we can forget about the absorption and write

$$n_{el}^2 = 1 - \frac{N_e e^2}{\epsilon_0 m \omega^2} = 1 - \frac{N_e e^2}{4\pi^2 \epsilon_0 m \nu^2} = 1 - \frac{\nu_p^2}{\nu^2} \qquad (11.4)$$

As ν decreases from infinity to ν_p, n_{el} goes from 1 to 0. The phase velocity v ($= c/n$) in a free electron gas is therefore greater than c when $\nu > \nu_p$, it approaches infinity as $\nu \to \nu_p$, and it is imaginary for $\nu < \nu_p$. The propagation of electromagnetic energy is determined by the group velocity u. This is most easily derived from Equation 11.4 by writing the phase velocity as $v = \omega/k$, where $k = 2\pi/\lambda$. Then

$$v = \frac{\omega}{k} = \frac{c}{n_{el}} = \frac{c}{\sqrt{[1 - (\omega_p/\omega)^2]}}$$

or

$$k = \frac{\omega}{c} \sqrt{\left(1 - \frac{\omega_p^2}{\omega^2}\right)} = \frac{1}{c} \sqrt{(\omega^2 - \omega_p^2)} \qquad (11.5)$$

As ω decreases, k tends to zero at the plasma frequency, and at lower frequencies it is imaginary. Differentiating Equation 11.5,

$$u = \frac{d\omega}{dk} = \frac{c}{\omega} \sqrt{(\omega^2 - \omega_p^2)} = c \sqrt{\left[1 - \left(\frac{\omega_p}{\omega}\right)^2\right]} \qquad (11.6)$$

u, of course, is always less than c, and it tends to zero as $\omega \to \omega_p$. The plasma frequency therefore acts as a cut-off frequency for the propagation of electromagnetic waves.

At some distance from the cut-off, when $\omega \gg \omega_p$, n_{el} is fairly close to 1, and Equation 11.4 can be expanded to give

$$n_{el} - 1 = -\frac{1}{2}\left(\frac{\nu_p}{\nu}\right)^2 = -\frac{N_e e^2}{8\pi^2 \epsilon_0 m \nu^2} = -\frac{N_e e^2}{8\pi^2 \epsilon_0 m c^2}\lambda^2 \quad (11.7)$$

The comparable expression for bound electrons in atoms and ions can be found from Equation 9.18 with the damping term again neglected. For $n \approx 1$, that equation becomes

$$n_{at} = \sqrt{\frac{\epsilon}{\epsilon_0}} \approx 1 + \frac{\mathcal{N}e^2}{2\epsilon_0 m(\omega_0^2 - \omega^2)} = 1 + \frac{\mathcal{N}e^2}{8\pi^2 \epsilon_0 m(\nu_0^2 - \nu^2)}$$

This equation is not identical with Equation 9.20 because in deriving the latter it was assumed that ν was in the neighbourhood of an absorption line, $|\nu_0 - \nu| \ll \nu_0$. This assumption is not justified in the infra-red, since most strong absorption lines fall in the visible and ultra-violet. Going from classical oscillators to real atoms, one replaces $\mathcal{N}$ by the product of the number density in the lower level of a transition, N_j, and the appropriate oscillator strength, f_j, as in Section 9.6, and sums over all transitions j:

$$n_{at} - 1 = \frac{e^2}{8\pi^2 \epsilon_0 m} \sum_j \frac{N_j f_j}{(\nu_j^2 - \nu^2)} \qquad (11.8)$$

In the infra-red it is reasonable to assume $\nu_j \gg \nu$, and Equation 11.8 can be expanded as

$$n_{at} - 1 = \frac{e^2}{8\pi^2\epsilon_0 m} \sum_j \frac{N_j f_j}{\nu_j^2} \left(1 + \frac{\nu^2}{\nu_j^2} + \cdots\right)$$

$$= \frac{e^2}{8\pi^2\epsilon_0 mc^2} \sum_j N_j f_j \lambda_j^2 \left(1 + \frac{\lambda_j^2}{\lambda^2} + \cdots\right) \qquad (11.8a)$$

This expression varies very slowly with λ in contrast to the λ^2 dependence of the electron refractive index (Equation 11.7). This very different wavelength dependence makes it possible to separate the free electronic from the atomic/ionic contributions to the refractive index by measuring at two different wavelengths. When n_{el} has been measured, ν_p and hence N_e are known.

The propagation of electromagnetic waves through a real plasma is considerably more complicated than this simple treatment indicates. Thermal motion and collisions tend to smear out the cut-off frequency, and electrons in highly excited states occupy a position intermediate between free and bound. Moreover, we have not treated at all the effects of external magnetic fields, which are important in astrophysical plasmas and in many laboratory plasmas such as pinch discharges. Magnetic fields give rise to a different type of collective motion, the electrons spiralling round the lines of magnetic induction with the cyclotron frequency $\omega_c = eB/m$ and the ions with the considerably lower frequency $(\omega_c)_i = eB/M$.

11.5 Distribution of energy – dissociation equilibrium

As the temperature of a gas is increased, molecules tend to dissociate into atoms, and atoms into ions plus electrons; some of the molecules, atoms and ions are excited to higher energy states; and the kinetic energy of all these particles and of the free electrons increases. We have already used the relations describing the distribution of velocity among particles (Maxwell) and the distribution of population among discrete energy states (Boltzmann). For conditions of equilibrium it is also possible to write down equations relating the number densities of dissociation products. The equation governing ionization equilibrium, Saha's equation, is simply an application of the general equilibrium relation for molecular dissociation. Like the other distributions, this is obtained by the methods of statistical mechanics (see, for example, [1], [3]) and will not be derived here. Before writing it down and discussing it, however, it should be helpful to look again at the Maxwell and Boltzmann distributions, and especially at the connection between them. For convenience, these equations are re-written here.

Equation 8.12 gave Maxwell's expression for the fraction of particles having a given velocity in one direction. Extending this to three dimensions, we have for the number with velocity components between v_x and $v_x + \mathrm{d}v_x$, v_y and $v_y + \mathrm{d}v_y$, v_z and $v_z + \mathrm{d}v_z$:

$$\mathrm{d}N(v_x, v_y, v_z) = N \left(\frac{m}{2\pi kT} \right)^{3/2} \exp \left(- \frac{m(v_x^2 + v_y^2 + v_z^2)}{2kT} \right) \mathrm{d}v_x \, \mathrm{d}v_y \, \mathrm{d}v_z$$

$$(11.9a)$$

If one is interested only in the total speed and not in its direction, one can rewrite this as

$$\mathrm{d}N(v) = N \left(\frac{m}{2\pi kT} \right)^{3/2} \exp \left(- \frac{\tfrac{1}{2}mv^2}{kT} \right) 4\pi v^2 \, \mathrm{d}v \qquad (11.9b)$$

The Boltzmann distribution, Equation 9.2, is

$$N_j = N \frac{g_j}{U(T)} \exp \left(- \frac{E_j}{kT} \right) \qquad (11.10)$$

where g_j is the statistical weight of the jth level, having energy E_j above the ground state, and $U(T)$ is the state sum or partition function

$$\sum_j g_j \exp \left(- \frac{E_j}{kT} \right).$$

Both these distributions have the same exponential $(-E/kT)$, and the partition function in the second plays the same role as the normalizing factor $(m/2\pi kT)^{3/2}$ in the first in ensuring that the sum over all possible states adds up to N. The analogy can be made more obvious by rewriting Equation 11.9a in terms of the momenta, $mv_x = p_x$, etc.:

$$\mathrm{d}N(p_x, p_y, p_z) = N \left(\frac{1}{2\pi mkT} \right)^{3/2} \exp \left(- \frac{p_x^2 + p_y^2 + p_z^2}{2mkT} \right) \mathrm{d}p_x \, \mathrm{d}p_y \, \mathrm{d}p_z$$

The uncertainty principle does not allow the momentum and the position of a particle to be specified simultaneously with a precision greater than that given by $\Delta p \Delta x = h$. In three dimensions, therefore, $\Delta p_x \Delta p_y \Delta p_z \Delta x \Delta y \Delta z = h^3$. Imagine now a six-dimensional momentum-space carved into cells of volume h^3. A particle cannot be specified more precisely than by assigning it to one such cell; but it is quite possible for the particle to have the same energy in any one of several cells, just as a bound electron may be found in several (degenerate) states of the same energy. So we take the statistical weight of the free particle to correspond to the number of cells in which it may be found: i.e., $g = dp_x\, dp_y\, dp_x\, V/h^3$, where V is the available volume. Comparing the two equations with this in mind, it is seen that they become identical if the term $(1/2\pi mkT)^{3/2} h^3/V$ corresponds to $1/U(T)$, which is to say that the partition function for translational motion of a free particle is given by

$$\frac{(2\pi mkT)^{3/2}}{h^3} V$$

We now come to dissociation and ionization. The process $A^+ + e^- \rightleftharpoons A$ is a special case of the general chemical equilibrium reaction $X + Y \rightleftharpoons Z$. By minimizing the free energy of the system it can be shown ([3], for example) that the equilibrium numbers of the three species are given by $XY/Z = Q_X Q_Y / Q_Z$. Each Q here is the *total* partition function of the species – that is, the product of the internal partition function $U(T)$ and the free particle translational partition function that we arrived at just above. Applying this equilibrium condition first to molecular dissociation, we have

$$\frac{N_X N_Y}{N_Z} = \frac{(2\pi \mu kT)^{3/2}}{h^3} \frac{U_X U_Y}{U_Z} e^{-D/kT}$$

V has been set equal to unit volume, so that N_X, etc., represent number *densities*. The first term on the right is obtained from the three translational partition functions, μ being a sort of reduced mass given by

$$\mu = \frac{M_X M_Y}{M_Z} = \frac{M_X M_Y}{M_X + M_Y}$$

The U's are the internal partition functions or state sums, the zero of energy being in each case the ground state of the atom or molecule. D is dissociation energy, and the factor $e^{-D/kT}$ appears because U_z has to be multiplied by $e^{D/kT}$ to give the state sum of the molecule the same zero of energy as that of the separated atoms.

In many atoms and ions the first excited state is several eV above the ground state, and only the first term in the state sum is significant – i.e., $U(T) \approx g_0$ – at any rate for temperatures up to say 10 000 K. g_0 for most simple atoms and ions is close to one (it is exactly one for closed shells and sub-shells, as in the inert gases and the alkaline earths, two for the alkali metals, etc.). On the other hand, $U(T)$ is very large for molecules, because even at moderate temperatures a large number of rotational states and even vibrational states are occupied. Consequently an appreciable fraction of molecules may remain undissociated even when T is high enough that $kT \to D$.

11.6   Saha's equation for ionization equilibrium

When the equilibrium condition is applied to ionization, again setting $V = 1$ and using number densities, we have

$$\frac{N_e N_i}{N_0} = \frac{2(2\pi m k T)^{3/2}}{h^3} \frac{U_i(T)}{U_0(T)} e^{-\chi/kT} \qquad (11.11a)$$

The first term on the right is the total partition function for the electrons, and the factor 2 incorporated in it is the state sum for the two possible spin states. The translational partition functions of the atom and ion cancel out, their masses being nearly enough equal, leaving only their internal partition functions $U_0(T)$ and $U_i(T)$. χ is the ionization energy, and the factor $\exp(-\chi/kT)$ accompanies $U_i(T)$ to give the state sum of the ion the same zero of energy as the state sum of the atom.

Similar equations hold for higher degrees of ionization. The number densities in the zth and $(z + 1)$th stages are related by

$$\frac{N_e N_{z+1}}{N_z} = \frac{2(2\pi m k T)^{3/2}}{h^3} \frac{U_{z+1}(T)}{U_z(T)} e^{-\chi/kT} \qquad (11.11b)$$

In general, by the time an appreciable number of ions have reached $z + 1$, there will be a negligible number of $z - 1$ present, so that it is adequate to consider only two stages of ionization for any one species.

The number density N_i in Saha's equation refers to the total number of ions per unit volume, regardless of the particular energy levels they may occupy, and the same is of course true for N_0. To find the number densities for the ground state or any particular

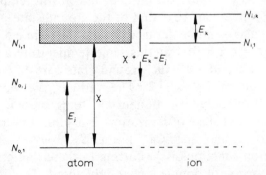

Fig. 11.3 Energy levels of atom and ion. The k'th level of the ion lies a distance $\chi + E_k - E_j$ above the j'th level of the atom, where χ is the ionization energy.

excited state it is necessary to combine Boltzmann's and Saha's relations. For example the number density of neutral atoms in state j is related to the number density of ions in state k by (see Fig. 11.3):

$$\frac{N_e N_{i,k}}{N_{0,j}} = \frac{2(2\pi m k T)^{3/2}}{h^3} \frac{g_{i,k}}{g_{0,j}} \exp\left(-\frac{\chi + E_k - E_j}{kT}\right) \quad (11.12)$$

Since the internal partition functions of both atom and ion are likely to be small, as remarked in the last section, the ratio U_{z+1}/U_z is usually of the order of unity and is not very strongly temperature-dependent. Inserting numerical values in Equation 11.11a, we have

$$\frac{N_e N_i}{N_0} = 4.83 \times 10^{21} \, T^{3/2} \, \frac{U_i}{U_0} \exp\left(-\frac{1.16 \times 10^4 I}{T}\right)$$

where the number densities are in m^{-3} and I is the ionization potential in volts. It can be seen that there is an appreciable degree of ionization even when $I \ll kT$ so that the exponential term is very small. With $I = 10$ volts (which is somewhere between the metals and the non-metals) and $T = 10\,000$ K the exponential factor is only about 10^{-5}, but the right hand side of the equation works out to about 4.8×10^{22}. For an electron density of $10^{17}\,cm^{-3}$, or $10^{23}\,m^{-3}$, such as might be found in a high current arc, we have $N_i \approx \frac{1}{2} N_0$, or 30% ionization. It is worth contrasting this with the population of excited states of the neutral atom: an excited level 10 eV above the ground state at this temperature has a population only about 10^{-5} that of the ground state.

As a particular example, let us calculate the degree of ionization of calcium ($I = 6.1$ volts) at this same temperature and electron density. For Ca and Ca^+ the ground states are $^1S_0$ and $^2S_{1/2}$ respectively, giving $U_i/U_0 \approx 2$. The ratio N_i/N_0 works out to about 100. With only 1% of neutral atoms present, one must investigate the next stage of ionization. The ionization potential of Ca^+ is 11·9 volts, and since Ca^{++} has a $^1S_0$ ground state the ratio of the partition functions is $\frac{1}{2}$. Putting in these figures, one finds that the proportion of double ions is only about 2%.

In a laboratory plasma of relatively low density and high temperature such as a pinch discharge – $N_e \sim 10^{15}\,cm^{-3}$, $T \sim 10^6$ K – Equation 11.11b gives $N_{z+1}/N_z \approx 5 \times 10^9\,e^{-I/100}$. Appreciable ionization should be attained for $\exp(-I/100) \sim 10^{-9}$, which corresponds to $I \sim 2000$ volts, say 10 to 12 stages of ionization. Such high ionization is not in fact attained in most pinch discharges because the electron density is too low and the discharge time too short for the establishment of thermal equilibrium, but in the plasma focus device ionization stages up to 18 have been reached.

In stellar atmospheres, where the temperature is likely to be a few thousand degrees, the degree of ionization is higher than in laboratory plasmas of comparable temperature because of the much lower electron density. For example, in the sun's lower chromosphere $T \sim 7000$ K and $N_e \sim 10^{10}\,cm^{-3}$. Astrophysicists usually replace the electron density by the electron pressure, $P_e = N_e kT$, in using Saha's equation. In this example the appropriate electron pressure is 10^{-3} newton m^{-2} (10^{-2} dyne cm^{-2}). Substituting the calcium values in Saha's equation, one finds only about 1 in 10^7 neutral calcium atoms.

11.7 Depression of ionization potential and Inglis–Teller limit

In accurate calculations it is necessary to reduce the actual ionization energy by an amount $\Delta\chi$. One would expect the electron to be no longer bound to its parent nucleus when it is at a distance of approximately ρ_D, this being the limit to which the influence of that nucleus extends. Thus 'ionization' has effectively taken place when the potential energy of the electron is $-ze^2/4\pi\epsilon_0\rho_D$ instead of zero. For one degree of ionization ($z = 1$) one could expect $\Delta\chi$ to be of order $e^2/4\pi\epsilon_0\rho_D$

$$\text{i.e.,} \qquad \Delta\chi \sim 3 \times 10^{-11}(N_e/T)^{1/2} \text{ eV}$$

In a typical arc, $N_e \sim 10^{17}$ cm^{-3} = 10^{23} m^{-3}, $T \sim 5000$ K, $\Delta\chi$ is only about 0·1 eV, too small to measure accurately. Because of the experimental difficulties it is not possible at present to decide conclusively in favour of any one of the various more sophisticated expressions for $\Delta\chi$.

The depression of the ionization potential removes a difficulty about the divergence of the partition function at high temperatures: in any atom there are an infinite number of bound states crowding up towards the ionization limit, and unless the exponential function $e^{-E_i/kT}$ is vanishingly small for these states (i.e., $\chi \gg kT$) the state sum diverges. However, the sum extends to infinity only if the principal quantum number and hence also the 'radius' of the atom extends to infinity. If the size is limited to ρ_D, the discrete levels should come to an end at a principal quantum number n^* such that $R/n^{*2} \geqslant \Delta\chi$, where R is the Rydberg constant and the high levels are assumed hydrogen-like.

This 'depressed series limit' must be distinguished from the superficially similar Inglis-Teller limit. The latter describes the point at which the higher series members are sufficiently broadened to merge with one another. In the case of hydrogen, which can be applied to the higher series members of all atoms, the total Stark splitting of a state of principal quantum number n is proportional to the electric field F and to $n(n - 1)$ (see Fig. 2.16). On the other hand, the separation of adjacent states of high n is proportional to $1/n^3$. At some value n_m the broadening must catch up with the separation, and this value of n_m defines the Inglis-Teller limit. If F is regarded as the most probable value of the field due to a neighbouring charged particle in the quasi-static approximation, F should depend on $1/\bar{r}^2$, where $\bar{r}$ is the mean inter-particle distance. Since $\bar{r} \propto N_e^{-1/3}$, it follows that the line width should be proportional to $N_e^{2/3}$, leading to $n_m^5 \propto N_e^{-2/3}$. n_m is in general lower than $n^\star$ and has a different functional dependence on N_e: since $n^\star \propto (\Delta\chi)^{-1/2}$ and $\Delta\chi \propto N_e^{1/2}$, it follows that $n^\star \propto N_e^{-1/4}$. This is understandable in that n_m limits the radius of the atom to the inter-particle distance $\bar{r}$ while $n^\star$ limits it to ρ_D, which has already been assumed greater than $\bar{r}$.

11.8   Temperature and equilibrium

The equilibrium distribution of energy among the different states of an assembly of particles is determined by the parameter T defining the temperature for that particular form of energy. The distributions appropriate to kinetic, excitation and ionization energy, Maxwell, Boltzmann and Saha respectively, appeared in the last sections as Equations 11.9, 11.10 and 11.11. The distribution of radiative energy is of course governed by the Planck function, as given in Section 3.1. It is repeated here for convenience:

$$B_0(\nu, T) = \frac{2h\nu^3}{c^2} \frac{1}{e^{h\nu/kT} - 1}$$ (11.13a)

Two approximations to this function are often useful. The Rayleigh-Jeans formula is valid when $h\nu \ll kT$:

$$B_0^{RJ}(\nu, T) = \frac{2kT\nu^2}{c^2}$$ (11.13b)

and the Wien approximation when $h\nu \gg kT$

$$B_0^{Wien}(\nu, T) = \frac{2h\nu^3}{c^2} e^{-h\nu/kT}$$ (11.13c)

It often happens that there is an equilibrium distribution of one of these forms of energy but not of another, or that the temperature parameter for one form of energy is different from that for another. For example, in a low pressure discharge tube the gas kinetic temperature may be a couple of orders of magnitude smaller than the excitation energy, and the spectrum bears no resemblance whatsoever to a black body continuum at any temperature.

Complete thermodynamic equilibrium exists when all forms of energy distribution are described by the same temperature parameter. Thermodynamic arguments show that the principle of detailed balance or microscopic reversibility must then operate: each energy exchange process must be balanced by its exact inverse. Statistically, for every photon emitted, a photon of the same frequency must be absorbed, for every excitation by electron collision (collision of the first kind) there must be a de-excitation by electron collision (collision of the second kind), etc. In practice this situation cannot be fully realized. However large and dense the plasma, photons must leak out from the edges, and if they did not we should be unable to observe the plasma. A close approach to thermodynamic equilibrium requires that such losses be small compared to the total energy.

11.9 Local thermodynamic equilibrium

The form of energy most likely to be out of balance with the others is the radiation energy, since radiative equilibrium requires the plasma to be optically thick at *all* frequencies. Many plasmas can be described by a state known as local thermodynamic equilibrium, or LTE, in which it is possible to find a temperature parameter for every point that fits the Boltzmann and Saha relations for the populations of the excited and ionic states and the Maxwell distribution of velocities among the electrons. The criterion for LTE is that collisional processes must be much more important than radiative, so that the shortfall of radiative energy does not matter. More precisely, an excited state must have a much larger probability of de-excitation by an inelastic collision than by spontaneous radiation. This is tantamount to requiring a high electron density because, as shown in the discussion of inelastic collisions in Section 8.4.3, only electrons are likely to attain velocities high enough to satisfy $\Delta E\, t_p \sim \Delta E \cdot \rho/v \sim \hbar$ for reasonably large values of the impact parameter ρ.

The criterion for LTE can be put more quantitatively as follows. The rate of collisional excitation $1 \rightarrow 2$ for electrons of velocity v is the product of the number of collisions per second, $N_1\, dN_e(v)v$ and the cross-section $\sigma_{12}(v)$:

i.e.,   collisional excitation rate =

$$N_1 N_e \int_0^\infty v\sigma_{12}(v)f(v)\, dv = N_1 N_e \overline{v\sigma_{12}(v)} \qquad (11.14)$$

where $f(v)$ is the velocity distribution of the electrons, assumed to be Maxwellian (Equation 11.9b) and the rule represents the average weighted with this distribution.

Fig. 11.4 shows all the radiative and collisional processes connecting levels 1 and 2. In complete thermodynamic equilibrium the collisional and radiative processes must balance separately. The radiative balancing has already been used in Section 3.4 to derive the relation between the Einstein coefficients. The collisional balancing gives the relation between the cross-sections

$$N_1 N_e \overline{v\sigma_{12}} = N_2 N_e \overline{v\sigma_{21}}$$

$$\therefore \quad \overline{v\sigma_{12}}/\overline{v\sigma_{21}} = N_2/N_1 = g_2/g_1 \; e^{-h\nu/kT} \qquad (11.15)$$

since the population ratio in TE is given by Boltzmann's formula. The σ's are atomic parameters, and so this relation between them must hold even when the system is not in TE. If the radiation

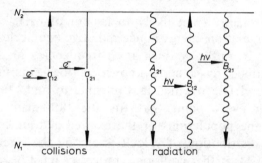

Fig. 11.4 Radiative and collisional processes connecting energy levels. The straight and wiggly lines represent collisional and radiative transitions respectively.

density is in fact much below the TE value, the two processes to the right of Fig. 11.4, absorption and induced emission, can be neglected. In any form of equilibrium the population and depopulation rates of level 2 are still equal, and if the problem is simplified by ignoring transitions to all levels other than level 1 we have

$$N_1 N_e \overline{v\sigma_{12}} = N_2 N_e \overline{v\sigma_{21}} + N_2 A_{21} \qquad (11.16)$$

The second term on the right can be dropped if N_e is made large enough. Then

$$N_2/N_1 = \overline{v\sigma_{12}}/\overline{v_{21}} = g_2/g_1 \; e^{-h\nu/kT}$$

$$= \text{Boltzmann ratio, from Equation 11.15.}$$

The criterion for LTE can therefore be written

$$N_e \gg \frac{A_{21}}{v\sigma_{21}} \qquad (11.17a)$$

The frequency and temperature dependence of this inequality can be estimated as follows. The criterion $\Delta E \cdot \rho/v \sim \hbar$ is a necessary but not a sufficient condition for a collisional transition; it is also necessary that the maximum value of the perturbation be of order ΔE. The term of lowest order in this perturbation is the interaction of the electric dipole transition moment of the atom, $R_{12}(er)$, with the electric field from the passing electron, $e/4\pi\epsilon_0\rho^2$. Combining these requirements we have

$$\frac{e}{4\pi\epsilon_0} \frac{R_{12}}{\rho v} \sim \hbar, \qquad \text{or} \qquad \rho \sim \frac{e}{4\pi\epsilon_0 \hbar} \frac{R_{12}}{v}$$

The excitation cross-section at threshold is

$$\sigma_{12}(v) \sim \pi\rho^2 \sim \pi \left(\frac{e}{4\pi\epsilon_0 \hbar}\right)^2 \frac{S}{v^2}$$

where S is the line strength. Putting this value of σ_{12} into Equation 11.17a makes the right hand side proportional to vA_{21}/S. Since $A_{21} \propto v^3 S$ and $v \propto T^{1/2}$, the critical value of N_e depends on $v^3 T^{1/2}$. In numerical form the criterion for LTE becomes (see, for example, [5]):

$$N_e \gg 1.6 \times 10^{12} T^{1/2} (\Delta E)^3 \text{ cm}^{-3} \qquad (11.17b)$$

where T is the electron temperature in K and ΔE is the energy difference in eV between the state in question and any neighbouring state to which it can make transitions.

This criterion is most difficult to satisfy for the low-lying states, where ΔE is large. At 10 000 K N_e must be greater than $10^{16}-10^{17}$ cm^{-3}. In a free-burning arc at atmospheric pressure running at low current (5–10 amp) N_e is 10^{15} cm^{-3} or less, and LTE is unlikely to exist in other than the highly excited states. In measurements of transition probabilities, when LTE has to be assumed in order to calculate the number densities of the excited states, some form of stabilized arc is generally used so as to attain currents, and hence electron densities, an order of magnitude greater.

The temperature dependence is actually greater than would appear from Equation 11.17b because at high temperatures one is concerned with ionic rather than atomic states, and ΔE scales as z^2 where z is the degree of ionization. The critical value of N_e therefore increases as z^6. Nevertheless, for any N_e and any degree of ionization it must be possible at high enough excitation to reach a limit where the states are crowded closely enough together for Equation 11.17b to hold; this is known as the 'thermal limit' for the particular values of N_e, T and z, and the plasma is said to be in partial LTE.

Many laboratory plasmas, particularly those at high temperatures, are transient, lasting a microsecond or less. In such cases it is necessary to estimate whether there is sufficient time for the establishment of collisional equilibrium. In a pinch discharge, for example, the particles are driven radially inwards with high velocity, but in so far as this is a unidirectional drift velocity it cannot be said to define a temperature. Elastic collisions between the electrons randomize their velocities fairly quickly, and it then becomes valid to speak of an electron temperature. The time required for the establishment of excitation and ionization equilibrium must vary inversely with N_e. A full discussion of all the factors involved may be found in references such as [4], [5], [6]. To give an example, full LTE in a plasma at 10 000 K would take about 1 μsec for $N_e \sim 10^{16}$ cm^{-3} (satisfying Equation 11.17b) if collisional processes only were involved, but the time may be significantly shortened by radiative excitation. The time required to establish partial LTE, among the higher states only, is of order 10^{-8} sec.

There may also be doubt as to whether the gas kinetic temperature determining the translational velocities of the atoms and ions is equal to the electron temperature. Elastic collisions between electrons and heavier particles are very inefficient in transferring energy because of the large mass difference. Only about $2m/M$ of the energy is transferred at each collision, so about 10^3 collisions are required for approximate equalization of electron and heavy particle energy. Again, the required time depends inversely on N_e. For the example given above, $T \sim 10\,000$ K and $N_e \sim 10^{16}$ cm^{-3}, it can be shown that kinetic equilibrium between electrons and hydrogen ions is established in a time comparable to the excitation equilibrium, $\sim 10^{-8}$ sec. The cross-sections for collisions with neutral atoms are about an order of magnitude smaller than those for ions, but as ion-neutral collisions are efficient energy exchangers and as also the odds are against a neutral atom remaining neutral for very long, these small cross-sections do not matter much. The equilibration time is proportional to the ion mass and to $T^{3/2}$ and is correspondingly longer for heavier ions and hotter plasmas.

11.10 Coronal equilibrium

In the absence of LTE it is still possible to estimate relative populations, but only if the relevant collisional cross-sections and radiative transition probabilities are known. The next most important approximation after LTE is described as coronal equilibrium, since it is applicable to the sun's corona where temperature is high ($\sim 10^6$ K) and electron density low ($\sim 10^8$ cm^{-3}). The radiation density is also low, and it can be assumed that level 2 in Fig. 11.4 is populated entirely by electron collisions and depopulated entirely by spontaneous radiation. The two-level atom of this figure is not a very realistic model, but it will serve to show the considerations involved. For $A_{21} \gg N_e \overline{v\sigma_{21}}$, equation 11.16 becomes

$$N_1 N_e \overline{v\sigma_{12}} = N_2 A_{21}$$

$$\frac{N_2}{N_1} = \frac{N_e \overline{v\sigma_{12}}}{A_{21}} \ll \frac{\overline{v\sigma_{12}}}{\overline{v\sigma_{21}}}$$

By Equation 11.15 the right-hand side of this inequality is equal to the Boltzmann value for N_2/N_1. The populations of higher states are therefore much lower in coronal equilibrium than in LTE, and their determination requires a knowledge of σ and A.

The same arguments apply to ionization equilibrium in the coronal approximation. Collisional ionization and three-body collisional recombination may be represented by

$$[z] + e^- \rightleftharpoons [z+1] + e^- + e^-$$

at rates

$$N_z N_e \overline{v\sigma_{z,z+1}} \quad \text{and} \quad N_{z+1} N_e^2 \overline{v\sigma_{z+1,z}}$$

respectively. The other possible recombination process is radiative,

$$[z+1] + e^- \rightarrow [z] + h\nu$$

at a rate $N_{z+1}N_e v\sigma_\nu$. In a steady state,

$$\frac{N_{z+1}}{N_z} = \frac{\overline{v\sigma_{z,z+1}}}{N_e\overline{v\sigma_{z+1,z}} + \overline{v\sigma_\nu}}$$

If N_e is large, so that σ_ν can be neglected, the collisional processes are balanced and the ratio of the cross-sections gives the value of $N_{z+1}N_e/N_z$ appropriate to LTE – that is, Saha's equation. But in the coronal approximation it is the first term in the denominator that must be dropped, giving

$$\frac{N_{z+1}}{N_z} = \frac{\overline{v\sigma_{z,z+1}}}{\overline{v\sigma_\nu}}$$

This is independent of N_e and is lower than the Saha value and again requires a knowledge of the cross-sections. These can be calculated for hydrogen-like ions, and semi-empirical modifications of the H-like data are used for other ions.

Coronal equilibrium can hold only if N_e is below some critical value, and even then only for the lower excited states. It must eventually break down for the highly excited states because of the rapid decrease of A_{21} with ν^3. The critical value has the same $T^{1/2}$ and z^6 dependence as Equation 11.17b; for $T \sim 10^4$ K and low z it is of order $10^{12}\,\mathrm{cm}^{-3}$, four or five orders of magnitude below the LTE value. It is quite possible to have the populations of the higher states following LTE while those of the lower states follow coronal equilibrium.

11.11   Continuous emission and absorption

It was mentioned in Section 1.4 that the unbound states of an electron-ion system, for which the energy is positive, have a continuous spectrum of energies above the ionization limit. Transitions between a bound and an unbound, or free, state give rise in both emission and absorption to radiation continua extending in the short wavelength direction from the line series limits. In absorption a radiative transition from a bound state j to a free state ϵ is known as photo-ionization. It can be seen from Fig. 11.5 that

$$h\nu = \chi - E_j + \epsilon \tag{11.18}$$

where ϵ is the kinetic energy $\frac{1}{2}mv^2$ of the free electron. The inverse process of radiative recombination, which occurs when an ion captures an electron and makes a radiative transition to a bound state, has already featured in the discussion of coronal equilibrium. This so-called bound-free (bf) continuum is characterized by discontinuities, or edges, in both the absorption and emission coefficients whenever $h\nu$ becomes large enough to reach the next bound level (Fig. 11.7).

Free-free transitions (ff), also illustrated in Fig. 11.5, correspond to loss or gain of energy by an electron in the field of an ion. This type of radiation is to be expected classically from any charged particle constrained to follow a curved path, since the

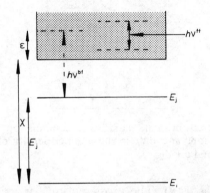

Fig. 11.5 Bound-free and free-free transitions. For explanation, see text.

particle is necessarily accelerated in the process; it is otherwise known as bremsstrahlung, meaning 'braking radiation'. There are no characteristic edges associated with it. For $h\nu \gg kT$ its importance increases with increasing wavelength, small energy changes being more probable than large ones, and at long wavelengths it approaches the black body function.

Both bound-free and free-free transitions are important in both laboratory and astrophysical plasmas. The principal features are discussed in this and the following sections, and references such as [1], [6] should be consulted for a fuller treatment.

In the case of bound-free radiation there must exist a general relation between the cross-sections for photo ionization and radiative recombination comparable with that between the Einstein coefficients for line radiation. Such a relation can be obtained by assuming the system to be in LTE so that Kirchhoff's Law (Equation 9.34) is valid:

$$j_\nu / k_\nu' = B_0(\nu, T)$$

k_ν' here is the absorption coefficient corrected for induced emission, and it can be shown that the relation

$$k_\nu' = k_\nu (1 - e^{-h\nu/kT}) \qquad (11.19)$$

used in Equation 9.34b for line radiation in LTE holds also for continuous radiation in LTE ([1], [6]). The photo-ionization cross-section from a given level j is analogous to the atomic absorption coefficient defined for line radiation in Section 9.5. It

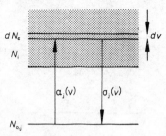

Fig. 11.6 Photo-ionization and radiative recombination. N_i is the total number density of ions and dN_e is the number density of electrons having velocity between v and $v + dv$.

may be written as $\alpha_j(\nu)$. The cross-section for radiative recombination to level j depends on the electron velocity and is designated by $\sigma_j(v)$ (Fig. 11.6). The absorption coefficient k_ν is then given by

$$k_\nu \, d\nu = N_{0,j} \alpha_j(\nu) \, d\nu \qquad (11.20)$$

The emission coefficient is the product of the rate of capture of electrons of velocity v, $N_i \, dN_e(v) \, v \, \sigma_j(v)$ and the energy radiated per unit solid angle with each recapture, $h\nu/4\pi$

$$\therefore \quad j_\nu \, d\nu = \frac{h\nu}{4\pi} N_i v \sigma_j(v) \, dN_e(v) \qquad (11.21)$$

In LTE $dN_e(v)$ is given by Maxwell's relation (Equation 11.9b); dv can be related to $d\nu$ by differentiating Equation 11.18 to give $h \, d\nu = mv \, dv$. Making these substitutions we have

$$\frac{j_\nu}{k_\nu'} = \frac{N_i N_e}{N_{0,j}} \frac{h^2 \nu v^2}{m} \left(\frac{m}{2\pi kT} \right)^{3/2} \frac{\sigma_j(v)}{\alpha_j(\nu)} \frac{e^{-mv^2/2kT}}{1 - e^{-h\nu/kT}} = B_0(\nu, T)$$

In LTE the term $N_i N_e / N_{0,j}$ can be evaluated from Saha's equation as in Equation 11.12, taking the state k of the ion to be the ground state. Substituting Equation 11.13a for B_0 and multiplying top and bottom by $h\nu/kT$, the equation becomes:

$$\frac{2m^2\nu v^2}{h} \frac{\sigma_j(v)}{\alpha_j(v)} \frac{g_{i,1}}{g_{0,j}} \frac{e^{-(x-E_j)/kT} e^{(h\nu-\epsilon)/kT}}{e^{h\nu/kT}-1} = \frac{2h\nu^3}{c^2} \frac{1}{e^{h\nu/kT}-1}$$

The exponentials in the numerator vanish by virtue of Equation 11.18, leaving

$$\frac{\sigma_j(v)}{\alpha_j(v)} = \frac{h^2\nu^2}{m^2c^2v^2} \frac{g_{i,1}}{g_{0,j}} \tag{11.22}$$

This is known as the Milne relation. Although it was derived for a system in LTE, the cross-sections are intrinsic atomic constants and the relation must remain valid in any conditions. It enables capture cross-sections, and hence emissivities, to be found from calculated or measured photo-ionization cross-sections. Electron densities can then be assessed from the absolute intensity of the continuous emission.

11.12 The continuous absorption coefficient

The absorption coefficient for a bound-free transition may be calculated in the same way as that for a bound-bound transition if the wave functions of the appropriate atomic or ionic state and of the free electron are known. Such calculations are difficult and somewhat unreliable for anything except hydrogen and the H-like ions. In practice they are usually based on a semi-classical expression that is obtained by extending the absorption coefficient for bound-bound transitions in H-like ions into the realm of imaginary quantum numbers beyond the series limit [1]. For absorption from the level of principal quantum number n the result is

$$\alpha_n(v) = \pi a_0^2 \frac{64}{3\sqrt{3}} \frac{e^2}{4\pi\epsilon_0 \hbar c} \left(\frac{\chi_H}{h\nu}\right)^3 \frac{z^4}{n^5} G_n^{bf}(v,z)$$

$$= \frac{64\pi^4}{3\sqrt{3}} \frac{m}{ch^6} \left(\frac{e^2}{4\pi\epsilon_0}\right)^5 \frac{z^4}{n^5 v^3} G_n^{bf}(v,z) \tag{11.23}$$

where χ_H is the ionization energy of neutral hydrogen and z is the effective nuclear charge. $G_n^{bf}(\nu, z)$ is the Gaunt correction factor, introduced to take care of the discrepancy between the semi-classical and quantum-mechanical formulae. Although it is a function of n, ν and z, it remains close to 1 for H-like ions. Since all atoms and ions become increasingly hydrogen-like as one goes to higher excited states, Equation 11.23 is a reasonably good approximation to the cross-section for any species except in the lower states. Calculations for the lower states of a number of atoms and ions of importance in laboratory and astrophysical plasmas have been carried out by the so-called quantum-defect method. Departures from the H-like behaviour are found even in the alkali spectra, where the continuous absorption passes through a minimum beyond the principal series limit, instead of decreasing steadily with $1/\nu^3$.

The actual absorption coefficient for photo-ionization from the state n is obtained from Equations 11.20 and 11.23, together with Boltzmann's relation in the form

$$N_{0,n} = \frac{N_0}{U_0} \, 2n^2 \, \exp\left(- z^2 \chi_H/kT\{1 - 1/n^2\}\right)$$

The total absorption coefficient at frequency ν for bound-free transitions must include the contributions from all bound levels lying within a distance $h\nu$ of the ionization limit:

$$k_\nu^{bf} = \frac{64\pi^4}{3\sqrt{3}} \frac{m}{ch^6} \left(\frac{e^2}{4\pi\epsilon_0}\right)^5 \frac{z^4}{\nu^3} \frac{2N_0}{U_0} e^{-z^2 \chi_H/kT} \sum \frac{1}{n^3} e^{z^2 \chi_H/n^2 kT} G_n^{bf}$$

$$(11.24)$$

The sum is to be taken over all levels for which $n \geqslant n_c$, where n_c is the cut-off value determined by $n_c^2 \geqslant z^2 \chi_H/h\nu$. A plot of log k_ν against log ν has the saw-tooth form shown in Fig. 11.7. Whenever ν gets large enough to reach a new bound level, the absorption rises sharply. It then falls off as $1/\nu^3$ until the next bound level is reached. The physical reason for the strong temperature-dependence is that at low temperatures the lower levels only are appreciably populated and the bound-free absorption is significant only at the very short wavelengths required to reach these levels. At all temperatures the edges get less pronounced as n is increased because of the decreasing population. For this reason the sum in Equation 11.24 can be replaced by an integral for the higher

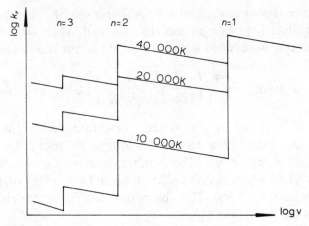

Fig. 11.7 Bound-free absorption coefficient as a function of frequency for a H-like atom

levels, $n \geqslant n'$, where n' is usually taken as 5. Taking the Gaunt factor as unity and writing $(z^2 \chi_H/kT) = C$,

$$\sum_{n=n'}^{n=\infty} \frac{1}{n^3} e^{C/n^2} \rightarrow \int_{n'}^{\infty} \frac{1}{n^3} e^{C/n^2} \, dn = -\frac{1}{2} \int_{n'}^{\infty} e^{C/n^2} \, d\left(\frac{1}{n^2}\right)$$

$$= -\frac{1}{2C} [e^{C/n^2}]_{n'}^{\infty} = \frac{1}{2C} (e^{C/n'^2} - 1)$$

If the cut-off value n_c is above n' - that is,

$$h\nu < \frac{z^2 \chi_H}{(n')^2}$$

or $\lambda > 25\ 000$ Å for $z = 1$ - the sum term vanishes altogether, leaving only the integral with n_c as its lower limit. The long wavelength approximation for k_ν is then:

$$k_\nu^{bf} = \frac{64\pi^4}{3\sqrt{3}} \frac{m}{ch^6} \left(\frac{e^2}{4\pi\epsilon_0}\right)^5 \frac{z^4}{\nu^3} \frac{2N_0}{U_0} e^{-(z^2\chi_H)/kT} \frac{kT}{2z^2\chi_H} (e^{h\nu/kT} - 1)$$

$$= \frac{16\pi^2}{3\sqrt{3}} \frac{k}{ch^4} \left(\frac{e^2}{4\pi\epsilon_0}\right)^3 \frac{z^2}{\nu^3} \frac{2N_0}{U_0} T e^{-(z^2\chi_H)/kT} (e^{h\nu/kT} - 1) \quad (11.25)$$

At high temperatures and long wavelengths free-free transitions become important. A semi-classical expression for the free-free cross-section of a H-like ion can be obtained by the same procedure (imaginary quantum numbers) used for the bound-free cross-section [1]. For an ion-electron pair when the velocity of the electron is between v and $v + dv$ the absorption coefficient is

$$\alpha_v(v)\, dv = \frac{4\pi}{3\sqrt{3}} \left(\frac{e^2}{4\pi\epsilon_0}\right)^3 \frac{1}{m^2 c^2 h} \frac{z^2}{v v^3} G^{ff}(v, z, T)\, dv$$

where $G^{ff}(v, z, T)$ is the Gaunt correction factor for free-free transitions. This is close to unity as long as kT does not exceed the ionization energy $z^2 \chi_H$. The number density of ion-electron pairs is $N_i\, dN_e(v)$, where $dN_e(v)$ is again given by the Maxwell distribution (Equation 11.9b). The absorption coefficient for electrons of velocity v is therefore

$$dk_v^{ff} = N_i N_e \frac{16\pi^2}{3\sqrt{3}} \frac{m}{ch(2\pi m k)^{3/2}} \left(\frac{e^2}{4\pi\epsilon_0}\right)^3 \frac{z^2}{v^3 T^{3/2}} G^{ff} v e^{-(m v^2)/2kT}\, dv$$

This expression has to be integrated over all v for a fixed value of v to obtain the contributions from all electrons. The result is:

$$k_v^{ff} = N_i N_e \frac{16\pi^2}{3\sqrt{3}} \frac{1}{ch(2\pi m k)^{3/2}} \left(\frac{e^2}{4\pi\epsilon_0}\right)^3 \frac{k}{T^{1/2}} \frac{z^2}{v^3} G^{ff}$$

$$= \frac{16\pi^2}{3\sqrt{3}} \frac{k}{ch^4} \left(\frac{e^2}{4\pi\epsilon_0}\right)^3 \frac{z^2}{v^3} \frac{2N_0}{U_0} T e^{-z^2 \chi_H/kT} G^{ff} \qquad (11.26)$$

The second expression is derived from the first by using Saha's equation (Equation 11.11). U_i in this equation has been put equal to one because the ion in the H-like case is a bare nucleus.

k_v^{bf} and k_v^{ff} have the same $1/v^3$ dependence. At long wavelengths, when the edge structure can be ignored, k_v^{bf} in Equation 11.25 is in fact identical with k_v^{ff} in Equation 11.26 except for the last factor. If G^{ff} in Equation 11.26 is put equal to unity, the combined absorption coefficient assumes the relatively simple form

$$k_v^{cont} = k_v^{bf} + k_v^{ff} = \frac{16\pi^2}{3\sqrt{3}} \frac{k}{ch^4} \left(\frac{e^2}{4\pi\epsilon_0}\right)^3 \frac{z^2}{v^3} \frac{2N_0}{U_0} T e^{-(z^2 \chi_H - hv)/kT}$$

$$(11.27)$$

It should be remembered that this expression is a long wavelength approximation, ignoring the edge structure in the visible and ultra-violet, and moreover that it is valid only for LTE.

11.13 Continuous emission

The emission coefficient is most easily obtained by taking advantage of the results for k_ν from the last section and assuming LTE, so that $j_\nu = B_0(\nu, T) k'_\nu$ and Equation 11.19 can be used to relate k_ν with k'_ν. This gives

$$j_\nu = \frac{2h\nu^3}{c^2} \frac{1}{e^{h\nu/kT} - 1} k_\nu(1 - e^{-h\nu/kT}) = \frac{2h\nu^3}{c^2} e^{-h\nu/kT} k_\nu$$

(11.28)

Using Equation 11.24 for k_ν^{bf},

$$j_\nu^{bf} = \frac{128\pi^4}{3\sqrt{3}} \frac{m}{c^3 h^5} \left(\frac{e^2}{4\pi\epsilon_0}\right)^5 \frac{2N_0}{U_0} z^4 e^{-(z^2 \chi_H - h\nu)/kT}$$

$$\times \sum_{n \geqslant n_c} \frac{1}{n^3} e^{z^2 \chi_H / n^2 kT} G_n^{bf}(\nu, z)$$

Since the emission is the result of a recombination process it is more logical to express it in terms of the number densities of ions and electrons rather than neutrals. Using Saha's equation for this purpose,

$$j_\nu^{bf} = \frac{128\pi^4}{3\sqrt{3}} \frac{m}{c^3 h^2 (2\pi mk)^{3/2}} \left(\frac{e^2}{4\pi\epsilon_0}\right)^5 \frac{N_e N_i}{U_i} \frac{z^4}{T^{3/2}} e^{-h\nu/kT}$$

$$\times \sum_{n \geqslant n_c} \frac{1}{n^3} e^{z^2 \chi_H / n^2 kT} G_n^{bf}$$

(11.29)

This expression can in fact be derived directly from Equation 11.21 for j_ν, together with the semi-classical expression (Equation 11.23) for $\alpha_n(\nu)$ and the Milne relation connecting $\alpha_n(\nu)$ with $\sigma_n(v)$. This derivation does *not* depend on the existence of LTE, except in so far as the electrons are assumed to obey the Maxwell distribution.

Evidently Equation 11.29 has the same edge structure as Equation 11.24, but the emission falls off exponentially between the edges instead of as $1/\nu^3$. This is shown on the right of Fig. 11.8. At wavelengths long enough to ignore the edge structure we have as the counterpart to Equation 11.25:

$$j_\nu^{bf} = \frac{16\pi}{3\sqrt{3}} \frac{1}{mc^3 (2\pi mk)^{1/2}} \left(\frac{e^2}{4\pi\epsilon_0}\right)^3 N_e N_i \frac{z^2}{T^{1/2}} (1 - e^{-h\nu/kT})$$

(11.30)

assuming as before that $G_n^{bf} \approx 1$. U_i has also been set equal to 1.

The free-free emission can be derived in the same way, using the first form of Equation 11.26 together with Equation 11.28 and again ignoring the Gaunt correction:

$$j_\nu^{ff} = \frac{16\pi}{3\sqrt{3}} \frac{1}{mc^3(2\pi mk)^{1/2}} \left(\frac{e^2}{4\pi\epsilon_0}\right)^3 N_e N_i \frac{z^2}{T^{1/2}} e^{-h\nu/kT} \quad (11.31)$$

Combining Equations 11.30 and 11.31 to find the total continuous emission at long wavelengths, we have

$$j_\nu^{cont} = j_\nu^{bf} + j_\nu^{ff} = \frac{16\pi}{3\sqrt{3}} \frac{1}{mc^3(2\pi mk)^{1/2}} \left(\frac{e^2}{4\pi\epsilon_0}\right)^3 N_e N_i \frac{z^2}{T^{1/2}} \quad (11.32)$$

This expression does *not* depend on the existence of LTE provided that the electron velocities follow a Maxwell distribution. It has the interesting property of being independent of the frequency.

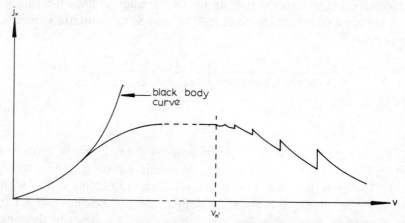

Fig. 11.8 Continuous emission coefficient as a function of frequency. ν_n' is the frequency above which the edge structure is significant. Below this frequency, j_ν is approximately constant until the plasma ceases to be optically thin.

The emissivity per unit frequency interval is constant for $\nu < \nu_{n'}$, the frequency at which the edge structure becomes significant (Fig. 11.8).

At very long wavelengths, $h\nu \ll kT$, the emission is predominantly free-free since the last term in Equation 11.30

$\rightarrow h\nu/kT$ while the last term in Equation 11.31 $\rightarrow 1$. However, j_ν cannot continue constant to indefinitely long wavelengths because, as shown in Fig. 11.8, it must eventually hit the black-body curve, represented in this region by the Rayleigh–Jeans approximation (Equation 11.13b). Thereafter it follows the black-body curve, so long as the electron velocity distribution is Maxwellian, and decreases as ν^2. Looked at another way, the absorption coefficient (Equation 11.27) behaves as $1/\nu^3$ when $h\nu \ll kT$. At sufficiently small ν the gas must become optically thick, and when this happens it radiates as a black body (Section 9.10). As a matter of fact it is quite possible for the plasma frequency to fall somewhere in the region where the gas is still optically thin, and in that case the emission falls abruptly to zero at $\nu \approx \nu_p$. This behaviour is further discussed in Section 11.16.5.

Energy losses through bremsstrahlung are a significant cause of cooling in laboratory plasmas. The z^2 factor in Equation 11.32 shows that it is important to eliminate as far as possible impurities of high atomic number from a hydrogen or helium plasma.

11.14 Other continua

Continuous emission and absorption coefficients for non-H-like ions have been calculated, but the results are of course less reliable. The deviations from H-like behaviour are most significant at short wavelengths when low-lying states are involved in the transitions. An additional complication may arise from auto-ionization, which causes anomalous bumps, or resonances, in the photo-ionization cross-section. In emission, the inverse process of di-electronic recombination may have a large effect on the electron capture cross-section.

The negative hydrogen ion is an important absorber in many stars, and other negative ions are important in some plasmas. Negative ions have only one stable state, and their emission and absorption is therefore all continuous, comprising both free-free and bound-free transitions, the latter connecting with a bound state of the neutral atom. The cross-sections are of the same order of magnitude as those involving positive ions, and the importance of the negative ion continua therefore depends on the number density of negative ions. This is given by the Saha equation in the form:

$$\frac{N_0 N_e}{N^-} = \frac{(2\pi m k T)^{3/2}}{h^3} \frac{2U_0(T)}{U^-(T)} e^{-\chi^-/kT} \tag{11.33}$$

χ^- is the detachment energy, which for H^- is $0\cdot75$ eV. Since the electron can be dislodged by radiation of energy greater than this, the absorption cross-section starts at 16,450 Å, rises to a peak at about 8500 Å and drops again towards shorter wavelengths. It can be seen from Equation 11.33 that for a large ratio N^-/N_0 a combination of large N_e and low T is required.

Another possible source of bremsstrahlung is the cyclotron radiation from electrons moving in magnetic fields, mentioned in Section 3.8 in connection with the electron synchrotron as a background source for the far ultra-violet. Cyclotron radiation may make a significant contribution to the bremsstrahlung from pinch discharges because of the magnetic fields needed to contain the plasma. Its importance increases as T^2, in contrast to the thermal bremsstrahlung which goes as $1/T^{1/2}$. Cyclotron radiation also occurs in the presence of strong magnetic fields in astrophysical plasmas and is held responsible for the strong continuous emission from certain supernovae.

11.15   Applications of plasma spectroscopy

Plasma spectroscopy developed initially as a diagnostic tool – that is, as a method of determining the number densities of the various particles, including electrons, in a plasma and their associated temperature parameters. Now that techniques such as laser scattering are available for diagnostic purposes (Section 11.18) the emphasis in laboratory plasma spectroscopy has switched in two rather different directions. First, the plasma may be regarded simply as a light source for the study of oscillator strengths and collisional cross-sections. In this context the electron temperature and number density may be determined either by the spectroscopic methods of the next two sections or by the non-spectroscopic methods of Section 11.18, as is most appropriate. For example, laser scattering methods are applicable to pinch discharges, but a determination of rotational temperature or reversal temperature is likely to be the best method for a flame, arc, or shock tube. Furthermore, the non-spectroscopic techniques have in many cases increased the reliability of the spectroscopic techniques; for example, the accuracy claimed for the determination of electron density from the width of the hydrogen β line in suitable conditions rests on the agreement of theory with independent methods of measurement. Secondly, in a plasma of known composition and temperature the interactions of the particles with each other and with the radiation can be investigated by studying such effects as the broadening and shifts of spectral lines and the appearance of satellite lines associated with plasma oscillations. In this context it is possible to work at temperatures and electron densities for which the simplifying assumptions necessary for spectroscopic diagnosis are no longer valid; for example, the density of charged particles may be so high that the interactions are dominated by collective effects, and the assumption of binary collisions is not even a good first approximation.

In astrophysics, spectroscopic methods remain, of course, the only ones available. The study of astrophysical plasmas has benefited greatly from the 'calibration' of spectroscopic diagnostic techniques in laboratory plasmas as well as from the greatly improved and increased data on oscillator strengths and collisional cross-sections stemming from laboratory work.

The next section describes the most widely used spectroscopic methods of temperature measurement. Of course, the 'temperatures' obtained from different methods – velocity distribution of ions, say, as compared with population distribution among excited states – will agree only in conditions of LTE. If LTE does not exist, the measurements must be regarded rather as providing information on population densities: any radiative transition can be used to measure the population of the initial state provided that the transition probability is known. The ratio of two transitions from different levels gives the ratio of the populations of these levels, and this information may still be needed even if, in the absence of LTE, the population ratio does not define a temperature. In coronal equilibrium, for example, it yields a value for the electron density if the collisional cross-sections are known.

11.16   Spectroscopic measurement of temperature

All but the first and last of the methods described here measure, by one means or another, the populations of excited states of atoms or ions and hence lead in the case of LTE to a 'Boltzmann' or 'Saha' temperature. The first method assumes a Maxwellian distribution of velocity among the emitting atoms or ions; only the last measures directly the Maxwellian distribution of electron velocities, or the electron temperature.

11.16.1   Gas kinetic temperature

It was shown in Section 8.3 that the half-value width of a Doppler-broadened line is determined entirely by the atomic or molecular weight of the emitter and its kinetic temperature, according to Equation 8.15:

$$\frac{\delta \nu_D}{\nu_0} = \frac{\delta \lambda_D}{\lambda_0} = \sqrt{\left(\frac{8kT \ln 2}{Mc^2}\right)} = 7.16 \times 10^{-7} \sqrt{\left(\frac{T}{M}\right)}$$

When using this relation to measure T, it is essential to make sure that instrumental width and pressure broadening have been allowed for. As explained in Chapter 8, pressure effects usually show up first in the line wings and may not greatly affect the half-width. It is sometimes possible to get rid of them altogether by using a transition between inner shells, sufficiently shielded by outer electrons to have an almost pure Doppler profile. In certain cases it is necessary to guard against bulk motions of the emitters – ions in an electric field, for example. Such drift velocities are not random and do not define a temperature. In astrophysical plasmas large Doppler broadening may be caused by macroscopic turbulence rather than by thermal motion of the emitters.

11.16.2   Population or excitation temperature

The temperature parameter in the Boltzmann distribution may be measured in a variety of ways, depending on the temperature, the plasma constituents, the availability of background sources for absorption, etc. Emission methods are the most generally applicable, and the discussion here is confined to emission. The techniques and problems are basically the same as those described in Section 10.4 in connection with the measurement of transition probabilities. If a sufficiently bright background source is available, the absorption methods described in Section 10.5 may be used in an analogous way to obtain population densities of the lower levels of transitions of known f-value, but weak absorption lines are more difficult to measure than weak emission lines because of the signal/noise problem.

The brightness of an optically thin source of thickness l is $1/4\pi\, N_2 A_{21} lh\nu_{12}$ (Equation 9.32). Using the Boltzmann formula (Equation 11.10),

$$I_{21} = \text{const}\,\frac{1}{U(T)}g_2 A_{21}\nu_{21}\,e^{-E_2/kT} \qquad (11.34)$$

In principle, T can be found by measuring the absolute intensity of the emission line, provided A_{21} is also known absolutely. In practice, the errors associated with absolute intensity measurements and absolute values of oscillator strengths make this method very unreliable. If ionization is appreciable, it also requires a knowledge of N_e. It is much better to work with relative measurements, for which purpose Equation 11.34 can conveniently be written

$$\log\frac{I_{21}\lambda_{21}}{g_2 A_{21}} = \text{const} - \frac{E_2}{kT} \qquad (11.35)$$

A plot of $\log I\lambda/gA$ against E for several spectral lines should be a straight line of slope $1/kT$. Deviations from the Boltzmann distribution should show up as deviations from the straight line. Apart from uncertainties about optical depth and inaccuracy of measurement, the main errors associated with this method are inaccuracies in the A-values and the limited range of E values that can be used. It is usually difficult to find suitable lines with a wide spread of upper states, and if they can be found they tend to fall in very different spectral regions, making the intensity comparison difficult.

Vibrational and rotational levels of molecules can be used instead of atomic levels in plasmas cooler than say 5000 K, when an appreciable fraction of molecules remains undissociated. The advantage of using molecules is that the relative vibrational and rotational transition probabilities, particularly the latter, can in many cases be calculated accurately. From Equation 9.7, $A_{21} \propto 1/g_2\, \nu^3 S_{12}$, where S is the line strength. Using Equation

9.39a, the emission line strengths within a band can be written as const. $\mathscr{S}_{J'J''}/(2J' + 1)$ where J' is the upper rotational level and $\mathscr{S}_{J'J''}$ is the Hönl-London factor, assumed calculable. The rotational degeneracy of the upper level is represented by the factor $(2J' + 1)$. Equation 11.34 applied to relative intensities of rotational lines becomes

$$I = \text{const.} \; (2J' + 1)\nu^3 \; \frac{\mathscr{S}_{J'J''}}{2J' + 1} \; \nu \, e^{-E_{J'}/kT}$$

leading to

$$\log I \, \lambda^4/\mathscr{S}_{J'J''} = \text{const} - E_{J'}/kT \approx \text{const} - BJ'(J' + 1)/kT$$

$$(11.36)$$

A plot of $\log I \, \lambda^4/\mathscr{S}$ against $J'(J' + 1)$ should yield a straight line whose slope determines T. A modification of this method is based on the fact that as J increases in any given branch of a band the intensity first increases because of the increasing Hönl-London factors, which go approximately as $(2J + 1)$, and then decreases because of the exponential factor. For two lines of equal intensity on either side of the maximum J_p and J_q, it follows from Equation 11.36 that

$$\ln \mathscr{S}_p \nu_p^4 - \ln \mathscr{S}_q \nu_q^4 = B/kT \{J_p(J_{p+1}) - J_q(J_{q+1})\}$$

from which T can be obtained by matching rather than measuring intensities. Since the lines have the same intensity and approximately the same frequency, the method is applicable even when the plasma is not quite optically thin, but it is not easy to check its accuracy. The rotational lines of certain molecules whose spectra are nearly always observed as impurities in arcs and flames, notably OH and CN, are frequently used for thermometric purposes. High resolution is required to resolve the rotational structure, especially if the molecule is not a hydride.

The method is applied to vibrational levels by measuring the relative integrated intensities of several bands. In this case the Hönl-London factor must be replaced by the band strength, which cannot usually be calculated with such certainty. If the electronic transition moment is constant over the band system, the relative band strengths are given by the Franck-Condon factors $q_{v'v''}$ (Equations 9.38 and 9.44). Equation 11.36 then takes the form:

$$\log I \, \lambda^4/q_{v'v''} = \text{const} - E_{v'}/kT$$

If the temperature of the plasma varies either spatially (inhomogeneous plasma) or temporally (transient plasma) it is possible to keep a rather crude track of temperature by observing the position or time at which any given spectral line has its maximum intensity. As T increases, the exponential factor in the Boltzmann formula also increases, and eventually approaches unity. On the other hand, the factor $N/U(T)$ decreases with T, partly because U increases as the higher states become populated and partly because N decreases with decreasing density and increasing ionization. The temperature for maximum intensity can be found by differentiating with respect to T. The method is not very sensitive, because the maximum is not usually very sharp; moreover, it is a function of N_e as well as T because of the ionization.

11.16.3 Ionization temperature

A much wider effective range of upper levels can be obtained by using different stages of ionization, together with Saha's equation to relate the number densities. The relative populations of excited levels of different stages of ionization are given by the combined

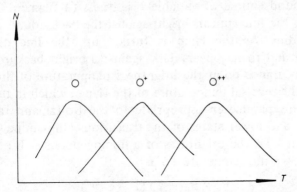

Fig. 11.9 Lines from successive stages of ionization as a function of temperature. The ordinate N is the number density of each stage of ionization.

Saha-Boltzmann equation (Equation 11.12). In an optically thin plasma the intensity $I_{i,k}$ of a line from the kth level of the ion relative to the intensity of a line $I_{0,j}$ from the jth level of the atom is given by

$$\frac{I_{i,k}}{I_{0,j}} = \frac{A_{i,k}}{A_{0,j}} \frac{2g_{i,k}}{g_{0,j}} \frac{\lambda_{i,k}}{\lambda_{0,j}} \frac{(2\pi mkT)^{3/2}}{h^3} \frac{1}{N_e} e^{-(\chi + E_k - E_j)/kT} \quad (11.37)$$

It is of course necessary to know N_e, and the accuracy is usually limited by uncertainties in the A values.

It is possible to assess the spatial or temporal variation of T from the appearance of different stages of ionization in a way analogous to that described for neutral atoms. As T increases, each stage climbs to a maximum and then declines because of further ionization. Comparison of the intensities of one suitable line from each stage enables the temperature to be estimated from a plot such as that in Fig. 11.9. To compute the curve for any one ionization stage, one has to know the relevant partition function, and this may present difficulties. This method is not in any case capable of great precision.

11.16.4   Reversal measurements

The reversal method much used for flame temperatures is a particularly accurate way of combining emission and absorption methods. Essentially it compares directly the populations of the upper and lower states of a transition; in LTE this population ratio defines the temperature. If a given spectral line is observed against a background source of variable brightness (a filament lamp for example), the line appears bright against the background when the lamp is dim. As the lamp is turned up, the line disappears altogether and then appears dark against a bright background. At the disappearance point the brightness temperature of the lamp is equal to the reversal temperature of the flame, which in turn is the temperature parameter appropriate to the Boltzmann factor for the upper and lower states of the transition. This can be shown as follows. In LTE the brightness of a homogeneous slab of hot gas of thickness l and temperature T is

$$B_\nu = B_0(\nu, T) (1 - e^{-k_\nu l})$$

by Equation 9.35. If this same slab is illuminated by a background source of brightness $B'_\nu(0)$, the energy absorbed (per unit of area, solid angle, frequency and time) is $B'_\nu(0) - B'_\nu(l)$, which, using Equation 9.10, becomes

$$B'_\nu(0) (1 - e^{-k_\nu l})$$

At the reversal point these two quantities must be equal, and we have

$$B'_\nu(0) = B_0(\nu, T)$$

If the background source is itself a black body at temperature T_r, then

$$B_\nu'(0) \equiv B_0(\nu, T_r) = B_0(\nu, T)$$

i.e., the temperature T of the plasma is equal to the brightness temperature T_r of the background. It is apparent from the derivation of Equations 9.34 and 9.35 that T describes the equilibrium distribution of population between the two states of the line considered.

The reversal method in its direct form is limited to plasmas of relatively low temperature for which a suitable adjustable background is available. It can, however, be extended to much hotter plasmas by comparing the total emission with the equivalent width in a number of spectral lines, using a background of fixed temperature for the latter. In the absence of the background, the brightness of the plasma integrated over the line profile is:

$$B_p = \int_{line} B_\nu \, d\nu = B_0(\nu, T) \int_{line} (1 - e^{-k_\nu l}) \, d\nu$$

where the variation of $B_0(\nu, T)$ over the line is assumed negligible. In absorption against a much brighter background source, such that B_p can be ignored by comparison with the background, the equivalent width of the line is given by Equation 9.28:

$$W_\nu = \int_{line} (1 - e^{-k_\nu l}) \, d\nu$$

$$\therefore \quad B_p / W_\nu = B_0(\nu, T)$$

If stimulated emission can be neglected so that Wien's approximation is valid,

$$\frac{B_p}{W_\nu} = \frac{2h}{c^2} \nu^3 \, e^{-h\nu/kT}$$

If B_p and W_ν are measured for several lines, a plot of log $B_p/(W_\nu \nu^3)$ against ν should give a straight line of slope $-h/kT$. As with relative intensity measurements, departures from LTE show up as deviations from the straight line, and the sensitivity of the method depends on the spread of wavelength. Reversal methods have, however, two substantial advantages. It is not necessary to know the transition probabilities, and the plasma need not be optically thin.

11.16.5   Black body temperature

It was shown in Section 9.10 that as the optical depth of a plasma in LTE is increased, the peak height of any emission line increases according to $B_\nu = B_0(\nu, T)(1 - e^{-k_\nu l})$ until it hits the black body

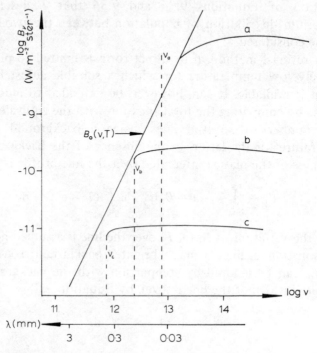

Fig. 11.10 Approach to black body radiation at different electron densities. The curves show the brightness of hydrogen plasma of depth 1 cm at 10^5 K as a function of frequency for different electron densities:

(a) 10^{18} cm^{-3} (10^{24} m^{-3})
(b) 10^{17} cm^{-3}
(c) 10^{16} cm^{-3}

The dashed lines show the corresponding plasma frequencies in each case and the solid line is the black body emission for 10^5 K (Rayleigh-Jeans approximation).

ceiling as $k_\nu l \to \infty$. The black body temperature can therefore be determined from the peak intensity of an optically thick line. Whether or not a line is optically thick can be checked by seeing if the intensity increases with multiple reflections or with end-on rather than side-on observation. The method is not usually a particularly accurate one because for $T > 10\,000$ K the peak of the black body distribution is in the ultra-violet, and the intensity varies rather slowly with T in the visible and near ultra-violet where absolute intensity measurements are most easily made. It is also necessary to resolve the line properly in order to pick out the optically thick section at the centre.

An alternative method is to measure the absolute intensity of the continuous emission at a wavelength sufficiently long that the plasma is optically thick (Fig. 11.8). This method is also dependent on LTE, and homogeneity of the plasma. Moreover, it is only feasible if the plasma becomes optically thick at a frequency higher than the plasma frequency. Since k_ν increases with $N_e N_i$ — i.e., with N_e^2 — while the plasma frequency ν_p increases with $N_e^{1/2}$ (Equation 11.3), the electron density must be above a critical value, which depends on the temperature and the geometrical thickness of the plasma, if this requirement is to be satisfied. For a hydrogen plasma this is given approximately by [6] : $N_e l > 6 \times 10^9 \ T^{3/2} \ \mathrm{cm}^{-2}$. Fig. 11.10 illustrates the effect of increasing the electron density in a hydrogen plasma of depth 1 cm at 10^5 K. The dashed lines mark the plasma frequency in each case. The critical density here is about $10^{17} \ \mathrm{cm}^{-3}$.

11.16.6 *Intensity of continuous radiation*

Equations 11.29 and 11.31 for the bound-free and free-free emission of an optically *thin* hydrogen-like plasma show that the intensity depends exponentially on the frequency according to $e^{-h\nu/kT}$. In the regions between the absorption edges a plot of $\log j_\nu$ versus ν should give a straight line of slope $-h/kT$. This is independent of LTE and requires only that the electrons should have a Maxwellian velocity distribution. Alternatively, the size of the discontinuity at an edge is a sensitive function of T, as shown for the case of the absorption in Fig. 11.7, and it may be possible to use this to measure T.

In cases where the absolute value of the continuous emissivity can be calculated with sufficient accuracy, one can use the ratio of the continuum intensity, which depends on $N_i N_e$, to the intensity of a neutral emission line, which depends on N_0, to find T from Saha's equation. Since only a reasonably high state of excitation of the neutral is likely to be involved, this method is valid so long as partial LTE holds.

11.17 Spectroscopic determination of electron density

Again, the methods described in this section are those most generally applicable, but a more specialized account should be consulted for a full treatment.

11.17.1 Line width

If inelastic collisions can be ignored, line broadening by charged particles is attributable to the linear Stark effect (C_2/r^2) in the case of hydrogen and to the quadratic Stark effect (C_4/r^4) for other atoms. The half-width of the line depends on the electron and ion densities and may be used to measure N_e. Such measurements are performed where possible on hydrogen or a H-like ion, introduced as an impurity if necessary, partly because the broadening is much greater and partly because the line profile can be calculated with much more certainty. As always, accurate calculations are very difficult for atoms with more than one electron.

For quasi-static (ionic) broadening, the line width is proportional to $N_e^{2/3}$, while for electron impact broadening it is proportional to N_e. The hydrogen lines are well described by a quasi--static model in which the smearing out from electron impacts changes the line profile but has little effect on the actual half-value width. N_e can be obtained from the relation

$$N_e = C(N_e, T) \, (\delta\lambda)^{3/2}$$

where C depends only weakly on N_e and T. For H_β, the second line of the Balmer series and the most overworked line in the spectrum in this connection, $\delta\lambda \approx 10$ Å for $N_e \approx 10^{16}$ cm^{-3}, and the determination of N_e should be good to about 5%.

Atoms that are not hydrogen-like are subject to the quadratic Stark effect, and, except in the line wings, the broadening is dominated by electron impacts. The half-value width should therefore be proportional to N_e. The constant of proportionality is usually subject to some uncertainty, and the accuracy with which N_e can be determined from the width is unlikely to be better than 20–30%.

11.17.2 Absolute intensity of continuous radiation

The absolute value of both bound-free and free-free emissivities depends on N_e. Calculations are difficult except for hydrogen, and measurements of absolute intensity are unlikely to be reliable to better than about 20%. Moreover, the z^2 factor in Equations 11.30 and 11.31 means that any impurities present are likely to contribute disproportionately to the intensity. Measurements are best made in the infra-red, where the combined intensity of bound-free and free-free radiation is frequency-independent (Equation 11.32) and is proportional to $N_e^2 T^{-1/2}$ if one takes $N_e = N_i$. The equations hold only if the plasma is optically thin, and it may be necessary to correct for self-absorption, particularly at long wavelengths.

11.17.3 Plasma frequency

The plasma frequency is related to N_e by Equation 11.3:

$$\nu_p = \sqrt{\frac{e^2 N_e}{4\pi^2 \epsilon_0 m}} = 9 N_e^{1/2} \ \text{sec}^{-1}$$

ν_p acts as a cut-off frequency for the transmission of electromagnetic waves, as was shown in Section 11.4. At frequencies below ν_p the plasma becomes reflecting. The emissivity j_ν must likewise cut off at ν_p, as illustrated in Fig. 11.10. An experimental determination of the cut-off frequency gives a value of N_e independent of temperature. This cannot be done with great precision, and in the case of laboratory plasmas it is preferable to measure the refractive index of the plasma at frequencies greater than ν_p and apply Equation 11.7 as described in the next section.

11.17.4 Inglis–Teller limit

In the case of Stark broadening, the quantum number of the last distinguishable line of a series is related to the electron density as described at the end of Section 11.7, and a crude estimate of N_e may be obtained from this relation [3]. Although far better methods can be applied to laboratory plasmas, the Inglis-Teller limit for the Balmer series of hydrogen has been used to estimate N_e in stellar atmospheres. It must be realized, however, that *any* line-broadening process will contribute to the lowering of the apparent series limit.

11.18   Non-spectroscopic methods

For laboratory plasmas spectroscopic diagnosis has been largely superseded by various microwave and, more importantly, laser techniques. Discussion of these methods is beyond the scope of this book, and a brief list only is given here. Full accounts may be found in books on plasma diagnostics, for example [5], [6], [9], [10]. It must be remembered that these new and powerful methods cannot be applied to astrophysical plasmas, for which the spectroscopic methods must still be used.

11.18.1 *Refractive index*

The refractive index of free electrons is given by Equation 11.7:

$$n_{el} - 1 = -\frac{e^2 N_e}{2\pi m c^2} \lambda^2$$

Comparing this with the atomic refractivity, Equation 11.8a, we see that the ratio is of order $N_e/N . \lambda^2/\lambda_0^2$ where N is the atomic number density and λ_0 the wavelength of the atomic resonance line. Most elements other than the metals have their resonance lines in the ultra-violet (for hydrogen and the inert gases, the most common constituents of laboratory plasmas, $\lambda_0 \sim 1000$ Å). For infra-red measurements the electron contribution therefore dominates when the ionization is only a few percent; in the visible one may require fairly high temperatures – say $T > 20\,000$ K – to achieve this. In any case, the atomic and ionic contributions, being virtually independent of wavelength except near an absorption line, may easily be assessed by measuring the refractive index at two different wavelengths.

Refractive index measurements on hot plasmas are normally possible only with lasers, because any other light source is swamped by the emission from the plasma itself. The refractive index is usually determined directly by incorporating the plasma in one arm of a Jamin-type interferometer. Since the optical path difference scales as λ^2 and the fringe shift for given optical path difference as $1/\lambda$, the sensitivity of the method increases as λ. The alternative methods of Schlieren and shadowgraph photography give respectively the gradient and the derivative of the gradient of the refractive index.

The so-called Ashby-Jephcott laser interferometer is an interesting variant of the usual refractive index method. In this the plasma is made virtually a part of the laser by putting a mirror beyond the plasma so that laser light transmitted through the plasma is reflected back through the laser. The behaviour of the laser is influenced by the phase of the reflected signal, and this shows up as an amplitude modulation depending on the refractive index of the plasma.

11.18.2 Microwave techniques

In low density plasmas the electron density may be found from the attenuation and phase shift of transmitted microwaves. As in the optical case, the sensitivity increases with λ, so that these methods come into their own when the electron density is small. They are in any case inapplicable for values of N_e greater than about 10^{13} cm^{-3}, because this corresponds to a plasma frequency cut-off at about 1 cm.

11.18.3 Scattering of laser light

Measurements of the incoherent scattering of light by a plasma give values for both electron density and temperature. The scattering cross-sections are small, so that the use of laser light is essential for sufficient intensity. Only the electrons contribute appreciably to the scattering, and the intensity of the scattered light is proportional to N_e. The thermal motion of the electrons introduces Doppler shifts into the scattered light, which, for large scattering angles, has an ordinary Doppler profile of half-width proportional to the square root of the electron temperature. In principle, laser scattering experiments at these large angles allow N_e and T to be determined at particular points in the plasma rather than from integration along the line of sight as in all other methods. At small scattering angles (forward scattering) correlation effects become important, and the form of the scattered radiation changes. In particular, the correlation between electron and ion motions allows the ionic kinetic temperature to be determined.

References

General

1. Aller, L. H. 'The Atmospheres of the Sun and Stars', Ronald Press, 1963
2. Boley, F. I. 'Plasmas – Laboratory and Cosmic', Van Nostrand, 1966
3. Cowley, C. R. 'The Theory of Stellar Spectra', Gordon and Breach, 1970
4. Griem, H. R. 'Plasma Spectroscopy', McGraw-Hill, 1964

5. Huddlestone, R. H. and Leonard, S. L. (eds.), 'Plasma Diagnostic Techniques', Academic Press, 1968
6. Lochte-Holtgreven, W. (ed.) 'Plasma Diagnostics', North Holland, 1968

Review Articles

7. Cooper, J. Plasma Spectroscopy, *Rep. Prog. Phys.* **22**, 35, 1966
8. Burgess, D. D. Spectroscopy of Laboratory Plasmas, *Space Science Reviews* **13**, 493, 1972

Diagnostic methods (mainly non-spectroscopic)

9. Rye, B. J. and Taylor, J. C. (eds.), 'Physics of Hot Plasmas', Oliver and Boyd, 1970
10. Griem, H. R. and Lovberg, R. H. (eds.), Plasma Physics, 'Methods of Experimental Physics', vols. 9A and 9B, Academic Press, 1970

Review articles on astrophysical applications may be found in the *Annual Reviews of Astronomy and Astrophysics* and in the Proceedings of various Symposiums of the I.A.U.

Appendix

Physical constants

Most of the constants listed below are known to eight significant figures. The four figures quoted here for the general constants give sufficient accuracy for most practical purposes, but the higher accuracy is frequently required for the spectroscopic constants.

General		*S.I. units*	*cgs units*
electron charge	e	$1 \cdot 602 \times 10^{-19}$ C	$4 \cdot 803 \times 10^{-10}$ e.s.u.
electron mass	m	$0 \cdot 9107 \times 10^{-30}$ kg	$0 \cdot 9107 \times 10^{-27}$ gm
electron charge/ mass ratio	e/m	$1 \cdot 759 \times 10^{11}$ C kg^{-1}	$5 \cdot 274 \times 10^{17}$ esu gm^{-1}
proton mass	M_{p}	$1 \cdot 672 \times 10^{-27}$ kg	$1 \cdot 672 \times 10^{-24}$ gm
proton/ electron mass ratio	M_{p}/m	1836·1	
Planck's constant	h $\hbar$	$\left.\begin{matrix} 6 \cdot 626 \\ 1 \cdot 055 \end{matrix}\right\} \times 10^{-34}$ J sec	$\left.\begin{matrix} 6 \cdot 626 \\ 1 \cdot 055 \end{matrix}\right\} \times 10^{-27}$ erg sec
Boltzmann's constant	k	$1 \cdot 380 \times 10^{-23}$ J/deg. K	$1 \cdot 380 \times 10^{-16}$ erg/deg. K

speed of
light c $2 \cdot 998 \times 10^8$ m sec^{-1} $2 \cdot 998 \times 10^{10}$ cm sec^{-1}

Avagadro's
number N_{AV} $6 \cdot 025 \times 10^{23}$ mole^{-1}

Loschmidt's
number N_L $2 \cdot 69 \times 10^{25}$ mol m^{-3} $2 \cdot 69 \times 10^{19}$ mol cm^{-3}

permittivity of vacuum $\epsilon_0 = 8 \cdot 854 \times 10^{-12}$ C^2(N m^2)$^{-1}$ or F m^{-1}

$1/4\pi\epsilon_0 = 9 \times 10^9$ N m^2 C^{-2} or m F^{-1}

permeability of vacuum $\mu_0 = 1 \cdot 257 \times 10^{-6}$ Wb(amp m)$^{-1}$ or H m^{-1}

$\mu_0/4\pi = 10^{-7}$ Wb(amp m)$^{-1}$ or H m^{-1}

Spectroscopic

Rydberg constant (S.I.)$R_\infty = \dfrac{me^4}{8\epsilon_0^2 ch^3} = 1 \cdot 09737312 \times 10^7$ m^{-1}

(cgs) $R_\infty = \dfrac{2\pi^2 me^4}{ch^3} = 1 \cdot 09737312 \times 10^5$ m^{-1}

(eV) $R_\infty = 13 \cdot 60$ eV

radius of first Bohr orbit (S.I.)$a_0 = \dfrac{4\pi\epsilon_0 \hbar^2}{me^2} = 5 \cdot 292 \times 10^{-11}$ m

(cgs) $a_0 = \hbar^2/me^2 = 5 \cdot 292 \times 10^{-9}$ cm

fine structure constant (S.I.)$\alpha = \dfrac{e^2}{4\pi\epsilon_0 \hbar c}$

(cgs) $\alpha = e^2/\hbar c$ $\Bigg\} = 7 \cdot 297351 \times 10^{-3}$ (dimensionless)

$1/\alpha$ $= 137 \cdot 03602$

Bohr magneton (S.I.) $\mu_B = \dfrac{e\hbar}{2m} = 9 \cdot 274096 \times 10^{-24}$ J(Wb m^{-2})$^{-1}$

(cgs) $\mu_B = \dfrac{e\hbar}{2mc} = 9 \cdot 274096 \times 10^{-21}$ erg gauss^{-1}

(cm^{-1}) $\mu_B = 0 \cdot 4666$ cm^{-1} (Wb m^{-2})$^{-1}$

$= 4 \cdot 66 \times 10^{-5}$ cm^{-1} gauss^{-1}

Energy conversion

	joule	eV	erg	cm^{-1}
1 joule =	1	$6 \cdot 242 \times 10^{18}$	10^7	$\equiv 5 \cdot 034 \times 10^{22}$
1 eV =	$1 \cdot 602 \times 10^{-19}$	1	$1 \cdot 602 \times 10^{-12}$	$\equiv 8065$
1 erg =	10^{-7}	$6 \cdot 242 \times 10^{11}$	1	$\equiv 5 \cdot 034 \times 10^{15}$
$1\ cm^{-1} \equiv$	$1 \cdot 986 \times 10^{-23}$	$1 \cdot 240 \times 10^{-4}$	$1 \cdot 986 \times 10^{-16}$	$\equiv 1$

$kT = 1 \cdot 380 \times 10^{-23}\ T$ joule $= 0 \cdot 861 \times 10^{-4}\ T$ eV and corresponds to $0 \cdot 695\ T\ cm^{-1}$

Index